Intelligent Image Processing in Prolog

AI – Its Appearance and Habits

The AI is about 2 feet long and is rather variable in colour, usually being brownish-grey but slightly variegated with differently tinted hairs. The head and face are darker than the body and limbs. The head is short and round, the eyes are deeply sunk in the head and the nose is large and moist. The hair has a curious hay-like appearance, being coarse and flat at the ends, but very fine closer to the root. The AI can hardly be distinguished from the branches of the trees, where it prefers to spend its time. This habit contributes to its survival, despite the fact that this unfortunate creature is dreadfully persecuted by man. The cry of the AI is low and plaintive. The young cling to their mother as soon as they are born, and are carried about by her. On trees, the AI is full of life and animation and traverses the branches at a speed that is anything but slothful. The AI travels best in windy weather, when it can pass from one tree to another without touching the ground. The only manner in which an AI can advance on the ground is to hitch its claws into any depression that it can find and so drag itself slowly and painfully forward.

Several related species are known to exist and are similar in appearance and habits.

Adapted from *The Illustrated Natural History: Mammalia,*
J. G. Wood, Routledge, London, 1876.

Bruce G. Batchelor

INTELLIGENT IMAGE PROCESSING IN PROLOG

With 131 Figures

Springer-Verlag London Ltd.

Bruce G. Batchelor, BSc, PhD, CEng, MIEE, FRSA, MSPIE
Professor, School of Electric, Electronic and Systems Engineering,
PO Box 904, King Edward VII Avenue, Cardiff CF1 3YH, UK

ISBN 978-1-4471-1131-3

British Library Cataloguing in Publication Data
Batchelor, Bruce G. (Bruce Godfrey) *1943–*
Intelligent image processing in Prolog.
1. Machine vision
I. Title
006.37
ISBN 978-1-4471-1131-3

Library of Congress Cataloging-in-Publication Data
Batchelor, Bruce G.
Intelligent image processing in Prolog/Bruce G. Batchelor.
p. cm.
Includes index.
ISBN 978-1-4471-1131-3 ISBN 978-1-4471-0401-8 (eBook)
DOI 10.1007/978-1-4471-0401-8
1. Prolog (Computer program language) 2. Image processing.
I. Title.
QA76.73.P76B37 1991
006.3--dc20 90-47488
CIP

Originally published by Springer-Verlag Berlin Heidelberg New York in 1991
Softcover reprint of the hardcover 1st edition 1991

Typeset by Fox Design, Surbiton, Surrey

69/3830-543210 Printed on acid-free paper

Do all the good you can,
By all the means you can,
In all the ways you can,
In all the places you can,
At all the times you can,
To all the people you can,
As long as ever you can.

John Wesley

PREFACE

After a slow and somewhat tentative beginning, machine vision systems are now finding widespread use in industry. So far, there have been four clearly discernible phases in their development, based upon the types of images processed and how that processing is performed:

(1) Binary (two level) images, processing in software
(2) Grey-scale images, processing in software
(3) Binary or grey-scale images processed in fast, special-purpose hardware
(4) Coloured/multi-spectral images

Third-generation vision systems are now commonplace, although a large number of binary and software-based grey-scale processing systems are still being sold. At the moment, colour image processing is commercially much less significant than the other three and this situation may well remain for some time, since many industrial artifacts are nearly monochrome and the use of colour increases the cost of the equipment significantly. A great deal of colour image processing is a straightforward extension of standard grey-scale methods.

Industrial applications of machine vision systems can also be subdivided, this time into two main areas, which have largely retained distinct identities:

(i) Automated Visual Inspection (AVI)
(ii) Robot Vision (RV)

This book is about a fifth generation of industrial vision systems, in which this distinction, based on applications, is blurred and the processing is marked by being much smarter (i.e. more "intelligent") than in the other four generations.

How did the distinction between RV and AVI become blurred? The simple answer is that there are many inspection tasks which require manipulation. For example, we pick up an object to examine it from a number of directions and with different lighting; one view is simply not sufficient to allow all of the important features to be seen properly. If you were asked to examine an electrical cable harness, you would need to separate the end connections in order to examine each one in turn. This requires both manipulation and viewing. You would naturally want to place a small lamp inside each hole on a complex casting, such as a car engine block, to investigate whether it has any debris (e.g. swarf) in it. In order to examine a cup, jug or teapot, you would need to look at it from a number of different directions. These and a myriad of other tasks require an integrated approach to image processing and robot control.

Why is intelligence needed for industrial vision systems? Suppose that you were instructed to inspect every solder joint on a printed circuit board. You would not need to be told where those joints are; you could find them for yourself and then examine each one in turn. Another example is detecting cracks in white ceramic objects (e.g. plain china cups). Of course, you would immediately realise that a crack is likely to be evident as a very thin dark streak; you do not need to be "programmed" in a more precise way. A machine should be able to accept a similar "high-level" definition of the task required. Another example is taken from the food-processing industry, where we might want to put a vision system in a feedback loop, where it controls the mixing of ingredients, or baking conditions. After inspecting cakes or loaves to determine the degree of baking and other visually discernible features/defects, the vision system might provide a feedback signal which will alter the manufacturing parameters, such as water content of the dough, or oven temperature. A high level of intelligence would also be needed in a visually controlled robot performing the brick-layer's task of building a wall. The bricks and the wall must both be examined, while the motions required by the manipulator are highly dependent upon the conditions discovered.

We find ourselves in the position of needing to integrate two technologies that have hitherto been almost disparate, namely Image Processing (IP) and Artificial Intelligence (AI). A cursory look at textbooks on these two subjects suggests that there is a closer linkage than has actually been realised in practice. Many books on AI include chapters on IP, although there is usually no *detailed* explanation as to how AI techniques could be applied to image understanding in practice. Books on programming for AI contain very little, if any, detail on image processing, although they frequently contain a chapter or two on processing picture-like data structures. There has been no really concerted effort to merge image processing ideas into either of the two main AI languages, Lisp and Prolog. This book attempts to fill the gap that exists here.

Several languages are currently in use for work in AI, of which Lisp and Prolog are by far the most popular. Lisp is still dominant in the U.S.A., while Prolog finds greater favour in Europe and Japan. Pop11 and Smalltalk have only a relatively small following at present. C is used mainly by people who cannot be bothered to learn one of the more specialised AI languages, or who wish to integrate AI calculations with other types of processing task.

Why was Prolog, rather than Lisp, chosen to be the basis for our work integrating vision and AI? The reason is partly historical, partly technical. I had been developing interactive image processing languages for about ten years, when I published a paper, in 1986, which described how the benefits of that approach to designing image processing algorithms could be retained by integrating image processing operations into Prolog. The result was a new language, called *ProVision* which was described originally as an amalgam of the *Autoview* interactive image processing language and Prolog. (Autoview was a product of British Robotic Systems Ltd and was also sold under the name *System 77* by The 3M Company.) A closely related language, called *SuperVision* was then defined which, in effect, merged the *VCS* image processing language (Vision Dynamics Ltd) with Prolog. It should be noted that VCS and Autoview are very closely related and owe their origins to the same progenitor, called *Susie*. Later, it was realised that the distinction between ProVision and SuperVision was largely cosmetic and the latter name was dropped. As time progressed, the name ProVision was discarded, since it seemed unnecessary to retain a name that was not universally understood. In order to acknowledge the fact that standard Prolog was enhanced by the image processing predicates, the name Prolog+ was coined. ProVision, SuperVision and Prolog+ are all dialects of the same basic language. Prolog+ is the name of the language which we shall describe here. It is a superset of standard Edinburgh Prolog.

There have been several different implementations of the new language, whatever name we give it. We shall describe these later. However, the latest, and simplest, is called VSP (Very Simple Prolog+) and has only two additional features, compared to standard Prolog. The more significant one for us is the # operator, which allows Prolog to issue commands to an image processor and to retrieve results from it. (The other new operator, ¶, performs a similar function for robot control.) Despite its *apparent* simplicity, VSP is able to simulate the fully developed language, Prolog+. This is a direct result of Prolog's great expressional power. Moreover, it does not matter to VSP which image processor is being used. It is possible to control image processors, from different manufacturers, with exactly the same VSP software. It is even possible to imagine a system in which VSP controls several different types of image processor, *at the same time*. The author has been using Prolog+ and its relatives for several years. Both he, his students and other associates have studied it intensively during that time. At each stage in its development, the language has shown itself to be very versatile. It has proved itself to be capable of expanding to meet the demands of new and previously unforeseen tasks; Prolog+ is still growing and has, in the author's opinion fully justified his faith in it.

This monograph represents a drawing together and, I hope, a rationalisation of a number of ideas which have been buzzing around in my head for several years. The ideas described in these pages are not yet accepted as conventional wisdom. Even Prolog does not yet enjoy that status, yet its popularity is growing rapidly. There are two types of people who, I hope will enjoy reading about these ideas. One of them consists of image processing specialists who might be persuaded to learn Prolog. The other comprises Prolog enthusiasts who might discover new things that this distinctive language can do. One of the main reasons for writing this book was to try to explain that there is a smooth path, which reaches from pixel-level data and processing, up to abstract, symbolic manipulation, even as far as Expert Systems. As I write this book, I am mindful of the fact that there are many excellent researchers in the same area of study whose work I have not acknowledged or cited. I apologise to them and hope that they will understand that the approach that I have adopted does not ignore their work. Instead, I have tried to integrate many of the ideas which have come from other minds into the computational framework represented by Prolog+. There are several guiding principles that have guided my own research and the writing of this book:

(i) If a technique works, use it. If not, find one that does.

(ii) Do not worry about the fine tuning, since the broad concept is more important. For example, arguing whether, say, Filter Method A is better or worse, on a theoretical basis, than Filter Method B is quite meaningless, if one is far superior to the other, when judged subjectively, or if one of them is far easier to implement than the other.

(iii) The engineering effort required to design a vision system for a novel application is far more expensive than the cost of electronic hardware. There are far too few *good* vision systems engineers in the world to meet the demand that could arise if our long vaunted claims of having a flexible technology were actually to be accepted by the very people who we hope will be our "customers". We need to equip ourselves with the very best design methods for industrial vision systems. Prolog+ was designed to be one of those tools.

(iv) There is no single image processing algorithm that stands above all others, in terms of its importance. In other words, there is no single algorithmic, nor heuristic method which could be said to be the heart of image processing, so that all other procedures have to be subjugated to its special requirements. Practical image processing procedures (e.g. for use in factory inspection systems) are very variable and are quite complex, often consisting of a sequence of computational steps. We require an operating framework in which to place modules for image processing, whether they be implemented in software or hardware. This framework should ideally allow easy prototype design and be fully compatible with the hardware that will eventually be placed on the factory floor.

(v) There remains a very large part of manufacturing that is closed to vision systems. Of special note are the inspection and manipulation of:

    (a) Complex objects
    (b) Assemblies of components
    (c) Hand-made products
    (d) Very variable products
    (e) Flexible objects and articulated assemblies of components
    (f) Products made in very small batch sizes

In other words, vision systems have only just touched the tip of a very large iceberg. The one characteristic which these new applications all have is that they cannot be performed without the use of AI techniques. (Whether they can be accomplished using the range of AI techniques that is available today, is quite a different question. However, I hope to show in these pages that considerable progress can be made using established AI methods.)

(vi) Do not rely on one technology too much. For example, we should never use AI or complex image processing methods to get us out of trouble that is caused by a casual approach to the design of the lighting sub-system. Whilst this monograph is about Prolog-style software, we should be just as concerned about the choice of lamps, optics, sensor, etc. The way that the ideas discussed here can be integrated into a wider systems approach to machine vision design is explained elsewhere.

(vii) Occam's Razor states that "It is vain to do with more what can be done with less". The colloquial version, "Keep it simple, stupid", is just as dismissive of excessive complexity, yet this wisdom is very often ignored or forgotten. Many designers use very complex procedures, when simpler ones would suffice.

(viii) Complexity, if it is needed can often be obtained by combining together quite simple operations. (Arithmetic is a good example. DNA is another.)

(ix) Sufficiency, not optimality, is required. A corollary of this is that machine vision systems need not, in any way, attempt to model natural vision.

Following this line of thought, I then concluded that image processing algorithm design/selection should, if possible, be regarded as the task of putting together sequences of "atomic" operations, to form complete procedures. One suggested software framework for doing this is embodied in Prolog+.

It is a pleasure to acknowledge the great contribution that other people have made to my understanding of the subject of Intelligent Image Processing. In particular, I should like to mention the following people with whom I have thoroughly enjoyed many fruitful hours of discussion: Dr John Parks, Dr Jonathon Brumfitt, Dr David Mott, Mr Michael Snyder, Dr Frederick Waltz, Dr Graham Page, Mr David Upcott, Dr Barry Marlow, Dr Simon Cotter, Dr Christopher Bowman, Mr Wayne Jones, Dr Anthony McCollum, Dr Derek Kelly, Mr Ian Harris and Mr John Chan. The importance of certain ideas and techniques does not become apparent until talking to enthusiastic and well-informed colleagues. The friends who I have just mentioned have taught me the importance of numerous techniques and ideas, in a subject, which like many other others has a "folklore" all of its own. My gratitude is also extended to Mr Stephen Paynter and Mr Ian Harris, both of whom greatly assisted me by proofreading the manuscript. The plant dissection program given in

Chapter 7 is based upon one written by Ian Harris who has kindly given his consent to my including it here.

My gratitude is also extended to Mr Kipling (Manor House Bakeries) who kindly supplied the exceedingly good Bakewell tarts, some of which used in our research.

I have made very extensive use of the *MORE*, *WORD*, *MacDraw* and *MacProlog* software packages in the preparation of the manuscript. These products have transformed the tedious process of writing into one of great fun. This would not have been possible without the harmonious operating environment of the Apple Macintosh family of computers. This is simply delightful to use. I have been able to use the excellent version of Prolog written by Logic Programming Associates Ltd, London. I am most grateful to Miss Nicky Johns, who has provided invaluable help, allowing me to understand some of its finer points.

I have made extensive use of the VCS image processing language in the pages that follow. I am very grateful to the directors of Vision Dynamics Ltd who have always supported and encouraged my work.

I dedicate this book to my darling wife, Eleanor. She has always totally supported me in my work. She has never faltered in her belief in me and is a source of unending encouragement and love. My Mother and late Father too have always encouraged me. I cannot let this opportunity pass by without thanking them for giving me such a good start in life. Although Helen and David came along later, I thank them for all that thay have taught me.

This book is not a treatise on the Christian faith, nor does it attempt to answer the great questions that have puzzled philosophers for centuries. However, I feel compelled to tell you that I undertook its writing because I was called by God to do so. Although I don't know why He wanted me to write this book, I trust that His purpose will be served by it.

Bruce Batchelor

Cardiff, December 1990

CONTENTS

CHAPTER 1

VISION AND INTELLIGENCE

"It is a foolish thing to make a long prolog and to be short in the story itself." The Second Book of Maccabees, chapter 2, verse 32.

1.1 Vision and Intelligence

Sight is, without doubt, the most precious and wonderful of the senses. The value of vision to the survival of both people and animals is obvious: food can be located, and a mate or a potential predator can be identified. All of this can be done at a safe distance from the object that is being observed. Detailed information about size, shape, colour and motion can be collected and then integrated with stored data and that derived from other sensors, in order to establish an understanding about the surroundings of the organism. The survival and safety of nearly all of the higher animals is critically dependent upon sight and human beings are no exception. Apart from the basic animal needs to which we have referred, people have come to rely upon vision for a multitude of uses, including manufacturing, navigation, commerce, education, communication, and entertainment. Blindness is so terrible a prospect that most of us would prefer a host of physical disabilities instead. We do not need to emphasise the practical and aesthetic importance of vision, because it is immediately apparent to us all, throughout our everyday lives. From the moment we wake up, to the very last second when we put the lights out before going to sleep at night, we use our eyes.

You are using your eyes very effectively at this very moment! Understanding a page of printed text, such as this one, clearly requires a high level of intelligence, although the role of intelligence in lower-level image interpretation (e.g. recognising individual typed/printed letters) is less obvious. However, a little thought soon shows that intelligence is also an integral part of this process; the enormous variability of letters in different type fonts provides evidence for this (see Figure 1.1). It is impossible to define an archetypal letter "A". Nor can we specify the limits of acceptability for

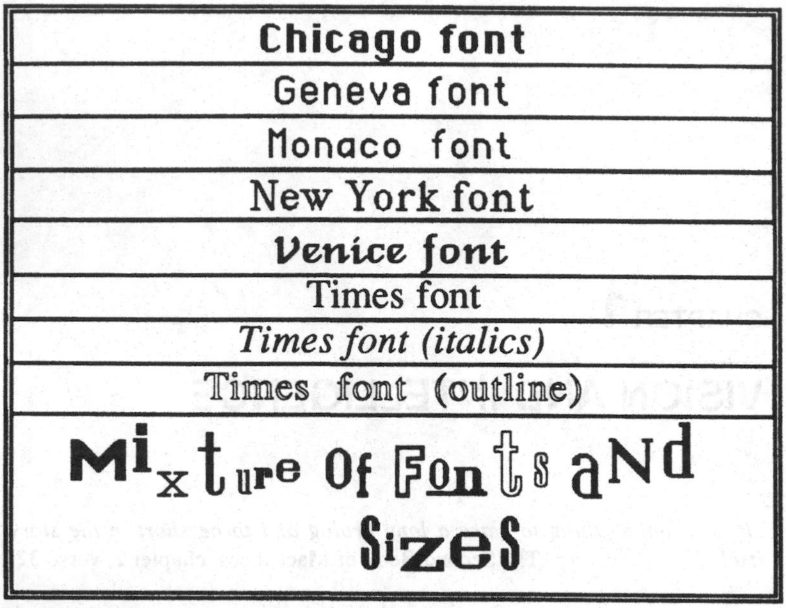

Chicago font
Geneva font
Monaco font
New York font
Venice font
Times font
Times font (italics)
Times font (outline)
Mixture Of Fonts aNd Sizes

Figure 1.1. Intelligence is required to read different typefaces because we cannot use template matching when the form of characters is so variable and unpredictable.

the parameters representing the class of objects that we call "A". We cannot list all examples of an acceptable letter "A". Despite this, neither you nor I have difficulty in understanding Figure 1.1. Clearly, you do not recognise the letters and words there by some trivial process of matching letters to mental templates. If you did, then you could not possibly cope with previously unseen type fonts, or handwriting. Hofstadter [HOF-85] presents an excellent discussion of what constitutes the "A-ness" of a printed letter "A" and, in the process, clearly demonstrates that vision and intelligence are inextricably inter-twined and inter-dependent (see Figure 1.2).

So far, we have argued that intelligent processing is an integral part of visual perception and the converse is also true. To illustrate this last point, consider the task of driving to a defined address in an unknown town, with the help of a street plan. Vision is used to locate both the present location and the desired address on the map. Then, a route is formulated and converted from a spatial form (i.e. a line drawn on the map) into a series of instructions, which are expressed in ordinary English. These might take the form "take the third left turn", "continue straight on at

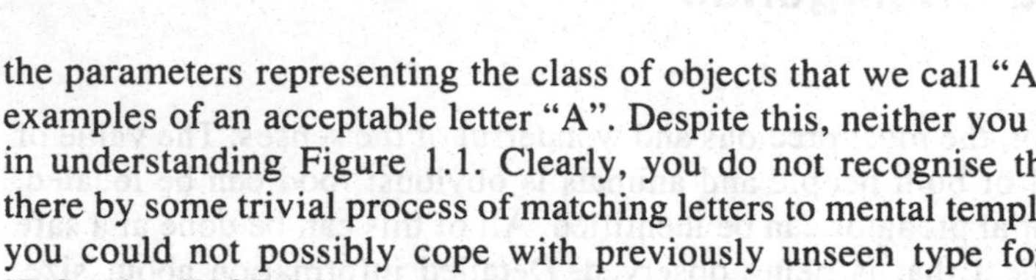

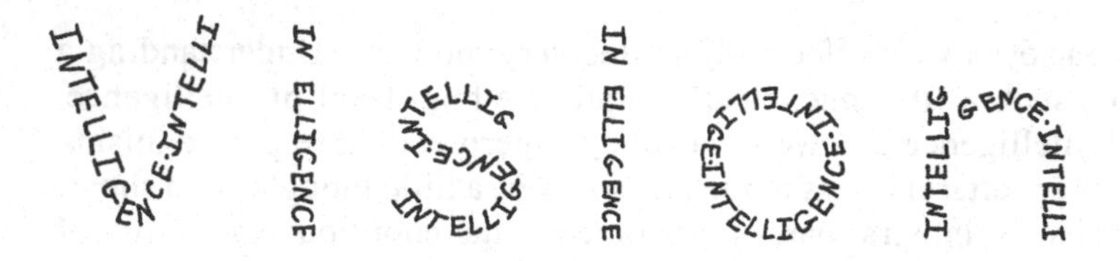

Figure 1.2. Further evidence that intelligence is an integral part of vision.

the traffic lights", etc. This particular task is one that we shall discuss again later (Chapter 7). The general point is that, once an image (or a series of images), has been analysed, it may be necessary to perform a large amount of intelligent planning, in which vision plays no further role. In this type of situation, vision "merely" provides the input to the intelligent process (i.e. thinking). There is a very long list of tasks which possess this characteristic, including:

1. Driving a car
2. Playing chess
3. Recognising a friend in a crowd
4. Appreciating a painting
5. Playing the piano
6. Decorating a cake
7. Choosing which dress to wear
8. Brick-laying
9. Operating a computer, etc.

So, we have shown that vision is an integral part of intelligence and vice versa:[1] *vision and intelligence are inter-twined and inter-dependent.* If you do not agree, think about it! By doing so, you are actually proving the point. First of all, you read the sentence set in italics, which is a visual task. Then, you *thought*, which is an intelligent activity. Neither process would have been meaningful without the other.

Several quite different approaches to studying the inter-dependence of intelligence and vision are possible: psychologists, physiologists, mathematicians, engineers, computer programmers, philosophers and painters all have different views on this subject. (See, for example, [ARB-87, HOF-85, GOM-88, GRE-90, MAR-82, PYL-88].)

Since vision is so useful to both people and animals, it has long been a dream that machines might be provided with an ability to sense their environment visually. This has been a reality, in a rather limited sense, for some years. Machines equipped with visual sensing have, for example, been in use for some years for inspecting industrial artifacts. However, the tasks which have been performed by machine vision systems have, to date, been rather limited in their versatility, because they have had virtually no ability to reason intelligently about what they "saw". There has been a considerable research effort to build intelligent machine vision systems but, so far, this has not resulted in useful factory-floor equipment. This book describes one attempt to add intelligence to industrial vision systems, in a way that will be both practical and useful.

1.2 Machine Vision Systems for Industry

During the last two decades, machines that can "see" have been developed for a variety of uses. Of special interest to us are the tasks of inspecting and manipulating industrial artifacts. Of course, there are numerous other actual and potential uses for machine vision, but our concern in this monograph is to discuss techniques that are specifically relevant to industrial vision systems. A large part of the fun which comes from working in this area is the exceedingly broad and diverse range of applications that it presents. Here is a much abbreviated list of some of the more unusual applications of machine vision, which the author has *personally* encountered in the course of his work.

Sorting shoe components [BRO-86]
Dating trees by counting growth rings
Detecting fragments of plastic in pipe tobacco
Inspecting cakes
Inspecting castings for toy cars
Calibrating spirit levels
Counting worm holes in a sample of soil
Measuring magnetic domains in a ferromagnetic material
Counting stones in a sample of concrete
Monitoring the winding of thread onto a bobbin
Locating the filling hole on an oil drum
Measuring aerosol sprays for central heating boilers [BAT-85a]
Inspecting coins [BAT-85a]
Counting the holes in bread wrappers
Identifying and counting seeds
Inspecting front panels of television tubes
Locating the silver "button" on an electrical switch leaf
Counting tea bags in a packet
Measuring the honing of razor blades
Inspecting munitions shells
Inspecting tampons for oil stains
Grading corks [BAT-85a]
Inspecting the rubber tube which is attached to a cow's udder during milking
Inspecting cornflakes
Inspecting surgical instruments
Inspecting photographic film optically
Harvesting mushrooms
Detecting foreign bodies in food using x-rays
Dissecting very small plants [BAT-88b]
Inspecting newly made bottles which are still red-hot [BAT-85a]
Detecting cracks in automobile connecting rods (con-rods) [BAT-85a]
Finding scratches in automobile brake hydraulics cylinders [BAT-85a]
Identifying guns and explosives in aircraft baggage
Measuring plaques of bacteria in a Petri dish

Such a list indicates the truly enormous range of tasks to which industrial machine vision systems might be applied. (Not all of those tasks in the above list have resulted in practical vision systems being built.) However, it should be appreciated that these are just a few of the many thousands of applications that have been proposed and studied to date. Further applications are listed by Parks [PAR-78], Hollingam [HOL-84], and Chin and Harlow [CHI-82].

To date, most industrial vision systems that have been installed have been notable in not having any ability to interpret images intelligently. Several years ago, the author published an article explaining why this was to be expected, when mass-produced, close-tolerance articles are to be examined [BAT-86b]. The predictions in that article have been upheld; the primary role of AI techniques in industrial vision systems lies elsewhere:

1. *Inspecting objects that are very complex*, compared to those that have been examined by AVI systems installed so far. (Examples are car engine blocks, complicated mouldings/castings, populated printed circuit boards, car body panels, etc.)
2. *Inspecting assemblies of objects*. (Examples are hair dryer, automobile carburettor, etc.)
3. *Inspecting objects which are very variable in form.* (The main examples are processed food items, such as cakes, loaves, pizza, pies, etc.)
4. *Inspecting non-rigid objects* and those which are composed of articulated levers. (Cable harnesses, leather and fabrics also present similar problems.)
5. *Aiding in the design process* for both AVI and RV systems. (Knowledge-based systems are finding their way into such tasks as choosing the camera, lens, and lighting arrangement for an industrial vision system.)
6. *Inspecting objects that are made in very small batches*. (It has been estimated that 75% of manufactured goods are made in batches of 50 or fewer items. Customised products are finding a ready market in a large number of areas, from leather-ware to cars.)
7. *Providing an operating/programming environment which is easier to use* than those available so far. (The user interface might use pull-down menus, well-structured human–computer dialogues, speech synthesis and speech recognition, natural language understanding, etc.)
8. *Analysing scenes in which objects can touch or overlap*. (Microscopic particles, wool, cotton and other fibres are difficult to separate physically and are the most obvious examples in this category.)
9. *Manipulating objects which are very variable in form*, or are to be found in places where there is a very large degree of variability. Tasks in this category include:

   Grading, sorting, processing and harvesting fruit/vegetables
   Butchery
   Manipulating plant/animal products
   Collision avoidance
   Calculation of safe, reliable gripping points
   Electrical wiring

Plumbing
Brick-laying
Laying floor tiles

How do we examine and measure "visual appearance" using a machine? For example, how do we design a machine that can automatically assess the visual appearance of a decorated cake, a box of chocolates, an in-flight meal tray, a potted plant, or decorative woodwork? These are tasks that typically exhibit a high degree of variability and it should be noted that they are fundamentally different from those that are amenable to the use of conventional, algorithmic (i.e. non-AI) methods.[2] How do we build a machine that can sense how a piece of crumpled cloth should be picked up by a robot, prior to stitching, folding or ironing? Can a machine be built that can arrange flowers in a vase? Of course, this is a problem that people may prefer to do themselves but it does exemplify a broad class of delicate manipulative tasks which require the integration of visual sensing, intelligent planning and fine, sensitive handling. Can visually guided machines be used to perform such tasks as mowing a lawn, painting a ship, operating a vacuum cleaner, or clearing a table of glass-ware, china and cutlery? A robot which automatically shears sheep has already been developed, although the author is not yet aware of the existence of an automated hairdresser! A visually guided robot has been used to decorate chocolates in a way that makes them appear to be hand made. Several machines for guiding robotic welders have been designed. Even in these tasks further research would be beneficial; the problems have not yet been totally solved!

In the past, Automated Visual Inspection and Robot Vision developed almost independently and in parallel. There has been a degree of cross-fertilisation between them more recently. In the past, their requirements have been regarded by some writers, including the present one, as being essentially different [BAT-85a]. An RV system requires short-delay and safe decision making, linked with software to control the manipulator arm. Optimal illumination is rarely achieved. On the other hand, "conventional AVI" was required to provide a low-cost system capable of a high throughput rate and could make use of carefully optimised lighting to achieve that goal. However, if a flexible inspection device is to be built which can manipulate and examine complex objects/assemblies, then it requires a multi-axis manipulator and computer controlled lighting (see Figure 1.3). In other words, inspection is now moving towards Robot Vision. Meanwhile, RV is also developing to encompass inspection, in order to make movement of the manipulator arm safer. Trying to pick up a malformed object could be dangerous, for the robot, the object being handled and for people standing nearby. This blurring of the distinction between RV and AVI is likely to become more pronounced as machine vision is used more and more for inspecting complex and highly variable objects.

1.3 Technologies and Techniques

It is important that we first establish the range of technologies that we must bring together in order to build an intelligent vision system for industrial use. Figure 1.4

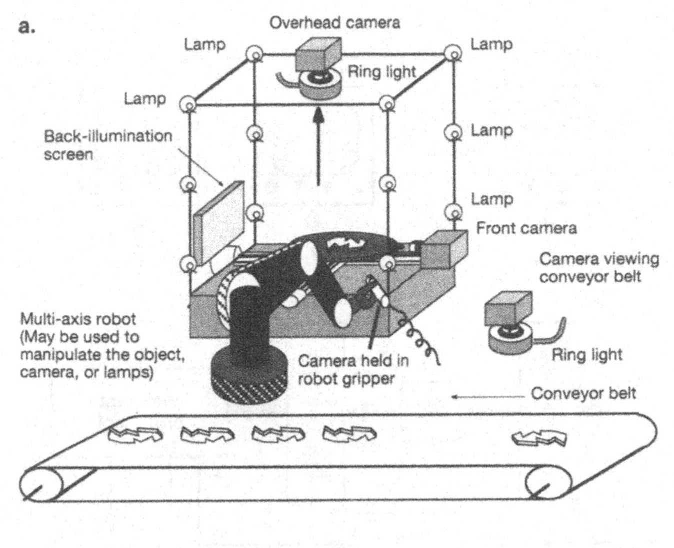

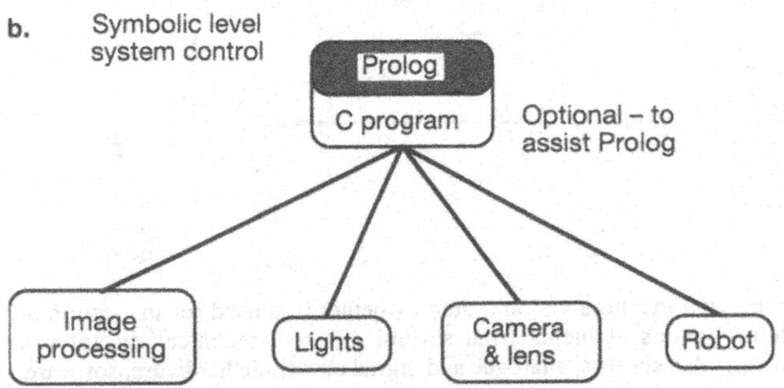

Figure 1.3. **Flexible Inspection Cell. (a) Physical layout. The cell consists of a frame which holds a set of lamps, each under computer control. At the centre of the frame is an (X,Y,θ)-table, which presents the object for examination in different postures. There is a need for some kind of mechanism for feeding objects to the cell. This may be a simple pick-and-place device, or a multi-axis robot. Notice that there are several cameras, which are intended for different purposes. One of the cameras is able to view the objects against a bright background, while another sees a dark background. There is an overhead camera and possibly more than one camera looking horizontally. (The "front camera", referred to later.) (b) Organisation of the control system for the Flexible Inspection Cell. Notice that a program written in Prolog acts as the top-level controller, although it may receive considerable help from a program written in a conventional, imperative language, such as C or Pascal. The important point to note is that symbolic processing is possible at the highest level.**

shows the system block diagram of an archetypal machine vision system, whether it be intended for AVI or RV. Notice that it indicates the requirement for a multi-disciplinary approach, involving:

1. Mechanical handling[3]
2. Lighting
3. Optics
4. Image sensing
5. Analogue (video) signal processing

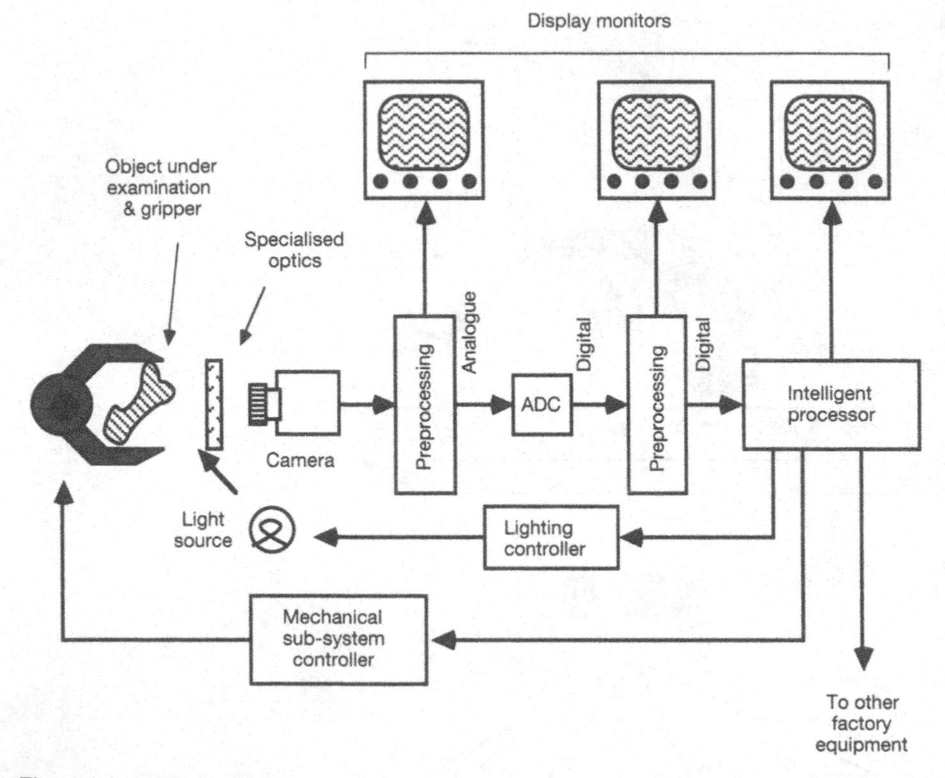

Figure 1.4. Block diagram of an archetypal machine vision system, whether it is used for inspection or robot control. Notice that the system contains elements from several different technical disciplines: mechanical engineering, lighting, optics, video sensors, analogue and digital electronic hardware, software, algorithms, quality assurance techniques, production engineering, ergonomics. Designing such systems is difficult and time consuming and there are very few competent vision engineers. As a result, the cost of an industrial machine vision system is dominated by that of the design process. The ideas expressed in this book form part of the industry-wide attempt to provide a set of tools for designing systems of this type. However, it should be understood that intelligent image processing systems has other uses too.

6. Digital signal processing
7. Computer systems architecture
8. Software
9. Heuristics and algorithms for image understanding
10. Industrial engineering, notably equipment protection and maintenance techniques
11. Human–computer dialogue engineering
12. Existing manufacturing and inspection practices

Since this monograph is primarily concerned with (8) and (9) in this list, it may be imagined that the author is ignoring the other items but this is not so. As evidence of this, we point out that the Prolog+ language, which we shall describe in these pages, contains facilities for controlling a robot (an (X,Y,θ)-table, plus pick-and-place arm), camera, lens, and lighting. In addition, the author has already developed an expert system which gives advice about lighting and is at an advanced stage in developing another advisor, for assisting with lens and camera selection. These are both written in the Prolog+ language.

There is an outstanding need for a computer language that is able to express a very broad range of ideas and operations needed for intelligent vision and the high-level control of a robotic manipulator. It was not by accident, that the word "ideas" was placed before "operations", in the previous sentence; we require a powerful language in which to represent concepts and to code operations.

Our experience has shown that there is a need for a very wide range of image processing operations in both AVI and RV systems. It was discovered by the author some years ago that algorithms for a very wide range of industrial vision applications could be expressed in a special-purpose language called Susie [BAT-79]. This has since been developed commercially and has resulted in a number of clones.

There are two major AI languages in use in the world and a number of less popular ones. Lisp and Prolog are, by far, the most frequently used AI languages. Together, they dominate AI research. Many people regard them as competitors, but it is perhaps better to view them as offering different strengths. The one feature of Prolog which renders it ideal for our use is its *declarative* nature. More will be said about this later.[4]

1.4 Prolog

Our objective in this monograph is to describe how image processing facilities can be added to Prolog. This language was not originally intended for image processing, but its use in this role has been discussed by several authors [BRU-84, MOT-85, FAI, GAR-89, HAN, POW-88, TOL]. Several good books describe the Prolog language [BRA-86, COE-88, CLO-81, FOR-89, GAZ-89, KLU-85, MAR-86, MER-89, ROS-89, STE-86]. The basic or "core" language has many limitations[5] but has been greatly extended in modern implementations.

The particular version of Prolog on which this monograph is based is that called MacProlog, written and sold by Logic Programming Associates Ltd.[6] It is simple to use and it makes good use of the superb operating environment of the Macintosh family of computers. MacProlog has many enhancements and extensions compared to "core" Prolog.

(i) Numerous additional Built-In Predicates (BIPs). (The user has full access to the Macintosh tool-box and, as a result, can program dialogues using pull-down menus, scroll menus, yes/no question boxes, etc.)
(ii) Properties
(iii) A comprehensive collection of graphics operators
(iv) Forward chaining
(v) A set of tools for building expert systems[7]
(vi) An enhanced set of tools for handling natural language
(vii) Object-oriented programming
(viii) Speech synthesis[8]
(ix) Ability to interface C and Pascal programs to MacProlog

All of these features make LPA MacProlog a formidable programming language and one that is very well suited for our purposes.

Another factor influencing the choice of the Prolog software was the need to communicate with the external world, via a serial port. LPA MacProlog provides an interface to the Macintosh I/O ports (RS-423). While the author will endeavour to highlight those areas where features specific to MacProlog are being used, it should be understood that many of the facilities offered by Prolog+ might not be transferable to other implementations.

1.5 Image Processing

It will be taken for granted that the reader is already familiar with, at least, the basic philosophy, objectives and concepts of digital image processing. There are many excellent textbooks which describe the fundamentals of this subject available for those readers who wish to revise this subject. A particularly good tutorial introduction is provided by Joyce Loebl [JOY-85], although this particular book is not directed specifically towards industrial image processing. An article which addresses this topic is given by Batchelor [BAT-85a]. Also see Batchelor and Waltz [BAT-91]. The books by Rosenfeld and Kak [ROS-82], Hall [HAL-79], Castleman [CAS-78], Ballard and Brown [BAL-82], Gonzalez and Wintz [GON-77] provide detailed descriptions of wider range of techniques. The book by Pratt [PRA-78] provides a very thorough description of image processing, for a wide range of applications.

In addition, there exists a wide range of commercial software, much of which can run on a standard desk-top computer. Dedicated image processing hardware, though rather less versatile than software of comparable price, operates much faster. There are numerous manufacturers who supply equipment for high-speed image processing [SNY-88].

The requirement for our present purpose is rather specific: a comprehensive and versatile tool-kit for image processing is essential. Several software packages for image processing, which have the desired characteristics, have been developed in the past. The list includes Susie [BAT-79, BAT-85a], Autoview [BAT-82] and VCS [VCS]. (There are many others.) Each of these offers an integrated set of software routines, which together implement a broad range of image processing operations. Any of these three languages would be suitable for our purpose of developing image processing within the framework of Prolog.[9] The VCS language, which we shall merge with Prolog, was inspired and influenced by Autoview, which in turn was descended from Susie. The present author was involved in the initial development of Susie and has been associated with the development of it, and its close relatives, Autoview and VCS, ever since. One reason why VCS was chosen as the basis for Prolog+ is that the author had the opportunity to work in close cooperation with the authors of the VCS package and could therefore continue to influence its development. Very recently, the author has begun to look at the interface between Prolog and the so-called *Intelligent Camera* (Image Inspection Ltd), which also owes its origins to Autoview. Susie and Autoview have together inspired many other languages.

Susie, Autoview and VCS are examples of interactive image processing languages. These have proved to be invaluable for prototyping machine vision systems, for use in both AVI and RV. The arguments in favour of an experimental try-it-and-see approach to the design/selection of image processing procedures have been rehearsed several times elsewhere [BAT-85a]. Like its progenitors, VCS was designed specifically for the role of prototyping machine vision systems. None of these interactive image processing languages was ever intended to be used in a factory-floor machine vision system, because until very recently, it was not possible to implement them at a speed and at a cost which would make them attractive in this role. This situation is now changing. There are hints in the literature that certain very flexible image processing systems of moderate cost are being developed. For example, one of the author's associates[10] has shown that it is possible to design a system, using standard image processing modules, that provides high-speed operation at a moderate cost. He predicts that it will be possible, within the very near future, to build a system which is more flexible than VCS, which operates at a speed of over

```
100   for i = 1 to 100                  % Begin outer loop
200   grb                               % Digitise an image
300   for j = 1 to 3                    % Begin inner loop
400   lpf                               % Low pass filter
500   next j                            % Terminate inner loop
600   gli min max                       % Max. and min. intensities
700   print "Maximum intensity =", max
800   next i                            % Terminate outer loop
900   stop                              % Program finished
```

Figure 1.5. VCS permits image processing commands to be incorporated into a Basic-like language. This program digitises a set of images, filters each one to reduce noise and then locates the point of highest intensity.

```
macro isophotes       % Define a macro called "isophotes"
lpf                   % Blur – low pass filter
lpf                   % More filtering ...
lpf                   % ... and still more of the same
sca 3                 % Reduce intensity to 3 bits (eight levels)
gra                   % Simple gradient operator
thr 1 255             % Threshold
endm                  % Complete definition of the macro
```

Figure 1.6. VCS macro, which draws intensity contours (isophotes).

```
isophotes :-          % Define the predicate "isophotes"
3·lpf,                % Low-pass filter, applied three times
sca(3),               % Reduce intensity to 3 bits (eight levels)
gra,                  % Simple gradient operator
thr(1, 255).          % Threshold
```

Figure 1.7. Prolog+ program. This program performs the same function as that in Figure 1.6.

100 times that of VCS (running on a Motorola 68000 processor) and which costs under £20,000.[11] Such a hardware system would, almost certainly, be fast enough and cheap enough to find application in a wide range of industrial inspection and robot vision tasks. Thus, there is the prospect that languages like VCS, which were originally intended for interactive prototyping of industrial vision systems, will be employed on the factory floor. Any improvements which can be made to these languages would have a benefit there and in the design laboratory.

At the time of writing, VCS provides almost 200 image processing operators. These can be written into Basic-like programs, or incorporated into macros. Figures 1.5 and 1.6 give examples of these two modes of programming. However, it is the image processing primitives, rather than the language control structures, which really concern us here. Of course, the command repertoire is increasing as new modules are added. On a simple count of primitives, VCS might not appear to be as attractive as some other languages. However, this can be a misleading comparison, since the primitive image processing operations in VCS are intended to be used as building blocks, for use in larger programs and macros. (This is the same role which they have within Prolog+, the language which we shall develop in this monograph. See Figure 1.7.) If we were to include the number of macros available in VCS, the comparison would be very much more favourable. Several hundred macros have been written to date. However, we emphasise that such numbers are not to be taken too seriously. Are there more valid sentences possible in VCS than in Image Processing Language X? Are there more valid sentences possible in French than in English? Are there more valid C programs than Pascal programs? A better basis for the comparison of image processing languages is their ability to express concepts and/ or algorithms and this is essentially non-quantitative in nature.

1.6 Outline of this Monograph

The next chapter reviews the basic ideas of image processing using VCS and Prolog. It should be possible for a reader who has not previously studied image processing to understand the programs in this monograph, since they are all annotated in detail. The review of Prolog is intended to refresh the memories of readers who have previously studied the language but who have forgotten its details. It is not practical in a book of this length to describe all of the features of this fascinating and powerful language. For this reason, readers who have no prior knowledge of Prolog should first consult one of the excellent tutorial books already available. Chapter 2 then describes how image processing predicates may be added to Prolog and we present some simple programs, in order to illustrate the nature and use of the new language, which is called Prolog+. The ability to program declaratively in Prolog+ is then discussed and the use of this facility is illustrated by showing how the language can be extended, how default values of parameters can be redefined and how Prolog-type operators can be created and used. The recognition of printed and hand-written letters is then considered. Finally, primitive predicates for controlling various electro-mechanical devices are described. These include an (X,Y,θ)-table, a pick-and-place arm, computer controlled lights (both on/off and continuously variable), camera, video multiplexer, lens and general I/O ports. We then show how higher level predicates for controlling external devices may be built up.

In Chapter 3, we discuss the operating environment for Prolog+. This makes use of the extensions to the core language offered by LPA MacProlog. The design and use of pull-down menus is an important feature of the environment. Command keys are also provided, in order to augment the pull-down menu system and provide short cuts for an experienced user of the Prolog+ system. The so-called interactive, or transparent, mode of operation allows the user to work interactively, thereby preserving continuity with Prolog+'s progenitors. The language is able to control a speech synthesiser, which is useful during both debugging and the training of new programmers. A simplified software interface (called VSP, Very Simple Prolog+) is then discussed. Using VSP, it is possible to interface Prolog to virtually any image processor which has a suitable command language. It is a trivial, if somewhat tedious, matter to write Prolog+ in terms of VSP. Additional AI facilities may be added to Prolog+, by making use of the language extensions offered by LPA MacProlog.

Chapter 4 discusses software extensions to Prolog+, in the form of high-level predicates which together form a Library of utilities and which are listed in Appendix II. Ever-higher levels of abstraction are being represented in the Library which is continually being expanded. It is in the Library that the expressional power of Prolog+ is manifest most clearly. The Library predicates are provided with a facility for self-documentation, which is also described. There is a section on the use of Prolog-style operators, to facilitate programming and another on the use of Definite Clause Grammars (DCGs) for natural language understanding. The use of DCGs for robot and lighting control is the next topic and the chapter concludes with the description of a grammar for describing picture.

In Chapter 5, we analyse some of the techniques which have been developed for analysing so-called "line drawings". These are not hand-drawn pictures, but are binary images that have been derived from a camera and whose elementary features are straight line segments. The chapter discusses how this type of image can be generated and how quantitative data about it can be acquired using Prolog+. Simple geometric figures can be recognised and tree-like structures can be identified and analysed. The next topic is the construction of fully connected figures from a set of disconnected linear segments. An alternative representation for edges is then presented. The chapter concludes with a section on curve parsing.

The control of a robot and lighting equipment form the twin subjects of Chapter 6. The calibration of a Robot Vision system is discussed and a program for converting from one coordinate system to the other is presented. A set of predicates for controlling a robotic device are then described and two demonstration programs are listed. A Robot Vision system might employ cameras on the manipulator arm, overhead, or at the side of the work cell. These present different calibration and control problems. The use of structured lighting (triangulation) to obtain range maps is then discussed. The chapter ends with a discussion of the possibilities for the automatic control of the illumination in a robot work cell, where complex objects can be examined from a variety of viewing positions.

Chapter 7 describes several very varied applications, including recognising playing card suits, stacking blocks, packing laminate objects of irregular shape, shape recognition by matching skeletons[12], searching a simple maze, automatic dissection of plantlets and learning from examples (so-called Teaching by Showing).

Knowledge-based inspection is the subject in Chapter 8. The simplest form that this can take is to follow a pre-defined sequence of moves and image processing steps. The next topic is generating search patterns for use whilst looking for specific image features. Rule-based inspection follows, after which we discuss the possible role of Prolog+ in declarative programming. A simple English description of a class of images is provided. This is then translated into a Prolog+ program. Images from the camera are then analysed to see whether or not they conform to the given description.

Chapter 9 concludes this brief discussion of Prolog+ with a look towards the future. Of special note is the exciting possibility of using an extended version Prolog+ to control *several* image processors which operate concurrently. Whilst this is a distinct possibility, it is likely that a revised form of Prolog, specifically intended to operate on a concurrent processor (a network of transputers), will prove to be even more useful.

There are four appendices: Appendix I presents a catalogue of the low-level image processing predicates (primitives) which are currently incorporated into Prolog+. Appendix II presents the Library of higher level operations, while Appendix III lists the VSP software. Appendix IV describes a system which uses a very small image processing system interfaced to Prolog. The image processing hardware is housed within the camera enclosure and the Prolog software is hosted on a portable computer. The whole Prolog+ system, including the sensor and video monitor, fits into a small suitcase.

Notes

1. Of course, blind people can be very intelligent but there are very many mental processes that are made much easier through the integration of vision and intelligence and which blind people find more difficult than sighted people. This is the point that we wish to make. We do not wish to imply that intelligence cannot exist without vision, since clearly it can.
2. This raises a semantic question as to what AI is. Many workers regard the answer as essentially the recursive definition that AI is the subject that uses so-called "AI techniques". In essence, this means that AI is the subject concerned with back-track programming, recursion, list processing, heursitics, rather than algorithms, etc. Lisp, Poplog, Prolog are regarded as essential tools for AI workers to master.
3. This is no less important for AVI than it is for RV.
4. It should be noted that it is possible to write a Lisp compiler in Prolog, and a Prolog compiler in Lisp, so there is perhaps less need to justify the choice at the outset. We prefer, instead, to begin by describing ProVision, which is a superset of Prolog, and then justify it by demonstrating its power. The author has not found any undue encumbrance as a result of selecting Prolog as the basis for Prolog+. On the contrary, his experience has more than fulfilled the expectations.
5. Core Prolog is commonly taken to be the version of the language described by Clocksin and Mellish [CLO-81]. It is also referred to as Edinburgh Prolog.
6. Logic Programming Associates Ltd, Royal Victoria Patriotic Building, Prince Consort Road, London, U.K.
7. The expert system tool-box which interfaces to MacProlog is called Flex.
8. The speech synthesis driver software was added by Mr Terry Gritton, 65 Nunes Road, Watsonville, California 95076, U.S.A.
9. In fact, Autoview and VCS have both been interfaced successfully to Prolog.
10. For reasons of confidentiality, his identity cannot be revealed.
11. About $30,000.
12. The same program can be applied to isolated, touching, overlapping and semi-flexible objects.

CHAPTER 2

FUNDAMENTALS OF PROLOG+

"Лучше один раз увидеть, чем сто раз услышать."
Russian proverb. Often paraphrased in English as:
"A picture is worth ten thousand words"

"Murphy's Law was not formulated by Murphy, but by another person with the same name."
Anonymous. In the spirit of D. Hofstadter's book *Metamagical Themas*

2.1 Basic Principles of Digital Image Processing

The key theoretical issues of image representation and manipulation are discussed in this section. The description of image processing techniques below provides an introduction to *industrial* image processing and does not purport to define the limits of our knowledge about this topic. There are many excellent textbooks on the theoretical basis of image processing, covering a wider range of applications and some of these have been listed in the previous chapter.

There is an important philosophical point that must be discussed first. Complexity and variety can be achieved by using combinations of a very limited number of items. An outstanding example of this is to be found in nature. DNA achieves tremendous variety and provides the basis for building organisms of enormous complexity, even though it contains only four "symbols" in its "alphabet". Strings of the same four amino acids are able to code the protein structure of *Homo sapiens* and all of the other animals, plants and protists. In a similar way, ordinary arithmetic contains only a few basic types of operation, namely addition, subtraction, multiplication and division, yet we would not describe arithmetic as being unduly limiting. Theoretical studies of abstract "computers" have resulted in the realisation that only three types of operation are required to perform any finite computational task. Of course, it is convenient to build computers which have a much larger instruction repertoire than this. Turing machines provide yet another example, although they cannot really claim to be practical computing devices.

In contrast to these situations in which there is a small "alphabet", the VCS image processing language contains a relatively rich repertoire of commands. (There are currently about 200 primitive commands in VCS.) However, VCS permits the construction of some very complex image processing ("high-level") operations, using strings of commands drawn from its basic "alphabet". It is important to understand this point for three reasons:

(a) Many of the basic image processing operations defined in this chapter and in the remainder of this monograph do not, when considered alone, seem to provide any useful function. However, when they are combined with a number of other simple operators, they form part of a very versatile language, in which it is possible to express a large number of complex and useful image processing operations.
(b) The second point relates to human pride! Many people admire complexity for its own sake and many individuals, particularly academics, do not admire a treatise if it does not achieve a certain level of complexity. When this section is scrutinised using such a criterion, there is no doubt that much of it is trivial in content. Yet, it is the interaction between these simple image processing operators that is so very complex. It is only when we reach the later chapters that this point becomes apparent. Many of the algorithms and heuristic procedures discussed there are of considerable complexity, even though they are expressed in terms of simple operators.
(c) Certain image processing operations, can be implemented in VCS but in a round-about way. If such a situation exists and the resulting method is slow in operation, then it may well be worthwhile adding a new operator to the VCS command repertoire. In fact, this has been done many times.

We may summarise thus: the reader is warned against adopting the attitude that the basic commands of VCS are trivial. The interactions between these seemingly simple "atoms" are both subtle and complex.

Digital Representations of Images

Digital image processing necessarily involves the study of both the storage and manipulation of pictures inside a computer, or special purpose electronic hardware. It is important that we interest ourselves in image storage, because the ability to process data efficiently and purposefully depends upon the suitability of the encoding used to represent the original optical pattern. Images may arrive inside a computer in a variety of forms, depending upon the type of sensor (i.e. camera) and coding hardware used to compact the data. We shall discuss the following types of images in this chapter:

(a) Monochrome (multi-level grey-scale) images
(b) Binary images in which only two brightness levels are allowed

(c) Colour images, which can be represented by three separate monochrome component images
(d) Multi-spectral images, which require more than three monochrome images to represent them
(e) Stereoscopic image pairs, such as those arising from binocular vision
(f) Moving images (strictly, image sequences)
(g) A generalised image, combining (d), (e) and (f)

There are many different forms of representation for each of these image types. However, it is not our intention to provide a comprehensive review of image coding techniques. Instead, we shall concentrate upon those methods which are necessary to understand the rest of this book. VCS permits only one type of image representation. This seemingly severe constraint is imposed in order to achieve the greatest flexibility and is not, in fact, unduly restrictive. It does, however, sometimes result in the need to use a lot more data storage than would be required by a more compact image representation. Occasionally, sub-optimal image coding results in a procedure being slow in execution. The software which is used to implement VCS may use a variety of data forms to represent images. However, each of the VCS image processing operators is subject to the requirement that both its input and output images are of the same basic type. Inside the operator, however, other data types may be used for efficiency's sake. Communication between these operators uses a standard image form.

Monochrome (Grey-Scale) Images

Let i and j denote two integers where

$$1 \leq i \leq m$$

and

$$1 \leq j \leq n$$

In addition, let f(i,j) denote an integer function such that

$$0 \leq f(i,j) \leq W$$

The array F, where

$$F = \begin{bmatrix} f(1,1), f(1,2), \ldots\ldots\ldots\ldots, f(1,n) \\ f(2,1), f(2,2), \ldots\ldots\ldots\ldots, f(2,n) \\ \ldots\ldots\ldots\ldots\ldots\ldots\ldots\ldots\ldots\ldots\ldots \\ \\ f(m,1), f(m,2), \ldots\ldots\ldots, f(m,n) \end{bmatrix}$$

will be called a *digital image*. An address (i,j) defines a position in F, called a *pixel*, *pel* or *picture element*. The elements of F denote the intensities within a number of small rectangular regions within a real (i.e. optical) image (see Figure 2.1). Strictly

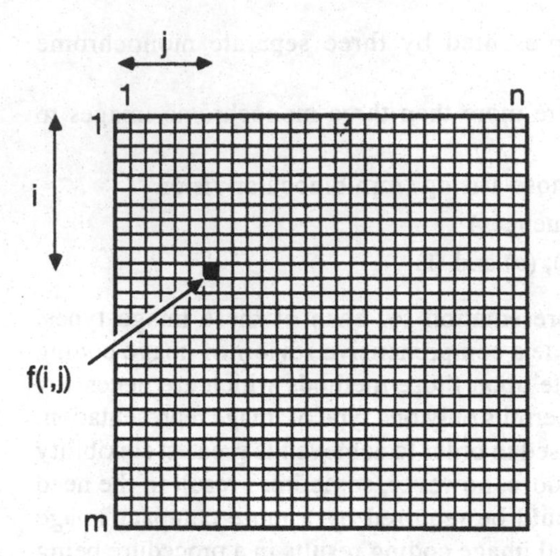

Figure 2.1. A digital image consisting of an array of m × n pixels. The pixel in the ith row and the jth column has an intensity equal to f(i,j).

speaking, f(i,j) measures the intensity at a single point but if the corresponding rectangular region is small enough, the approximation will be accurate enough for most purposes. The array F contains a total of m,n elements and this product is called the *spatial resolution* of F. We shall arbitrarily assign intensities according to the following scheme:

$f(i,j) = 0$	black
$0 < f(i,j) \leq 0.33W$	dark grey
$0.33W < f(i,j) \leq 0.67W$	mid-grey
$0.67W < f(i,j) < W$	light grey
$f(i,j) = W$	white

This simple coding scheme is used throughout this book, both for grey-scale and binary images. More will be said later about binary images, but let us first consider how much data is required to represent a grey-scale image in this form. Each pixel requires the storage of

$\log_2(1+W)$ bits

This assumes that (1+W) is an integer power of two. If it is not, then $\log_2(1+W)$ must be rounded up to the next integer. This can be represented using the ceiling function, $\lceil \ldots \rceil$. Thus, an image requires the storage of

$\lceil \log_2(1+W) \rceil$ bits/pixel

Since there are $m \times n$ pixels, the total data storage for the entire digital image F is equal to

$$m.n.\lceil \log_2(1+W) \rceil \text{ bits}$$

In VCS, the white level is usually represented by $W = 255$, thereby requiring the storage of

m.n bytes/image

If $m = n \geq 128$, and $W \geq 64$, we can obtain a good image of a human face. Nearly all of the images in this book use $m = n = 512$ and $W = 255$.

Binary Images

A binary image is one in which only two intensity levels are permitted. At first sight, the obvious way to represent such an image is to follow the convention defined above and simply set *W* equal to 1. This coding scheme requires a total of

m.n bits/image.

In VCS, however, it is customary to use the less efficient but much more useful scheme below:

$f(i,j) = 0$	black
$1 \leq f(i,j) \leq W$	white

Despite the fact that this is a redundant coding scheme, it does permit a lot of operations to be performed which would not otherwise be possible, if we separated binary from grey-scale images. We can apply grey-scale operators to binary images, in order to obtain some interesting and useful effects.

Colour and Multi-Spectral Images

Colour printing uses *four* imprints: black, red, yellow and blue. Notice that an impression of colour is conveyed to the eye by superimposing four separate images. Ciné film operates in a similar way, except that when different colours of light, rather than ink, are added together, *three* components are needed: red, green and blue. Television operates in the same way as film; the signal from a colour television camera may be represented using three components

$R = \{r(i,j)\}$
$G = \{g(i,j)\}$
$B = \{b(i,j)\}$

where R, G and B are defined in the same way as F above. The vector (r(i,j), g(i,j), b(i,j)) defines the intensity and colour at the point (i,j) in the colour image. Using this idea, VCS can represent and manipulate colour images without difficulty.

There is no reason, however, for us to be restricted to using only three images. For example, certain satellites capture pictures of the earth's surface using up to 11 image components. This is akin to having eyes with 11 colour sensors. Multi-spectral images can be represented using several monochrome images. The total amount of data required to code a colour image with r components is equal to

$m.n.r.\lceil \log_2(1+W) \rceil$ bits

Here, W is simply the maximum signal level on each of the channels, whereas white is represented by a colour vector

$(X_1, X_2,..., X_r)$, where all $X_i = W$.

Stereoscopic Image Pairs

Binocular vision is used by *Homo sapiens* and the Great Apes, owls and certain other animals to obtain range information about the scene being viewed. Algorithmic techniques are known which can derive range information from a pair of images, but they are computationally expensive. For this reason, we shall not consider stereo pairs further, except to state that the amount of data required to represent a (multi-spectral) stereo-image pair is

$2.m.n.r.\lceil \log_2(1+W) \rceil$ bits

Moving Images

Once again, we shall refer to ciné film and television in order to explain how moving scenes may be represented in digital form. A ciné film is, in effect, a time-sampled representation of the original moving scene. Each frame in the film is a standard colour, or monochrome image, and can be coded as such. Thus, a monochrome ciné film may be represented digitally as a sequence of two-dimensional arrays

$[F_1, F_2, F_3, F_4,...]$

Each F_i is an $m \times n$ array of integers as we defined above, when discussing the coding of grey-scale images. If the film is in colour, then each of the F_i has three components. In the general case, when we have a sequence of r-component colour images to code, we require

$m.n.p.r.\lceil \log_2(1+W) \rceil$ bits per image sequence

where

(a) The spatial resolution is $m \times n$ pixels
(b) The intensity scale along each spectral channel permits (1+W) levels
(c) There are r spectral channels
(d) p is the total number of "stills" in the image sequence

VCS can accommodate such image sequences. However, as far as both storage and processing are concerned, there may be practical problems, in view of the large amounts of data that moving colour imagery generates. In order to demonstrate this, let us use the formula just given to calculate the bandwidth required to code a standard broadcast quality television picture. We shall take the variables to have the following values:

m = n = 512 pixels

p = 3 channels

r = 25 images/second

W = 16

Then, a total of 78.6 Mbits/second is required. The human visual system can accommodate data at a very much higher rate than this, as the following calculations show. With

m = 5000

n = 5000

p = 3

r = 50

W = 255

we obtain a crude estimate of the bandwidth of the human visual system equal to 10^{10} bits/second.

We must emphasise that we have considered only those image representations which are relevant to the operation of VCS. There are many possible methods of coding images, but our brief review of image processing ideas in this chapter and VCS are both based upon these ideas.

Image Processing Functions

There are several types of image processing functions:

(a) Array–array mappings
(b) Array–vector and array–scalar mappings
(c) Vector–array and scalar–array mappings

Of these, (c) includes the image generation routines familiar in CAD and graphics, while we shall be concerned principally with type (a) and (b) functions. Array–vector and array–scalar mappings are data reduction techniques, and are used in deriving measurements from images. Many examples of these are to be found in VCS, although most of our attention in this chapter will be directed towards array–

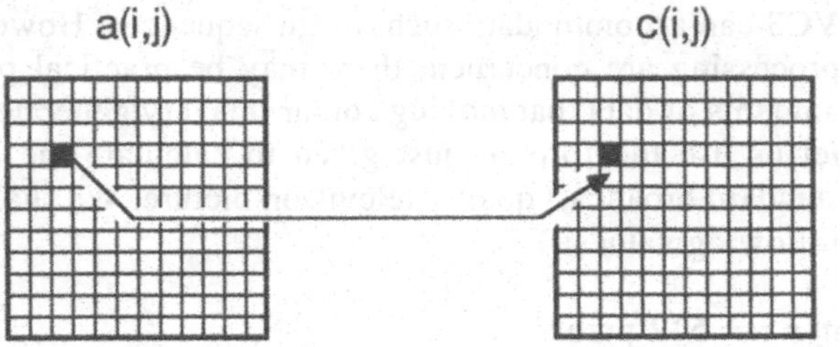

a(i,j) c(i,j)

Figure 2.2. Monadic point-by-point operator. The (i,j)th pixel in the input image has intensity a(i,j). This value is used to calculate the intensity, c(i,j), for the intensity of the corresponding pixel in the output image.

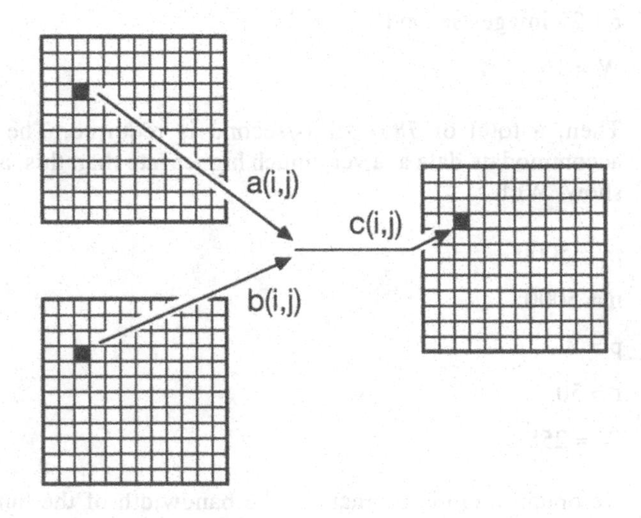

Figure 2.3. Dyadic point-by-point operator. The intensities of the (i,j)th pixels in the two input images (i.e. a(i,j) and b(i,j)) are combined to calculate the intensity, c(i,j), at the corresponding address in the output image.

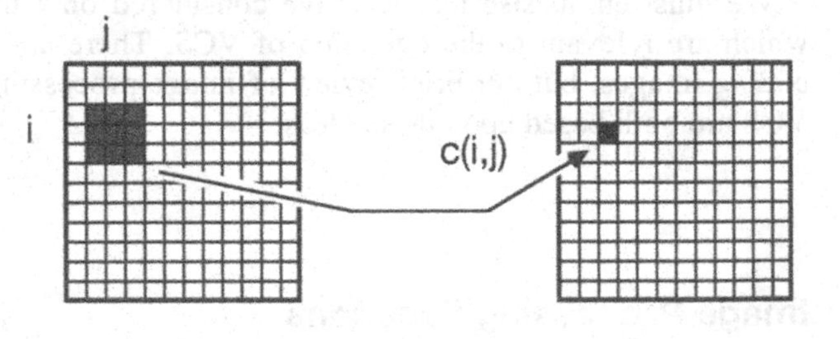

Figure 2.4. Local operator. In this instance, the intensities of nine pixels arranged in a 3 × 3 window are combined together. Local operators may be defined which uses other, possibly larger windows. The window may, or may not, be square and the calculation may involve linear or non-linear processes.

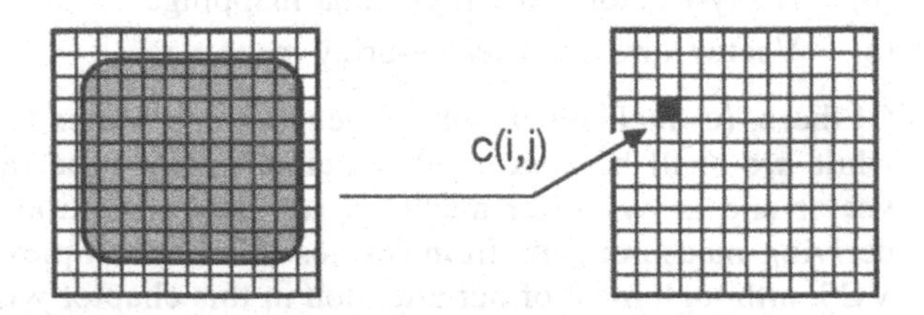

Figure 2.5. Global operators use a large proportion of the image, possibly all of it, in order to calculate just one intensity value.

array mappings. Such a function may be represented by the form of a *Mapping Function:*

Input Image(s) ⟶ Output Image

We shall find it convenient to divide array–array mappings according to the following schedule:

(i) *Monadic point-by-point operators*, in which there is only one input image and the intensity at each point in the output image depends upon that at only one point in the input image (Figure 2.2).

(ii) *Dyadic point-by-point operators*, in which the intensity at a certain point in the output image depends only upon the intensities at the corresponding points within the two input images (Figure 2.3).

(iii) *Local operators*, in which the intensity at one point in the output image depends upon the intensities at a compact, clustered group of points in the input image. There is only one input image (Figure 2.4).

(iv) *Global operators*, in which the intensity at a given point in the output image depends upon the intensities at a large number of points in the single input image (Figure 2.5).

(v) *Data dependent operations*, in which the function which is applied to an image is determined in part by the properties of the input image. It is common to find the data dependent operators calculating one or a few parameters from the given image and then using the value(s) so obtained in another procedure which then modifies the input image.

(vi) *Image transformations*, which result in some translation, rotation or warping of the input image.

We shall now consider each of these in turn, beginning with the grey-scale images and later turning our attention to binary images. We shall frequently indicate the equivalent VCS operators to the mathematical operations described in what follows and will enclose the names of any VCS commands inside square brackets. In certain cases, sequences of VCS commands are needed to perform an operation and these are similarly listed.

Notation

The following notation will be used in this chapter and in the mathematical descriptions of the basic functions listed in Appendix I.

(1) i and j are address variables and lie within the ranges

$$1 \leq i \leq m$$
$$1 \leq j \leq m$$

Notice that we shall use i to define the vertical position and j the horizontal position in the images (Figure 2.1).

(2) A = {a(i,j)}, B = {b(i,j)} and C = {c(i,j)}
(3) W denotes the white level
(4) g(X) is a function of a single independent variable X
(5) h(X,Y) is a function of two independent variables X and Y
(6) The assignment operator

$\longleftarrow$

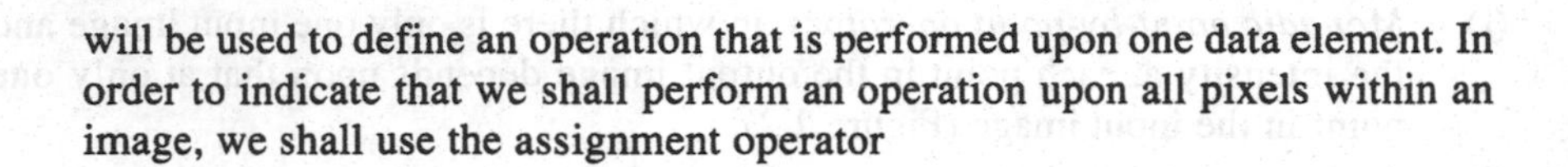
will be used to define an operation that is performed upon one data element. In order to indicate that we shall perform an operation upon all pixels within an image, we shall use the assignment operator

7. k, k1, k2, k3 are constants
8. N(i,j) is that set of pixels arranged around the pixel (i,j) in the following way:

(i–1, j–1)	(i–1, j)	(i–1, j+1)
(i, j–1)	(i, j)	(i, j+1)
(i+1, j–1)	(i+1, j)	(i+1, j+1)

Notice that N(i,j) forms a 3 × 3 set of pixels and is referred to as the 3 × 3 neighbourhood of (i,j). In order to simplify some of the definitions, we shall refer to the intensities of these pixels using the following notation:

A	B	C
D	E	F
G	H	I

(Ambiguities over the dual use of A, B and C should not be troublesome, as the context will make it clear which meaning is intended.)

The points

{(i–1, j–1), (i–1, j), (i–1, j+1), (i, j–1), (i,j+1), (i+1, j–1), (i+1, j), (i+1, j+1)}

are called the *8-neighbours* of (i,j) and are also said to be *8-connected* to (i,j). The points

{(i–1, j), (i, j–1), (i,j+1), (i+1, j)}

are called the *4-neighbours* of (i,j) and are said to be *4-connected* to (i,j).

Monadic Point-by-Point Operators

Monadic point-by-point operators have a characteristic equation of the form:

$$c(i,j) \Longleftarrow g(a(i,j))$$

or

$$E \Longleftarrow g(E)$$

Such an operation is performed for all (i,j) in the range [1,m] × [1,n] (see Figure 2.2).

Intensity Shift ***[acn]***

$$c(i,j) \Longleftarrow \begin{cases} 0, & 0 > a(i,j) + k \\ a(i,j)+k & 0 \leq a(i,j) + k \leq W \\ W & W < a(i,j) + k \end{cases}$$

k is a constant, set by the system user. Notice that this definition was carefully designed to maintain c(i,j) within the same range as the input, viz. [0,W]. This is an example of a process which we shall call *intensity normalisation* and is important because it permits the result of this operation to be processed using any of the other procedures. Normalisation will be used frequently throughout this chapter.

Intensity Multiply ***[mcn]***

$$c(i,j) \Longleftarrow \begin{cases} 0, & a(i,j).k < 0 \\ a(i,j).k & 0 \leq a(i,j).k \leq W \\ W & W < a(i,j).k \end{cases}$$

Once again notice the presence of normalisation within the mapping formula.

Logarithm ***[log]***

$$c(i,j) \Longleftarrow \begin{cases} 0 & a(i,j) = 0 \\ W.\log(a(i,j))/\log(W) & \text{otherwise} \end{cases}$$

This definition *arbitrarily* replaces the infinite value of log(0) by zero, and thereby avoids a difficult rescaling problem.

Antilogarithm (exponential) ***[exp]***

$$c(i,j) \Longleftarrow W.\exp(a(i,j))/\exp(W)$$

Negate ***[neg]***

$$c(i,j) \Longleftarrow W - a(i,j)$$

Gamma Correction. Gamma correction takes its name from a mathematically similar process which is used in both television and photographic work.

$$c(i,j) \Longleftarrow [a(i,j)]^n/W^{n-1}$$

For small integer values of N, Gamma correction can be performed using a combination of the VCS commands **mul** and **sqr**. Transformations of this type can, of course, be performed very easily using a look-up table.

tra is able to perform a general intensity transformation by using a look-up table.

*Threshold [**thr**]*

$$c(i,j) \Longleftarrow \begin{cases} W & k1 \le a(i,j) \le k2 \\ 0 & \text{otherwise} \end{cases}$$

*Highlight [**hil**]*

$$c(i,j) \Longleftarrow \begin{cases} k3 & k1 \le a(i,j) \le k2 \\ a(i,j) & \text{otherwise} \end{cases}$$

*Squaring [**sqr**]*

$$c(i,j) \Longleftarrow [\, a(i,j)\,]^2/W$$

*Modulus [**abs**]*

$$c(i,j) \Longleftarrow 2 \cdot |\, a(i,j) - W/2\, |$$

Dyadic Point-by-Point Operators

Dyadic point-by-point operators have a characteristic equation of the form:

$$c(i,j) \Longleftarrow h(a(i,j),b(i,j))$$

Notice that there are two input images,

$A = \{a(i,j)\}$

and

$B = \{b(i,j)\}$

while the output image is

$C = \{c(i,j)\}$

It is important to realise that c(i,j) depends upon only a(i,j) and b(i,j) (see Figure 2.3).

*Add [**add**]*

$$c(i,j) \Longleftarrow (a(i,j) + b(i,j))/2$$

Notice the presence of normalisation in this equation.

Subtract [sub]

$$c(i,j) \Longleftarrow (a(i,j) - b(i,j))/2 + W/2$$

Multiply [mul]

$$c(i,j) \Longleftarrow a(i,j).b(i,j)/W$$

Maximum [max]

$$c(i,j) \Longleftarrow \mathbf{MAX}(a(i,j), b(i,j))$$

When this operator is applied to a pair of binary images, the *union* (OR function) of their white areas is computed. This function may also be used to *superimpose* white writing onto a grey-scale image.

Minimum [min]

$$c(i,j) \Longleftarrow \mathbf{MIN}(a(i,j), b(i,j))$$

When A and B are both binary, the *intersection* (AND function) of their white areas is calculated.

Local Operators

Figure 2.4 illustrates the principle of operation of the so-called local operators. Notice that the intensities of several pixels are combined together in order to calculate the intensity of just one pixel. Amongst the simplest of the local operators are those which use a set of nine pixels arranged in a 3 × 3 square. These have a characteristic equation of the following form:

$$c(i,j) \Longleftarrow g(a(i-1,j-1), a(i-1,j), a(i-1,j+1), a(i,j-1), a(i,j), a(i,j+1), a(i+1,j-1), a(i+1,j), a(i+1,j+1))$$

where g(.) is a function of nine variables. This is an example of a local operator which uses a *3 × 3 processing window*. (That is, it computes the value for one pixel on the basis of the intensities within a region containing 3 × 3 pixels. Other local operators employ larger windows and we shall discuss these briefly later.) In the simplified notation which we introduced earlier, the above definition reduces to:

$$E \Longleftarrow g(A, B, C, D, E, F, G, H, I)$$

Linear Local Operators

An important sub-set of the local operators is that group which performs a linear weighted sum, and which are therefore known as *linear local operators*. For this group, the characteristic equation is:

$$E \Longleftarrow k1.(A.W1 + B.W2 + C.W3 + D.W4 + E.W5 + F.W6 + G.W7 + H.W8 + I.W9) + k2$$

where W1, W2,..., W9 are weights, which may be positive, negative or zero. Values for the normalisation constants, k1 and k2 are given later. The matrix

$$\begin{bmatrix} W1 & W2 & W3 \\ W4 & W5 & W6 \\ W7 & W8 & W9 \end{bmatrix}$$

is termed the *weight matrix* and is important, because it determines the properties of the linear local operator. The following rules summarise the properties of this type of operator:

(i) If all weights are either positive or zero, the operator will *blur* the input image. Blurring is referred to as *low-pass filtering*. Subtracting a blurred image from the original results in a highlighting of those points where the intensity is changing rapidly and is termed *high-pass filtering*. (Applying an electrical filter to a video signal produces similar effects to those obtained from the digital image.)

(ii) If W1 = W2 = W3 = W7 = W8 = W9 = 0, and W4, W5, W6 ≥ 0, then the operator blurs along the rows of the image; horizontal features, such as edges and streaks, are not affected.

(iii) If W1 = W4 = W7 = W3 = W6 = W9 =0, and W2, W5, W8 ≥ 0, then the operator blurs along the columns of the image; vertical features are not affected.

(iv) If W2 = W3 = W4 = W6 = W7 = W8 = 0, and W1, W5, W9 ≥ 0, then the operator blurs along the diagonal (top-left to bottom-right). There is no smearing along the orthogonal diagonal.

(v) If the weight matrix can be reduced to a matrix product of the form P × Q, where

$$P = \begin{bmatrix} 0 & 0 & 0 \\ V4 & V5 & V6 \\ 0 & 0 & 0 \end{bmatrix}$$

and

$$Q = \begin{bmatrix} 0 & V1 & 0 \\ 0 & V2 & 0 \\ 0 & V3 & 0 \end{bmatrix}$$

the operator is said to be of the "separable" type. The importance of this is that it is possible to apply two simpler operators in succession: with weight matrices P and Q, in order to obtain the same effect as that produced by the separable operator.

(vi) The successive application of linear local operators which use windows containing 3 × 3 pixels produces the same results as linear local operators with larger windows. For example, applying that operator which uses the following weight matrix:

$$\begin{bmatrix} 1 & 1 & 1 \\ 1 & 1 & 1 \\ 1 & 1 & 1 \end{bmatrix}$$

twice in succession results in the same image as that obtained from the 5 × 5 operator with the following weight matrix:

$$\begin{bmatrix} 1 & 2 & 3 & 2 & 1 \\ 2 & 4 & 6 & 4 & 2 \\ 3 & 6 & 9 & 6 & 3 \\ 2 & 4 & 6 & 4 & 2 \\ 1 & 2 & 3 & 2 & 1 \end{bmatrix}$$

Applying the same 3 × 3 operator, twice is equivalent to using the following 7 × 7 operator:

$$\begin{bmatrix} 1 & 3 & 6 & 7 & 6 & 3 & 1 \\ 3 & 9 & 18 & 21 & 18 & 9 & 3 \\ 6 & 18 & 36 & 42 & 36 & 18 & 6 \\ 7 & 21 & 42 & 49 & 42 & 21 & 7 \\ 6 & 18 & 36 & 42 & 36 & 18 & 6 \\ 3 & 9 & 18 & 21 & 18 & 9 & 3 \\ 1 & 3 & 6 & 7 & 6 & 3 & 1 \end{bmatrix}$$

(Notice that all of these operators are separable. Hence it would be possible to replace the last-mentioned 7 × 7 operator with four simpler operators: 3 × 1, 3 × 1, 1 × 3 and 1 × 3, applied in any order.) It is not always possible to replace a large-window operator with a small number of 3 × 3 operators. This becomes obvious when one considers, for example, that a 7 × 7 operator uses 49 weights and that three 3 × 3 operators provide only 27 degrees of freedom. Separation is often possible, however, when the larger operator has a weight matrix with some redundancy, for example when it is separable or symmetrical.

(vii) In order to perform normalisation, the following values are used for k1 and k2:

$$k1 \longleftarrow 1/\sum_{p,q} |W_{p,q}|$$

$$k2 = [1 - \sum_{p,q} W_{p,q} / \sum_{p,q} |W_{p,q}|].W/2$$

(viii) VCS commands **con, lpf, hpf, lap, hge, vgr** and **raf** implement local operators.

(ix) The command **raf(11,11)** performs a local averaging function over an 11 × 11 window. The weight matrix is

$$\begin{bmatrix} 1,1,1,1,1,1,1,1,1,1,1 \\ 1,1,1,1,1,1,1,1,1,1,1 \\ 1,1,1,1,1,1,1,1,1,1,1 \\ 1,1,1,1,1,1,1,1,1,1,1 \\ 1,1,1,1,1,1,1,1,1,1,1 \\ 1,1,1,1,1,1,1,1,1,1,1 \\ 1,1,1,1,1,1,1,1,1,1,1 \\ 1,1,1,1,1,1,1,1,1,1,1 \\ 1,1,1,1,1,1,1,1,1,1,1 \\ 1,1,1,1,1,1,1,1,1,1,1 \\ 1,1,1,1,1,1,1,1,1,1,1 \end{bmatrix}$$

This produces quite a severe two-directional blurring effect. Subtracting the effects of a blurring operation from the original image generates a picture in which spots, streaks and intensity steps are all emphasised. On the other hand, large areas of constant or slowly changing intensity become uniformly grey. This process is called *high-pass filtering* .

Non-Linear Local Operators

There are many useful non-linear local operators, the principle ones being those listed below. Notice that some are simple combinations of linear local operators:

*Largest Intensity Neighbourhood Function [**lnb**]*

E ⟸ **MAX** (A, B, C, D, E, F, G, H, I)

This operator has the effect of spreading bright regions and contracting dark ones.

*Edge Detector [VCS Command Sequence: **lnb,sub**]*

E ⟸ **MAX** (A, B, C, D, E, F, G, H, I) – E

This operator is able to highlight edges (i.e. points where the intensity is changing rapidly).

*Median Filter [**mdf(5)**]*

E ⟸ **FIFTH_LARGEST** (A, B, C, D, E, F, G, H, I)

This filter is particularly useful for reducing the level of noise in an image. (Noise arises in all types of camera and can be a nuisance if it is not eliminated by hardware or software filtering.)

Crack Detector. This operator is equivalent to applying the following sequence of operations:

Largest intensity neighbourhood function **[lnb]**
Largest intensity neighbourhood function **[lnb]**
Negate **[neg]**
Largest intensity neighbourhood function **[lnb]**
Largest intensity neighbourhood function **[lnb]**
Negate **[neg]**

and then subtracting the result from the original image (Recall that **lnb** is the largest intensity neighbourhood function and **neg** performs negation.) Trying to express such an operator in formal mathematical terms is not particularly helpful. The crack detection operator is able to detect thin *dark* streaks and small dark spots in a grey-scale image; it ignores other features, such as bright spots and streaks, edges (intensity steps) and broad dark streaks.

*Sobel Edge Detector [**sed**]*

$$c(i,j) \Longleftarrow \sqrt{[(A + 2.B + C) - (G + 2.H + I)]} + \sqrt{[(A + 2.D + G) - (C + 2.F + I)]}$$

This is one of the most popular edge detection methods, in view of its theoretical properties. The following approximation is simpler to implement in software and hardware and, fact defines the VCS operator **sed**.

$$c(i,j) \Longleftarrow \{|(A + 2.B + C) - (G + 2.H + I)| + |(A + 2.D + G) - (C + 2.F + I)|\} /6$$

*Roberts Edge Detector [**red**]*

$$c(i,j) \Longleftarrow (|A - I| + |C - G|)/2$$

*Rank Filters [**mdf**, **rid**].* The rank filters are apparently related to the linear local operators but differ in one important point. Let

A' = **LARGEST**(A, B, D, C, D, E, F, G, H, I)

B' = **SECOND_LARGEST**(A, B, C, D, E, F, G, H, I)

C' = **THIRD_LARGEST**(A, B, C, D, E, F, G, H, I)

.
.
.

I' = **NINTH_LARGEST**(A, B, C, D, E, F, G, H, I)

The generalised 3 × 3 rank filter is then obtained using the formula:

$$c(i,j) \Longleftarrow k1.(A'.W1 + B'.W2 + C'.W3 + D'.W4 + E'.W5 + F'.W6 + G'.W7 + H'.W8 + I'.W9) + k1$$

The normalisation constants, k1 and k2, have the following values:

$$k1 \Longleftarrow 1/\sum_{p,q} |W_{p,q}|$$

and

$$k2 = [1 - \sum_{p,q} W_{p,q} / \sum_{p,q} |W_{p,q}|\,].W/2$$

The rank filters are quite versatile as the following examples show.

Edge Detection

(W1, W2, W3, W4, W5, W6, W7, W8, W9) = (4, 3, 2, 1, 0, –1, –2, –3, –4)

Noise Reduction

(W1, W2, W3, W4, W5, W6, W7, W8, W9) = (0, 1, 2, 3, 4, 3, 2, 1, 0)

Edge Sharpening

(W1, W2, W3, W4, W5, W6, W7, W8, W9) = (–4, –3, –2, –1, 20, –1, –2, –3, –4)

Image Enhancement

(W1, W2, W3, W4, W5, W6, W7, W8, W9) = (0, –1, –2, –3, 7, –3, –2, –1, 0)

Using Direction Codes (lgr, dbn, rid)

A function **DIRECTION_CODE** is defined thus:

$$\textbf{DIRECTION\_CODE}(A, B, C, D, F, G, H, I) \Longleftarrow \begin{cases} 1, \text{ if } A \geq \textbf{MAX}\ (B, C, D, F, G, H, I) \\ 2, \text{ if } B \geq \textbf{MAX}\ (A, C, D, F, G, H, I) \\ 3, \text{ if } C \geq \textbf{MAX}\ (A, B, D, F, G, H, I) \\ 4, \text{ if } D \geq \textbf{MAX}\ (A, B, C, F, G, H, I) \\ 5, \text{ if } F \geq \textbf{MAX}\ (A, B, C, D, G, H, I) \\ 6, \text{ if } G \geq \textbf{MAX}\ (A, B, C, D, F, H, I) \\ 7, \text{ if } H \geq \textbf{MAX}\ (A, B, C, D, F, G, I) \\ 8, \text{ if } I \geq \textbf{MAX}\ (A, B, C, D, F, G, H) \end{cases}$$

Using this definition, an operator [**dbn**] may be defined thus

E $\Longleftarrow$ **DIRECTION_CODE**(A, B, C, D, F, G, H, I)

The value of this function is that it can detect the *direction* of the intensity gradient; a *ridge-shaped peak* in the intensity function of the input image, results in a step-like intensity change in the output image.

N-Tuple Operators

The N-tuple operators are closely related to the local operators and have a large number of (linear and non-linear) variations. N-tuple operators may be regarded as

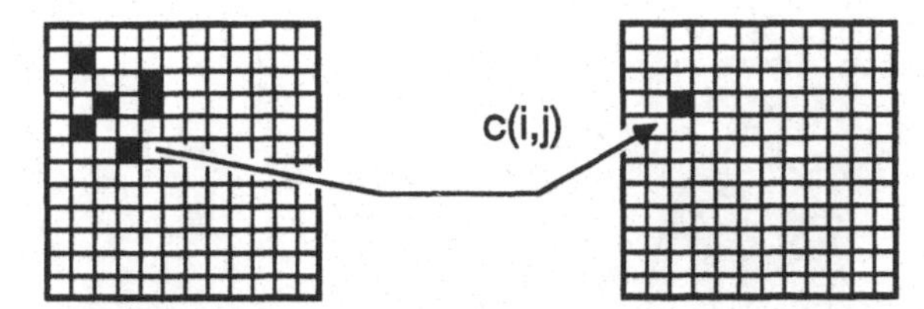

Figure 2.6. An N-tuple filter operates much like a local operator. The only difference is that the pixels whose intensities are combined together do not form a compact set. A linear N-tuple filter can be regarded as being equivalent to a local operator which uses a large window and in which many of the weights are zero.

generalised versions of local operators. In order to understand the N-tuple operators, let us first consider a *linear* local operator which uses a large processing window, (say $r \times s$ pixels) with most of its weights equal to zero. Only N of the weights are non-zero, where

$$N \ll r.s$$

This is an N-tuple filter. The N-tuple filters are usually designed to detect specific patterns (see Figure 2.6). In this role, they are able to locate a simple feature, such as a corner, annulus, letter "E", in any position. However, they are sensitive to changes of orientation and scale. An N-tuple can be regarded as a sloppy template, which is convolved with the input image.

Non-linear N-tuple operators may be defined in a fairly obvious way. For example, we may define operators which computed the maximum, minimum or median values of the intensities of the N pixels covered by the N-tuple. An important class of such functions is that containing the so-called morphological operators.

Morphological Operators [dil, ero]

Figure 2.7 illustrates the basic principle of a one-dimensional morphological operator. The kernel function is called a structuring element. One of the simplest morphological operators, with structuring element E(x) applied to an intensity function F(x), produces an output G(x) calculated thus:

$$\mathbf{MAX}_{y,z} \left(F(x-y) - E(x-z) + G(x)\right) = 0$$

Another operator can be defined with exactly the same definition, except that the **MIN** function replaces **MAX**. Two-dimensional morphological operators can also be defined.

The crack detector, whose VCS operator sequence is given above, is an example of a more complex morphological operator in which the structuring element, E(x), has the form of a pulse.[1]

Edge Effects

All local operators and N-tuple filters are susceptible to producing peculiar effects around the edges of an image. The reason is simply that in order to calculate the

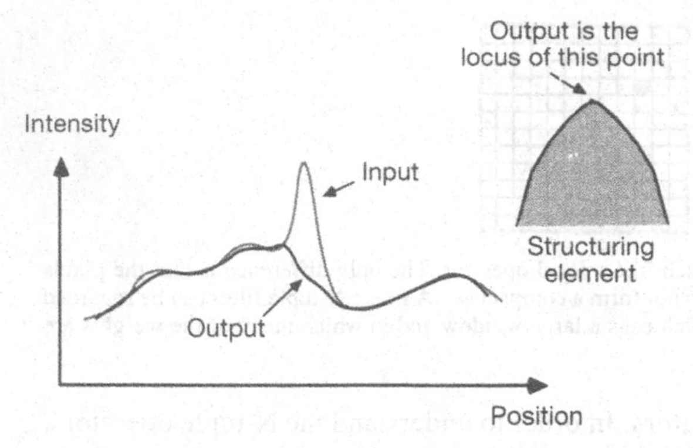

Figure 2.7. A one-dimensional morphological filter, operating on an analogue signal (equivalent to a grey-scale image). The input signal is represented by the grey curve and the output by the black curve. In this simple example, the structuring element has an approximately parabolic form. In order to calculate a value for the output signal, the structuring element is pushed upwards, from below the input curve. The height of the top of the structuring element is noted. This process is then repeated, by sliding the structuring element sideways. Notice how this particular operator smooths the intensity peak but follows the input signal quite accurately everywhere else. Subtracting the output signal from the input would produce a result in which the intensity peak is emphasised and all other variations would be eliminated. A similar two-dimensional morphological operator might use a structuring element which has a conical or bell-shaped form.

intensity of a point near the edge of an image we require information about pixels outside the image, which, of course, are simply not present. In order to make some attempt at calculating values for the edge pixels, it is necessary to make some assumptions, for example that all points outside the image are black, or have the same values as the border pixels. The strategy, whatever one we adopt, is perfectly arbitrary and there will be occasions when the edge effects are so pronounced that there is nothing that we can do but to remove them by masking [**edg**]. Edge effects are important because they require us to make special provisions for them when we try to patch several low-resolution images together.

Global Image Transforms

An important class of image processing operators is characterised by an equation of the form

$$B \longleftarrow f(A)$$

where A = {a(i,j)} and B = {b(p,q)}. Each element in the output picture, B, is calculated using all, or at least a large proportion of, the pixels in A. The output image, B, may well look quite different from the input image, A, or if it does, the resemblance will probably be small. Examples of this class of operator are given below.

*Integrate Intensities along Image Rows [**rin**]*

$$b(i,j) \Longleftarrow b(i,j-1) + a(i,j)/n$$

where

$b(0,0) = 0$

This operator is rarely of great value when it is used on its own, but is most useful when combined with other operators. For example, it can be used in a macro to detect *horizontal* streaks and edges.

*Row Maximum [**rox**]*

$$c(i,j) \Longleftarrow \mathbf{MAX}(a(i,j), c(i,j-1))$$

This function is often used to detect local intensity minima. This is one of those useful functions which is used in a wide variety of situations, although when considered on its own, it seemingly has little value.

*Copy RHS to All Other Columns [**csh**]*

$$c(i,j) \Longleftarrow a(i,n)$$

This is another of those useful functions which is used in a wide variety of situations, although when considered on its own, it seemingly has little value.

Geometric Transforms. Algorithms exist by which images can be shifted [**psh**], rotated [**tur**], magnified [**pex** and **psq**] and warped. The last three of these are quite complicated, for reasons which will become apparent later.

*Shifting without Wraparound [**psh**]*

$$c(i,j) \Longleftarrow \begin{cases} a(i+k1,j+k2) & 1 \leq i+k1 \leq m \text{ and } a \leq j+k2 \leq n \\ 0 & \text{otherwise} \end{cases}$$

Shifting with Wraparound

$$c(i,j) \Longleftarrow a(((i+k1) \bmod m), ((j+k2) \bmod n))$$

*Rotate by 90° [**yxt**, **roa**, **roc**]*

$$c(i,j) \Longleftarrow a(j,i)$$

This assumes that the image is square, i.e. $m = n$.

*Invert Axes [**lrt**, **tbt**]*

$$c(i,j) \Longleftarrow a(m+1-i,j)$$

and

$$c(i,j) \Longleftarrow a(i,m+1-j)$$

Notice that by using a combination of axis-inversion and 90° rotation, unidirectional operators such as **int** and **rox** can be generalised.

Rotate by an Arbitrary Angle [tur]. Orthogonal Cartesian coordinate axes can be rotated using the familiar transformation:

$$X' = X.\cos(\Psi) + Y.\sin(\Psi)$$

$$Y' = X.\sin(\Psi) + Y.\cos(\Psi)$$

where (X,Y) are the original axes (before rotation) (X′,Y′) are the new axes (after rotation). Ψ is the rotation angle. This equation is perfectly adequate as a means of describing the rotation of a continuous image but fails to take account of quantisation (see Figure 2.8). When a point (i,j), where i and j are both integers, is transformed using this equation, the result (i′,j′) is not necessarily a pair of integers, where

$$i' = i.\cos(\Psi) + j.\sin(\Psi)$$

$$j' = i.\sin(\Psi) + j.\cos(\Psi)$$

Furthermore, we shall define four variables which are derived from i′ and j′ by rounding up and down to the nearest integers. These will be denoted by i_{up}, i_{down}, j_{up} and j_{down}, where

$$i_{down} \le i' \le i_{up}$$

$$j_{down} \le j' \le j_{up}$$

Figure 2.8. Rotating a digital image causes some difficulties because pixels in the input image are not mapped exactly to pixels in the output image.

Once these four variables have been calculated we locate four pixels in the input image and find intensities $a(i_{up},j_{up})$, $a(i_{up},j_{down})$, $a(i_{down},j_{up})$ and $a(i_{down},j_{down})$. Using these values and the following linear interpolation formula, we can estimate the intensity c(i,j) in the output image

$$c(i,j) \Longleftarrow [a(i_{down},j_{down}).(j'-j_{down}) + a(i_{down},j_{up}).(j_{up}-j')].(i_{up}-i') + [a(i_{up},j_{down}).(j'-j_{down}) + a(i_{up},j_{up}).(j_{up}-j')].(i'-i_{down})$$

Points outside the border of the input image are assumed, quite arbitrarily, to be black (zero). This same interpolation procedure is used for a number of other geometric transformations, including magnification, warping and axis conversion.

Magnification and Aspect Adjustment [pex, psq, aad]

Image magnification, whether it is performed along one or two coordinate axes, requires interpolation. The address transformation formula for two-dimensional magnification is simply

$$(i',j') \Longleftarrow (i.k1,j.k2)$$

The variables i_{up}, i_{down}, j_{up} and j_{down} are then calculated from i′ and j′, and the interpolation formula defined above is used to complete the operation. Notice that the magnification parameters k1 and k2 may be chosen independently and that they may be greater than, equal to, or less than unity.

Axis Conversion [ctr, rtc]

There are numerous instances when we are confronted with the examination of circular objects, or those displaying a series of concentric circular arcs, or streaks radiating from a fixed point. Inspecting such objects is made very much easier if we first convert from Cartesian to polar coordinates. The following equation for axis conversion should be used in conjunction with the interpolation procedure given above:

$$(i',j') \Longleftarrow (k1+i.\cos(2\pi.j/n), k2+i.\sin(2\pi.j/n))$$

The reverse process of converting from polar to Cartesian coordinate axes uses the following address transformation

$$(i', j') \Longleftarrow (\sqrt{(i^2 + j^2)}, (n/\pi).\tan^{-1}(j/i))$$

Warping

Naive image warping can be achieved using the following formula, which ignores interpolation

$$c(i,j) \Longleftarrow a([i.s(i,j)+t(i,j)],[j.u(i,j)+v(i,j)]$$

where s(i,j), t(i,j), u(i,j) and v(i,j) are two-dimensional functions which control the warping process. A particularly simple example of warping is accomplished by slipping the rows of an image sideways by an amount which varies with the vertical position of that row:

$$c(i,j) \Longleftarrow a(i,j+q(i))$$

where q(i) is a function depending only on i. If all of the q(i) are integers, this formula is exact; no interpolation is needed. However, the result may be rather ragged, particularly if there are any values of i such that

$$q(i+1) - q(i) \gg 1$$

Interpolation is often able to make the output image appear to be smoother. Warping is useful in a variety of situations. For example, it is possible to compensate for *barrel* or *pin-cushion distortion* in a television camera, geometric distortions introduced by a wide-angle lens, or *trapezoidal distortion* due to viewing from a oblique angle. Another possibility is to convert simple curves of known shape into straight lines, in order to make subsequent analysis easier.

Intensity Histogram [hpi, hgi, hge, hgc]

The intensity histogram is defined in the following way:

(a) Let

$$s(p,i,j) \longleftarrow \begin{cases} 1, & a(i,j) = p \\ 0, & \text{otherwise} \end{cases}$$

(b) Let h(p) be defined thus

$$h(p) \longleftarrow \sum_{i,j} s(p,i,j)$$

It is not, in fact, necessary to store the s(p,i,j), since the calculation of the histogram can be performed as a serial process in which the estimate of h(p) is updated iteratively, as we scan through the input image. The *cumulative histogram*, H(p), can be calculated thus **[hgc]**:

$$H(p) = H(p-1) + h(p)$$

where

$$H(0) = h(0)$$

Both the cumulative and the ordinary histogram have a great many uses, as will become apparent later.

It is possible to calculate various intensity levels which indicate the occupancy of the intensity range **[pct]**. For example, it is a simple matter to determine that intensity level, p(k), which when used as a threshold parameter ensures that a proportion k of the output image is black; p(k) can be calculated using the fact that

$H(p(k)) = m.n.k$

The *mean intensity* [**avr**] is equal to

$$\sum_{p} (h(p).p)/(m.n)$$

while the *maximum intensity* [**gli**] is equal to

$$\mathbf{MAX}\,(p \mid h(p)) > 0)$$

and the *minimum intensity* [**gli**] is given by

$$\mathbf{MIN}\,(p \mid h(p)) > 0)$$

One of the principal uses of the histogram is in the selection of threshold parameters. It is useful to plot h(p) as a function of p. When we do this, it is commonly found that the resulting graph has one of the three forms shown in Figure 2.9. In these instances, it is often found that a suitable position for the threshold can be related directly to the position of the "foot of the hill", (b) and (c), or to the "valley" (a).

The standard deviation of the intensity, within a defined band, can be calculated using **gav**, while the threshold intensity which will segment the image such that the resulting image has a defined ratio of the number of black to white pixels can be computed using **pct**.

A particularly important operator for image enhancement is given by the transformation:

$$c(i,j) \Longleftarrow W.H(a(i,j))/(m.n)$$

This has the interesting property that the histogram of the output image {c(i,j)} is flat, giving rise to the name *histogram equalisation* for this operation [**heq**, also see

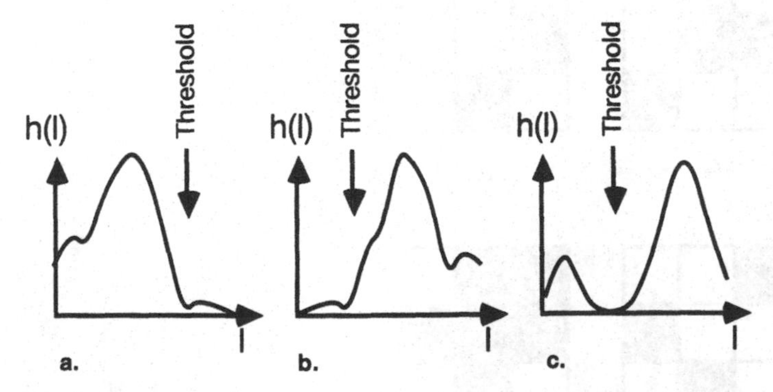

Figure 2.9. Three commonly occurring histogram types: (**a**) A few pixels have higher intensities than the background, whose histogram has the form of a steep hill. Place the threshold at the foot of the hill. (**b**) As (**a**), except that there are just a few pixels darker than the background. (**c**) The histogram is bimodal. Place the threshold at the centre of the valley. Sometimes, there are several hills and valleys. In this situation, several thresholds might be chosen to partition the image.

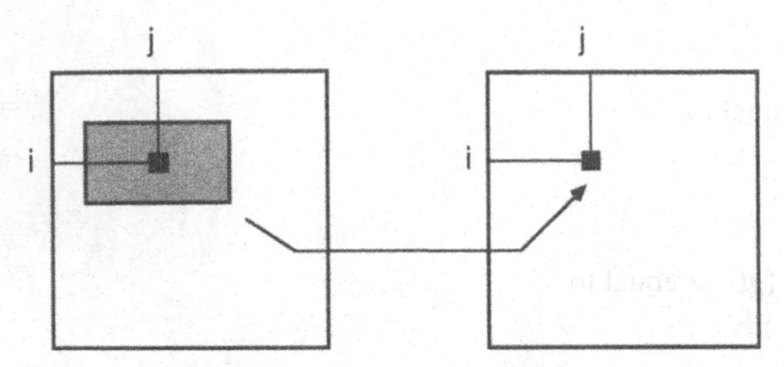

Figure 2.10. Local-area histogram equalisation. A window, represented by the stippled rectangle, is used to compute the intensity of a single point (i,j) in the output image. In this respect, this type of operator resembles other local operators. The number of pixels in the window that are darker than the central pixel are then counted. This number defines the intensity at the equivalent point in the output image. The operator takes its name from the fact that this process is equivalent to performing histogram equalisation within the stippled area but only one of the new intensity values is retained.

hge]. Notice that histogram equalisation is a data-dependent monadic, point-by-point operator. An image whose histogram has been flattened by this process can be transformed by further applications of other monadic operators; this forms one of the major techniques used in image enhancement.

An operation known as *local area histogram equalisation* is a local operator, which relies upon the application of histogram equalisation within a small window (Figure 2.10). This is a powerful filtering technique, which is particularly useful for texture analysis.

Binary Images

For the purposes of this description of binary image processing, it will be convenient to assume that a(i,j), b(i,j) can assume the values 0 (black) and 1 (white). (It may be

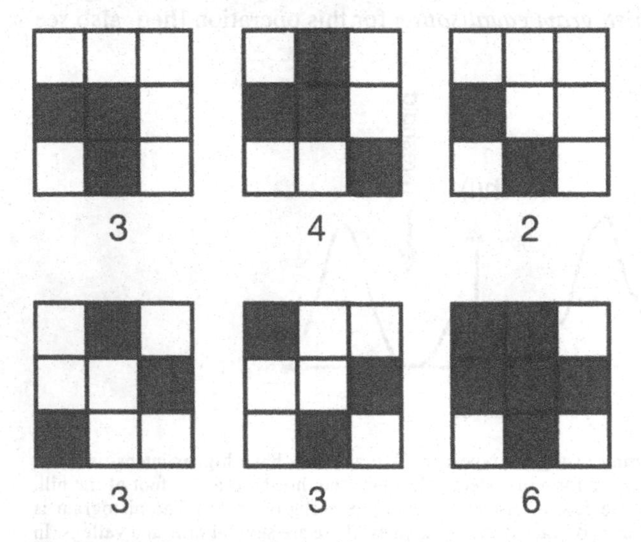

Figure 2.11. Counting white neighbours.

recalled that we indicated earlier that in VCS it is customary to use a different convention.) The operator "+" denotes OR and "." represents AND. Let #(i,j) denote the number of white points addressed by N(i,j), including (i,j) itself (Figure 2.11).

*Inverse [**not, neg**]*

$c(i,j) \Longleftarrow$ **NOT**(a(i,j))

*AND White Regions [**and, min**]*

$c(i,j) \Longleftarrow a(i,j).b(i,j)$

The binary operator "." denotes **AND**.

*OR [**ior, max**]*

$c(i,j) \Longleftarrow a(i,j) + b(i,j)$

The binary operator "+" denotes **OR**.

*Find Differences between White Regions [**xor**]*

$c(i,j) \Longleftarrow a(i,j) \sim b(i,j)$

The binary operator "~" denotes the **Exclusive OR** function.

*Expand White Areas [**exw**]*

$$c(i,j) \Longleftarrow a(i-1,j-1) + a(i,j-1) + a(i+1,j-1) + a(i-1,j) + a(i,j) + a(i+1,j) + a(i-1,j+1) + a(i,j+1) + a(i+1,j+1)$$

Notice that this is closely related to the local operators defined earlier. This equation may be expressed in the simplified notation;

$$c(i,j) \Longleftarrow A + B + C + D + E + F + G + H + I$$

*Shrink White Areas [**skw**]*

$$c(i,j) \Longleftarrow a(i-1,j-1).a(i,j-1).a(i+1,j-1).a(i-1,j).a(i,j).a(i+1,j).a(i-1,j+1).a(i,j+1).a(i+1,j+1)$$

or more simply

$$c(i,j) \Longleftarrow A.B.C.D.E.F.G.H.I$$

*Edge Detector [**bed**]*

$c(i,j) \Longleftarrow a(i,j).\mathbf{NOT}(a(i-1,j-1).a(i,j-1).a(i+1,j-1).$
$a(i-1,j).a(i+1,j).a(i-1,j+1).a(i,+1).a(i+1,j+1))$

or more simply

$c(i,j) \Longleftarrow E.\mathbf{NOT}(A.B.C.D.F.G.H.I)$

*Remove Isolated White Points [**wrm**].* This operator can be expressed in analytic form, but it is easier to understand if it is expressed in a different way.

$$c(i,j) \Longleftarrow \begin{cases} 1 & a(i,j).(\#(i,j) > 1) \\ 0 & \text{otherwise} \end{cases}$$

Noise reduction in binary images can also be performed using **brm** and **bfl**.

*Count White Neighbours [**cnw**]*

$c(i,j) \Longleftarrow \#(a(i,j) = 1)$

where #(Z) is the number of times that Z occurs. Notice that {c(i,j)} is a grey-scale image.

*Connectivity Detector [**cny**].* Consider the following pattern

1	0	1
1	X	1
1	0	0

If X = 1, then all of the 1's are 8-connected to each other. On the other hand, if X = 0, then they are not connected. In this sense, the point marked X is critical for connectivity, as are those given in the following examples:

1	0	0
0	X	1
0	0	0

1	1	0
0	X	0
0	0	1

0	0	1
1	X	0
1	0	1

However, those points marked X below are not critical for connectivity, since setting X = 0 rather than 1 has no effect on the connectivity of the 1's

1	1	1
1	X	1
0	0	1

0	1	1
1	X	0
1	1	1

0	1	1
1	X	0
0	1	1

A connectivity detector shades the output image with 1's to indicate the positions of those points which are critical for connectivity and which were white in the input image. Black points and those which are not critical for connectivity are mapped to black points in the output image. There is no analytic expression for the mapping function; this operator can be easily implemented, however, using a simple 512-element look-up table.

Euler Number [eul]. The Euler number represents a simple method of counting blobs in a binary input image, provided that they have no holes in them. Alternatively, it can be used to count holes, provided that they have no "islands" in them. The Euler number is actually equal to the number of blobs minus the number of holes. While this may seem to be a little awkward to use, compared to a simple count of the number of blobs, it should be borne in mind that the Euler number is very easy (and fast) to calculate, whereas the number of blobs is not. In effect, the Euler number is computed by using three local operators. Let us define three numbers N1, N2 and N3 as follows: N1 is the number of times that one of the following patterns occurs in the input image:

```
00   00   10   01
01   10   00   00
```

N2 is the number of times that one of the following patterns occurs in the input image:

```
01   10
10   01
```

N3 is the number of times that one of the following patterns occurs in the input image:

```
11   11   01   10
10   01   11   11
```

Then the Euler number is equal to

$$[N1 - 2.N2 - N3]/4$$

Figure 2.12. Shading blobs in a binary image (**a**) according to the order in which they are found during a raster scan (left to right; top to bottom); (**b**) according to their areas.

This formula calculates the *8-connected Euler number*; holes and blobs are both defined in terms of 8-connected figures. It is possible to calculate the *4-connected Euler number* in a similar way, but this parameter can give results which seem to be anomalous when we attempt to count the blobs and holes by eye.

The operator **ndo** gives a correct count of the number of blobs.

Filling Holes [blb]. Imagine a white blob-like figure, against a black background and that the figure has a hole in it. Applying a hole-filling operator will not alter the outer edge of the figure but will make the black region corresponding to the hole white. Despite its apparent simplicity an algorithm to accomplish this operation is quite difficult to define.

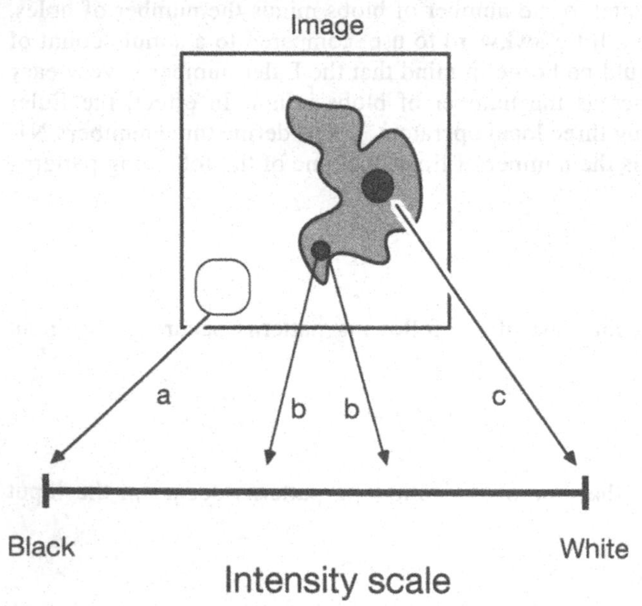

Figure 2.13. Using a grey-scale filter (such as the local averaging filter **raf**) to remove noise from a binary image. (**a**) Background points are mapped to black. (**b**) Edge points are mapped to the central part of the intensity range. Thresholding at mid-grey has the effect of smoothing the edge of large blobs. ("Hair" is removed and tiny crevices are filled.) Tiny blobs are also removed by filtering and thresholding (not shown). (**c**) Central areas of large white blobs are mapped to white.

*Region Labelling [**ndo**].* Imagine an image containing a number of separate blob-like figures. A region-labelling operator will shade the output image so that each blob is given a separate intensity value. We could shade the blobs according to the order in which they are found during a conventional raster scan of the input image (Figure 2.12). Alternatively, the blobs could be shaded according to their areas; the biggest blobs become the brightest. Despite the simplicity of its objective, the algorithm for blob shading is not easy to understand.

This is one of the most useful functions in the VCS repertoire, since it allows objects to be separated and analysed, one at a time. Small blobs can be eliminated from an image. In Chapter 7, we shall see how the third largest "bay" in a blob-like figure can be used to determine its orientation. This uses **ndo**.

Other Methods of Detecting/Removing Small Spots. As we explained earlier, a binary image can represented in the form of a *grey-scale* image in which we permit only two grey levels, 0 and W. Suppose we apply a conventional low-pass (blurring) filter to such an image. The result will be a grey-scale image in which there is a larger number of possible intensity values, than in the original binary image. Pixels which were well inside large white areas in the input image are mapped to very bright pixels in the output image. Those which were well within large black areas are mapped to very dark pixels in the output. However, pixels which were inside small white spots in the input image become mapped to pixels which are within the central part of the intensity range (Figure 2.13). Pixels on the edges of large white regions are also mapped to this central band of intensities. However, if there is a cluster of small spots, which are spaced close together, some of them may also disappear.

Using this knowledge the following procedure can be understood. It has been found to be effective in distinguishing between small spots and, at the same time, achieving a certain amount of edge smoothing of the large bright blobs which remain:

```
raf(11,11),   % Low-pass filter
thr(128).     % Threshold at the mid-grey level (W/2)
```

This technique is generally easier/faster to implement than the blob shading technique described above. Thus, although it may not achieve exactly the desired result, it can be performed at high speed.

An N-tuple filter having a weight matrix of the following form can be combined with simple thresholding to distinguish between large and small spots.

```
        -1 -1 -1
      -1        -1
   -1              -1
-1                    -1
-1        20          -1
-1                    -1
   -1              -1
      -1        -1
        -1 -1 -1
```

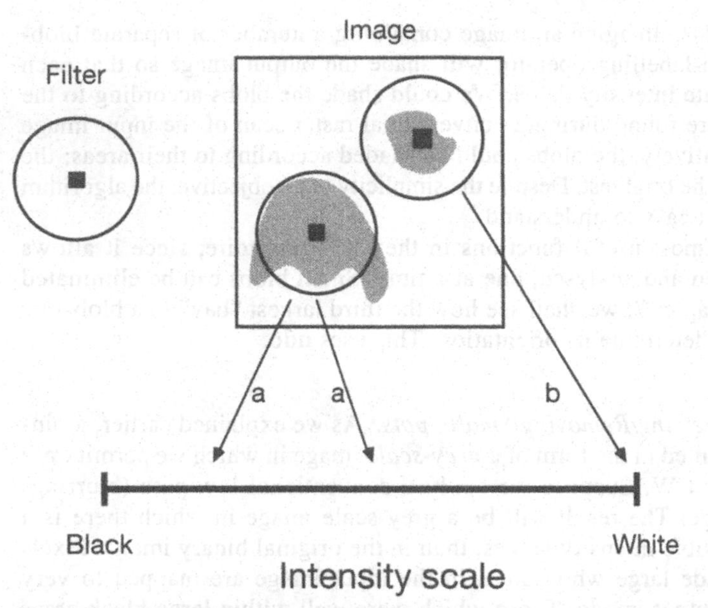

Figure 2.14. Distinguishing between small (isolated) spots and large ones, using a Laplacian-type filter. This type of filter is a linear local operator which uses a ring of negatively weighted pixels surrounding a small region with positive weights. (**a**) No point within a large blob will produce white pixels. (**b**) All pixels within a small blob will generate white pixels.

Assume that there are several small white spots within the input image and that they are spaced well apart. All pixels within a spot which can be contained within a circle of radius of three pixels will be mapped to white by this particular 21-tuple. Now, pixels within a larger spot will become darker than this (Figure 2.14). Hence, the following procedure will distinguish between large and small spots.

Apply an N-tuple filter which uses a ring-shaped weight matrix, plus a contribution from the central pixel
Threshold at the white level (W)

Grass-Fire Transform and Skeleton [gfa, mdl, mid]. Consider a binary image containing a single white blob. Imagine that this blob is made of some inflammable material and that a fire is ignited all around its outer edge and the edges of any holes that it may contain. All edge pixels are ignited at the same instant. The fire will burn inwards, until at some instant, two advancing fire lines meet. When this occurs, the fire becomes extinguished locally. An output image is generated and is shaded in proportion to the time it takes for the fire to reach each point. Background points in the input image are mapped to black.

The importance of the grass-fire transform is in the fact that it indicates distances to the nearest edge point in the input image. Thus, it is possible to distinguish thin and fat limbs of white blobs. Those points at which the fire lines meet are known as quench points. The set of *quench points* forms a figure usually known as a *skeleton*, although it is sometimes referred to as the *medial axis transform*. Very similar figures to the skeleton can be generated in a number of ways.

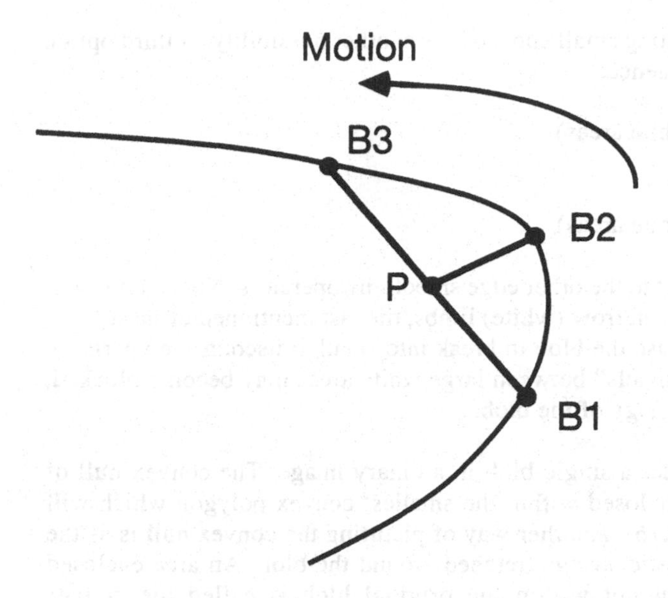

Figure 2.15. Edge smoothing and corner detection.

The one about to be described uses an algorithm known as *onion peeling*. Consider a single white blob in a binary image and a "bug" which runs around its outer edge, removing one pixel at a time. No edge pixel is removed if by so doing we would break the blob into two disconnected parts. In addition, no white pixel is removed if it has only one white neighbour amongst its 8-neighbours. This simple procedure leads to an undesirable effect in those instances when the input blob has holes in it; the skeleton which it produces has small loops in it which fit around the holes, like a noose pulled tight. A related algorithm has been developed which results in the skeleton lying mid-way between the outside edge and the edges of the holes. Its details are too complicated to describe here.

Edge Smoothing and Corner Detection. We shall restrict our attention to a single blob in a binary image. Consider three "bugs" (B1, B2, and B3) which are placed quite close together on the edge, while the spacing between B1 and B2 is equal to that between B2 and B3. That is, the same number of steps is needed to move around the edge from B1 to B2 as from B2 to B3. Define the point, P, to be that at the centre of the line joining B1 and B3 (see Figure 2.15). The three bugs now move around the edge of the blob, keeping the spacing between them constant. Now, the locus of P traces a smoother path than that followed by B2 as it moves around the edge. This simple procedure may be used for edge smoothing.

A related algorithm is that which shades the edge according to the distance between P and B2. This results in an image in which the corners are highlighted, while the smoother parts of the edge are much darker.

Many other methods of edge smoothing are possible. For example, we may map white pixels which have fewer than three white 8-neighbours to black. This has the effect of eliminating "hair" around the edge of a blob-like figure. One of the techniques described above for eliminating small spots offers another possibility. A third option is to use the processing sequence:

```
exw,        % Expand (white areas)
skw,        % Shrink
skw,        % Shrink
exw.        % Expand (white areas)
```

This achieves a similar effect to the other edge smoothing operators. Notice however, that if the blob contains very narrow (white) limbs, the last mentioned of these edge smoothing methods may cause the blob to break into smaller disconnected parts. In addition, narrow black "channels" between large white areas may become blocked, causing a change in the topology of the blob.

Convex Hull [chu]. Consider a single blob in a binary image. The convex hull of such a figure is that area enclosed within the smallest convex polygon which will enclose the shape (Figure 2.16). Another way of picturing the convex hull is as the area enclosed within an elastic string stretched around the blob. An area enclosed within the convex hull but not within the original blob is called the *convex deficiency*. It consists of a number of disconnected parts, including any holes which there may have been inside the original shape and any indentations. If we regard the blob as being like an *island*, we can understand the logic of referring to the former as *lakes* and the latter as *bays*. Both are examples of *concavities*.

Measurements on Binary Images

In order to simplify the explanation, it will be convenient to restrict our attention

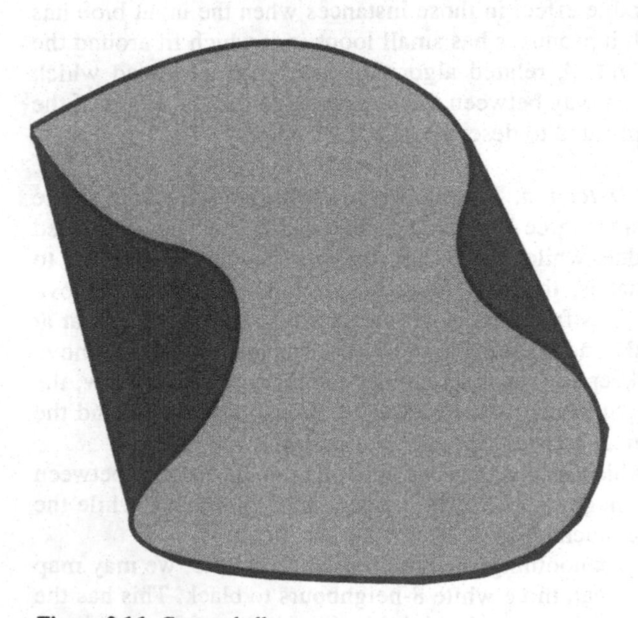

Figure 2.16. Convex hull.

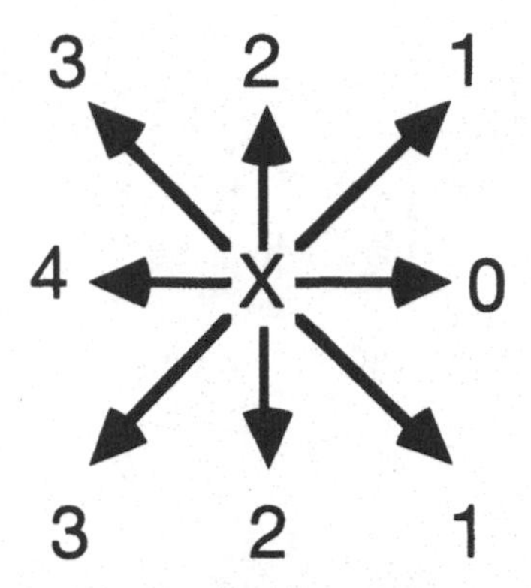

Figure 2.17. Chain coding.

again to the analysis of a binary image containing a single blob. The area of the blob can be measured by the total number of white pixels in the image. The perimeter of the blob can be measured approximately, as we shall show. First however, we must define two different types of edge points. Imagine a "bug" which runs around the edge of the blob. As it does so, the bug shouts a code to indicate in which direction it is moving, according to the scheme given in Figure 2.17.

The sequence of code values which indicates which moves the bug makes is called the *chain code* or *Freeman code* **[fcc]**. Let N_o indicate how many *odd-numbered* code values that the bug shouts in one complete journey around the blob and N_e how many *even-numbered* code values that it shouts. The *perimeter* of the blob is given *approximately* by the formula:

$$N_o + \sqrt{2}.N_e$$

This formula will normally suffice for use in those situations where the perimeter of a smooth object is to be measured. The *centroid* of a blob **[cgr]** determines its position within the image and is calculated using the formulae:

$$I \longleftarrow \sum_j \sum_i (a(i,j).i\,)/N_{i,j}$$

$$J \longleftarrow \sum_j \sum_i (a(i,j).j\,)/N_{i,j}$$

where

$$N_{i,j} \longleftarrow \sum_j \sum_i a(i,j)$$

This needs a little explanation. Although we are considering images in which the a(i,j) are equal to 0 (black) or 1 (white), it is convenient to use a(i,j) as ordinary arithmetic variable as well. This results in simpler equations for the centroid.

A convenient method of describing a blob-like shape is to find the "distribution of mass", as a function of distance measured from the centroid **[cgr(X,Y), hic(X,Y)]**. Using the centroid, it is possible to measure various features, such as the following:

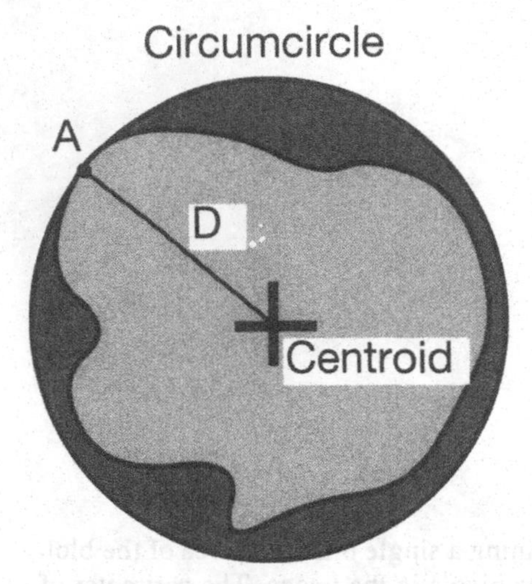

Figure 2.18. The distance of the furthest point from the centroid.

(a) The distance of the furthest point in the blob from the centroid (see Figure 2.18).

(b) The distance of the closest point on the edge of the blob, measured from the centroid.

(c) The number of protuberances as defined by that circle whose radius is equal to the average of the parameters measured in (a) and (b).

(d) The distance of the points on the edge of the blob from the centroid, as a function of angular position. This describes the silhouette in terms of polar coordinates.

A wide variety of shape parameters may be defined for a blob; just a few are listed below:

(e) The following ratio is a popular method of measuring shape:

Area/Perimeter^2

Notice that this ratio $\leq 0.25\pi$.

(f) The number of lakes. (Use **eul** or **ndo** to count them.)

(g) The number of bays.

(h) The ratio of the areas of the original blob to that of its convex hull.

(i) The ratio of the areas of the original blob to that of its circumcircle.-(Use **ccc** to draw the circumcircle.)

(j) The ratio of the area of the blob to the square of the total limb-length of its skeleton. (Use **mdl** or **mid** to draw the skeleton.)

(k) The ratio of the projections onto the major and minor axes. (Use **lmi** to find the orientation of the axis of minimum second moment.)

(l) Distances between joints/limb ends and similar features of the skeleton. Angles formed by joining such features together may also be measured. (Use **ang** to compute angles.)

Commentary

Within a relatively small number of pages, a large number of techniques have been defined and others described informally. It must be emphasised that we have not done much more than indicate the existence of a large corpus of knowledge about image processing techniques. It has already become apparent that we often need to combine several image processing techniques together. A natural corollary of this is that we require some formalised language in which to express such sequences of operations. VCS is one such language.

Notice too that we do not need to know exactly how certain operations are performed. For example, we do not require detailed knowledge about the method of labelling blobs according to their sizes in order to be able to use such an idea and incorporate it with other image processing operators. A comparison of the material in this chapter with that in the authoritative texts on image processing will show a marked difference: we have concentrated upon very simple operators, often without even justifying our interest in them. We have in effect explained the meaning of some of the "words" in a language, without bothering to state why we need to have them in our vocabulary. It will not become clear as to why some of these primitive image processing functions are needed until we have defined the syntax of our new language in more formal terms. We must repeat the remark made earlier to the effect that the image processing operators that we have described above may be regarded as the "atoms" from which practical image processing operations ("molecules") can be constructed .

Another general point that we must make is that some of the techniques perform a certain task quite satisfactorily but with some unwanted "side effects". The precise choice of procedure that we would use in practice often has to be a compromise between what we want to achieve and the "cost" of doing so. In many instances, a simple procedure may suffice, even though it is theoretically inferior to some other more complicated and expensive method. All engineering design is based on compromise and it should be clearly understood that, in practice, we cannot always afford to use the best behaved procedure. Many of the procedures we use in practice are heuristic, as distinct from algorithmic. Another question remains: how do we decide which technique to use when we are faced with a given image processing problem? The last thing that we should expect to do is to decide on the method of processing after simply viewing the given image. We need a set of tools so that we can try various options and see the results immediately. We also need a framework in which we can operate those tools. This is the point at which we should introduce Prolog+.

2.2 Introducing Prolog+

Prolog+ is an interactive language and is intended to provide a vehicle for developing intelligent image processing procedures. It is a superset of Prolog and also incorporates extensive image processing functions, as well as facilities for controlling various external electro-mechanical devices, such as an (X,Y,θ)-table, camera (pan, tilt, focus and zoom), relays, solenoids, computer-controlled lighting, etc. Prolog+ programs can also be written, edited and run, just as one would expect using normal Prolog programs. Apart from its video input, Prolog+ can sense and act upon data derived from a range of other sensors.

The breadth of the image processing command repertoire embedded within Prolog+ is evident from the discussion earlier in this chapter and also from Appendix I. The list of image processing operators is constantly growing, as new image processing primitives are being added. The exact names of these commands is unimportant, since it is a trivial exercise to define new ones, as we shall show later. Appendix I is based upon the command repertoire of the VCS image processing language existing on 1 April 1990. The list of VCS commands is constantly growing.

It should be noted that VCS provides various control features, very similar to those in the Basic language (Figures 2.19 and 2.20). These have not been incorporated into Prolog+, since Prolog's control facilities (i.e. recursion and back-tracking) are far more powerful and can easily provide equivalent functions.

VCS Macro

```
macro isophotes    % Define a macro called "isophotes"
frz                % Digitise an image from the camera
raf 11 11          % Blur the image, low-pass filtering
sca 3              % Reduce the number of grey levels to eight
gra                % Gradient
thr 1 255          % Threshold
endm               % End the macro "isophotes"
```

Equivalent Prolog+ program

```
isophotes :-
    frz,
    raf(11, 11),
    sca(3),
    gra,
    thr(1,255).
```

Equivalent VSP program

```
isophotes :-
    # frz,
    # raf(11, 11),
    # sca(3),
    # gra,
    # thr(1,255).
```

Figure 2.19. Defining a simple VCS macro.

Figure 2.20. *(opposite)* VCS contains control structures similar to those in BASIC. This simple program shows some of these facilities in use. It counts the number of "bays" on each blob in the binary input image and forms the cumulative sum.

Sample VCS Program

```
100   sof                          % Set auto-switching off
200   ndo 3 N                      % Count objects in the image. Shade according to size
300   S = 0                        % Set counter to zero
400   while i = 1 to N             % Loop counter
500   thr i i                      % Select one blob at a time
600   blb                          % Fill holes, if any
700   chu                          % Convex hull
800   max                          % Superimpose convex hull on blob
900   eul M                        % C = 1 – Number of bays in this blob
1000  S = S – M + 1                % Count the bays found so far in all blobs
1100  swi                          % Return to the original image
1200  next i                       % End of loop
1300  son                          % Put auto-switching on again
1400  print 'The total number of bays is',S
1500  stop 'Program finished'
```

Equivalent Prolog+ Program

```
count(bays,S) :-
        # sof,                     % Auto-switching off
        # ndo(3,N),                % Count objects in the image
        counter(N,0,S),            % See below
        # son,                     % Auto-switching on
        write('The total number of bays is'),
        write(S),
        nl,
        write('Program finished').

counter(0,S,S).

counter(N,P,Q) :-
        # thr(N,N),                % Select one blob at a time
        # blb,                     % Fill holes, if any
        # chu,                     % Convex hull
        # max,                     % Superimpose convex hull on blob
        # eul(M)                   % C = 1 – Number of bays in this blob
        R is P – M + 1             % Count the bays found so far in all blobs
        # swi,                     % Return to the original image
        N1 is N – 1,               % Decrement loop counter
        !,                         % Limit back-tracking
        counter(N1,R,Q).           % Repeat for all blobs in the input image
```

Equivalent VSP Program

```
count(bays,S) :-
        # sof,                     % Auto-switching off
        # ndo(3,N),                % Count objects in the image
        counter(N,0,S),            % See below
        # son,                     % Auto-switching on
        write('The total number of bays is'),
        write(S),
        nl,
        write('Program finished').

counter(0,S,S).

counter(N,P,Q) :-
        # thr(N,N),                % Select one blob at a time
        # blb,                     % Fill holes, if any
        # chu,                     % Convex hull
        # max,                     % Superimpose convex hull on blob
        # eul(M)                   % C = 1 – Number of bays in this blob
        R is P – M + 1             % Count the bays found so far in all blobs
        # swi,                     % Return to the original image
        N1 is N – 1,               % Decrement loop counter
        !,                         % Limit back-tracking
        counter(N1,R,Q).           % Repeat for all blobs in the input image
```

2.3 Review of Prolog

Let us now quickly review the major features of this fascinating language. However, the reader who has not previously encountered Prolog should stop here and consult one of the excellent textbooks on Prolog.

Prolog differs from most other computer languages, such as Pascal, C, Forth, APL, Occam, Lisp, Fortran and assembly code, in several very important ways.

(a) Firstly, a Prolog "program"[2] does *not* consist of a number of instructions, as routines written in these other languages do. Instead, it is a description of (part of) the world. For this reason, Prolog is referred to as a *declarative* language, whereas most other computer languages, military orders, knitting patterns, automobile repair manuals, musical scores and culinary recipes are all examples of *imperative* languages. This is a vital difference, which distinguishes Prolog (and a very small group of related languages) from the well-known conventional languages of Computer Science.

(b) The "flow of control" in a Prolog program does not follow the normal convention of running from top to bottom. We shall see later that the flow is just as likely to be in the reverse direction, through a control mechanism called *back-tracking*.

(c) Through the use of back-tracking, it is possible to revise temporary assignments of values to variables. This process is called *instantiation*. Instantiation is performed, in order to try and prove some postulate, theorem or statement, which may or may not be true. As far as Prolog is concerned, theorem proving is the equivalent process to running or executing an imperative language program.

(d) It is possible to make very general statements in Prolog. This is not nearly so convenient in most other languages, such as those listed above. We shall see much more of this feature later, but for the moment, let us illustrate the point with a simple example. In Prolog it is possible to define a relationship, called **right** in terms of another relationship, called **left**.

    In English

    "A is to the **right** of B if B is to the **left** of A."

    In Prolog

    right(A,B) :- left(B,A).

    (Read ":-" as "**is proved to be true if**".) Notice that neither A nor B have yet been defined. In other words, we do not need to know what A and B are in order to define the relationship **right**.[3] For example, A and B might be features of an image such as blob centres or corners, or any other objects.

(e) Prolog makes very extensive use of *recursion*. While, Pascal, C and certain other imperative languages also allow recursion, in Prolog it forms an essential control mechanism. Recursion is central to Prolog as it is to Lisp.

Prolog has been chosen by the Japanese computer industry as the language which will form the basis for programming their Fifth Generation machines. Surely, this reason alone makes it worthy of further study?

Sample Prolog program

The following program deals with the ancestry and ages of members of two families.

```
/* ____________________________________________________________
                     Start of the program

The following facts specify in which years certain people were born.
Interpretation:
            born(roger,1943).
means that "roger was born in 1943".
______________________________________________________________ */

born(roger,1943).
born(susan,1942).
born(pamela,1969).
born(graham,1972).
born(thomas,1953).
born(angela,1954).
born(elizabeth,1985).
born(john,1986).
born(marion,1912).
born(william1912).
born(patricia,1911).
born(gertrude,1870).
born(david,1868).

/* ____________________________________________________________
These facts describe the parent–child relationships which exist.
Interpretation:
            parent(X,Y).
means that "X is a parent of Y".
______________________________________________________________ */

parent(roger, pamela).
parent(roger, graham).
parent(patricia, roger).
parent(anne, patricia).
parent(david, patricia).
parent(marion, susan).
parent(susan, graham).
parent(susan, pamela).
```

```
parent(thomas, john).
parent(angela, john).
parent(thomas, elizabeth).
parent(angela, elizabeth).

/* ______________________________________________________________
Defining a relationship called "child". Read this as follows:

"A is a child of B if B is a parent of A."
______________________________________________________________ */

child(A,B) :- parent(B,A).

/* ______________________________________________________________
Defining a relationship called "older". Read this as follows:

A is older than B if
        the age of A is X AND
        the age of B is Y AND
        X > Y.
______________________________________________________________ */

older(A,B) :-
        age(A, X),
        age(B,Y),
        X > Y.

/* ______________________________________________________________
Defining a relationship "age". Read this as follows:
A has age B if
        A was born in year X AND
        it is now year Y AND
        X ≤ Y AND
        B is equal to Y – X.
______________________________________________________________ */

age(A, B) :-
        born(A, X),
        date(Y,_,_),
        X ≤ Y,
        B is Y – X.

/* ______________________________________________________________
The definition of "ancestor" has two clauses; always try to satisfy the top one first.
If this fails, then try to satisfy the second clause. Interpretation:
        ancestor(A, B).
means that "A is an ancestor of B".
```

```
The first clause should be interpreted as follows:
        "A is an ancestor of B if A is a parent of B."
________________________________________________________________ */

ancestor(A,B) :- parent(A, B).

/* ______________________________________________________________
The second clause should be interpreted as follows:
        "A is an ancestor of B if
                A is a parent of Z AND
                Z is an ancestor of B."
Notice the use of recursion here
________________________________________________________________ */

ancestor(A, B) :-
        parent(A, Z),
        ancestor(Z, B).

/* ______________________________________________________________
Definition of "print_descendants". This uses back-tracking to find all possible
solutions. The first clause always fails but in doing so it prints the descendants and their
dates of birth
________________________________________________________________ */

print_descendants(A) :-
        nl,                     % New line
        write('The known descendants of '),     % Print a message
        write(A),               % Print value of A
        write('are:'),          % Print a message
        ancestor(A, Z),         % Find Z such that A is ancestor of Z
        born(Z, Y),             % Z was born in year Y
        nl,                     % New line
        tab(10),                % 10 white spaces
        write(Z),               % Print value of Z
        write(', born'),        % Print a message
        write(Y),               % Print value of Y
        fail.                   % Force back-tracking

/* ______________________________________________________________
The second clause always succeeds and prints a new line
________________________________________________________________ */

print_descendents(_) :- nl.

/* ______________________________________________________________
                        End of the program.
________________________________________________________________ */
```

Sample Queries

Query: born(susan, 1942)
YES

Query: born(susan, X)
X = 1942
YES

Query: born(X, 1942)
X = susan
YES

Query: born(X, Y)
X = roger
Y = 1943

X = susan
Y = 1942

X = pamela
Y = 1969

X = graham
Y = 1972

X = thomas
Y = 1953

X = angela
Y = 1954

X = elizabeth
Y = 1985

X = john
Y = 1986

X = marion
Y = 1912

X = patricia
Y = 1911

X = gertrude
Y = 1870

X = david
Y = 1868

No more solutions
(Notice the alternative solutions generated by this general query.)

Query: age(marion, Z)
Z = 77
YES

Query: older(marion, susan)
YES

Query: older(susan, marion)
NO
(This really means NOT PROVEN)

Query: child(susan, Z)
Z = marion
YES

Query: ancestor(susan, Z)
Z = graham

Z = pamela

NO MORE SOLUTIONS
(Notice the alternative solutions.)

Query: ancestor(Z, graham)
Z = roger

Z = susan

Z = patricia

Z = anne

Z = david

Z = marion

NO MORE SOLUTIONS

Query: print_descendants(marion)

The known descendants of marion are:
susan, born 1942
graham, born 1972
pamela, born 1969
YES

Query: print_descendants(anne)

The known descendants of anne are:
patricia, born 1911
roger, born 1943
pamela, born 1969
graham, born 1972
YES

Query: print_descendants(wilfred)
The known descendants of wilfred are:
YES
(There are no known descendants of wilfred. The goal still succeeds.)

2.4 Sample Prolog+ Programs

The following Prolog+ program identifies table cutlery. In its present limited form, it merely recognises forks and knives; additional clauses for **object_is** are needed to identify other utensils, such as spoons, salad tongs, cheese knives, etc. Image processing primitives are set in bold type merely to help the reader to understand the program.

```
camera_sees(Z) :-
    frz,              % Digitise an image from the camera
    thr(120),         % Threshold image – naive segmentation method
                      % is used here for simplicity of illustration
    ndo,              % Shade image so each blob has different intensity
    wri(temp),        % Save image in disc file file "temp"
    repeat,           % Continue until all blobs have been analysed
    next_blob,        % Select a blob from the image saved in "temp"
    object_is(Z),     % Identify the blob
    finished.         % Succeeds only when no more blobs to analyse

next_blob :-
    rea(temp),        % Read image from disc
    gli(_,A),         % Identify next blob – i.e. brightest
    hil(A,A,0),       % Remove it from stored image
    wri(temp),        % Save remaining blobs
    swi,              % Revert to previous version of stored image
    thr(A,A).         % Select one blob
```

```
object_is(fork) :-
      mma(X,Y),        % Find lengths along major and minor axes(X, Y)
      X ≥ 150,         % Length must be ≥ 150 pixels
      X ≤ 450,         % Length must be ≤ 450 pixels
      Y ≥ 25,          % Width must be ≥ 25 pixels
      X ≤ 100,         % Width must be ≤ 100 pixels
      Z is Y/X,        % Calculate aspect ratio – whatever orientation
      Z ≤ 10,          % Aspect ratio must be ≤ 10
      Z ≥ 4,           % Aspect ratio must be ≥ 4
      count(N),        % Instantiate N to number of limb ends
      N ≥ 3,           % Skeleton of fork must have ≥ 3 limb ends
      N ≤ 5.           % Skeleton of fork must have ≤ 5 limb ends

object_is(knife) :-
      mma(X,Y),        % Find lengths along major and minor axes(X, Y)
      X ≥ 150,         % Length must be ≥ 150 pixels
      X ≤ 450,         % Length must be ≤ 450 pixels
      Y ≥ 25,          % Width must be ≥ 25 pixels
      X ≤ 100,         % Width must be ≤ 100 pixels
      Z is Y/X,        % Calculate aspect ratio – whatever orientation
      Z ≤ 12,          % Aspect ratio must be ≤ 12
      Z ≥ 6,           % Aspect ratio must be ≥ 6
      count(2).        % Skeleton of knife must have two limb ends

finished :-
      rea(temp),       % Read image from disc
      thr(1),          % Threshold stored image
      cwp(0).          % Succeeds if number of white points = 0

count(N) :-
      mdl,             % Generate skeleton of the blob
      cnw,             % Count white neighbours for 3 × 3 pixel window
      min,             % Ignore background points
      thr(2,2),        % Select limb ends
      eul(N).          % Instantiate N to number of limb ends
```

Commentary

(a) Prolog+ is a superset of Prolog, and provides a large range of Built-in Predicates (BIPs). Compared to most languages, standard or "core" Prolog, as described by Clocksin and Mellish [CLO-81], includes very few BIPs. A Prolog+ program consists of image processing predicates, for example **frz, thr, ndo, wri** etc., embedded within what is otherwise perfectly standard Prolog code.

(b) **camera_sees** is the top-level predicate in the above program. Its argument (Z) becomes instantiated to the type of table cutlery discovered by the program. In the present program, this is limited to either "fork" or "knife". Additional clauses for **object_is** are needed, if other types of object are to be recognised.

(c) The way that the image processing predicates operate follows the pattern established for printing in Prolog (c.f. **nl, write, tab**). That is, they always succeed but are never resatisfied on back-tracking. For example, as a "side effect" of trying to satisfy the goal

neg

Prolog+ performs the image-negation function on the current image. As in VCS, the operation is performed on the current image, a copy of which is saved and renamed as the "alternate" image. The result of the processing is placed in the current image. Similarly, the goal

thr(125,193)

performs thresholding, setting all pixels in the current image in the range [125,193] to white and all others to black.

(d) The goal

cwp(Z)

always succeeds and instantiates Z to the number of white pixels in the current image. However, the goal

cwp(15294)

will only succeed if there are exactly 15 294 white pixels in that image.

(e) While Prolog+ is trying to prove the compound goal

avr(Z), thr(Z)

Z is instantiated to the average intensity within the current image (operator **avr**). This value is then used to define the threshold parameter used by **thr**. Once the outcome of this compound goal is known (i.e. it either succeeds or fails) the value of Z is no longer defined.

(f) Consider the following endless loop

```
run :-
        repeat,
        frz,            % Digitise an image
        avr(Z),         % Calculate average intensity
        thr(Z),         % Threshold at average intensity level
        fail            % Back-track, to repeat
```

As it is trying to satisfy **run**, Prolog+ repeatedly back-tracks and instantiates/deinstantiates Z; the threshold parameter is calculated afresh for each newly digitised image. The effect is that an endless series of thresholded images is displayed.

Another Program and Some Further Comments

(a) The following Prolog+ program searches for a large difference between two successive images digitised from the camera. When such a pair of images has been found, the difference between them is measured and defines the value for the "output parameter", A.

```
big_changes(A) :-
    repeat,         % Always succeeds on back-tracking
    frz,            % Digitise an image from the camera
    lpf,            % Perform lpf (low-pass filter)
    lpf,            % Perform lpf (low-pass filter)
    lpf,            % Perform lpf (low-pass filter)
    sca(3),         % Retain only three bits of each intensity value
    rea(temp),      % Read image stored during previous cycle
    swi,            % Switch current and alternate images
    wri(temp),      % Save the image for the next cycle
    sub,            % Subtract images
    abs,            % Compute "absolute value" of intensity
    thr(1),         % Threshold at intensity level 1
    cwp(A),         % Instantiate A to number of white pixels in image
    A > 100.        % Test to see whether differences between images
                    % are large. If not, back-track to "repeat"
```

The operator **repeat** succeeds, as does each of the image processing operators, up to and including **cwp**. If the test

A > 100

then fails, the program back-tracks to **repeat**. Another image is then captured from the camera and the whole processing sequence is repeated. The loop terminates when A exceeds 100. A is then instantiated to the number of white pixels in the image.

(b) Although image processing commands, such as **frz**, **thr**, **cwp** etc. are always satisfied, errors will be signalled if arguments are incorrectly specified. Since **thr** requires either one or two numeric arguments,

thr(X)

will fail if X is uninstantiated. On the other hand, the following compound goal is satisfied

X is 197, thr(X)

The compound goal

```
X is 186, Y is 25, thr(X,Y)
```

fails, since **thr** fails when its second argument is less than the first one.[4] For a similar reason

```
X is 1587, thr(X)
```

fails, because the parameter (X) is outside the range of acceptable values, i.e. [0, 255].

(c) The disc commands, **rea** (read an image file) and **wri** (write an image to a disc file), are typical operators which require a string argument. This may be generated according to the usual Prolog conventions. For example, the symbol generator, **gensym**, may be used to create a series of file-names, image_file1, image_file2, ..., as the following illustration shows:

```
process_image_sequence :-
    frz,                          % Digitise an image
    process_image,                % Process the image
    gensym(image_file,X),         % Generate new symbol name
    wri(X),                       % Use it to define a new file
    process_image_sequence.       % Repeat processing
```

(d) A simpler, revised version of **big_changes** may be defined using a subsidiary predicate, **process**:

```
big_changes(A) :-
    process      % Specified at the VCS level. Listed below
    cwp(A),      % Instantiate A to number of white pixels in image
    A > 100.     % Test to see whether differences between images
                 % are large. If not, back-track to process
```

where **process** is defined thus:

```
process :-
    frz,
    lpf,
    lpf,
    lpf,
    sca(3),
    rea(temp),
    swi,
    wri(temp),
    sub,
    abs,
    thr(1).
```

The use of subsidiary predicates, such as **process**, allows the programmer to think at a higher conceptual level. In the same way, the command repertoire can be extended; a library of predicates for high-level image processing will be described later.

2.5 The Declarative Nature of Prolog+

Like standard Prolog, Prolog+ is a declarative language. This means that it is very easy to create new commands in terms of existing ones, change command names and alter default values. We shall now illustrate these points. Later, we shall explain how the declarative nature of the language alters the mode of thinking about programming.

Defining New Names for Image Processing Operators

The use of three-letter mnemonics may not be to everyone's taste. Suppose that we wish to define a new name for the image processing operator **neg**. This can be accomplished very simply:

```
negate :- neg.
```

Either **neg** or **negate** may now be used. In the same way, a two-letter mnemonic form may be defined.

```
ne :- neg.
```

When we wish to define a new name for an operator, such as **thr**, which has a variable arity,[5] the situation is slightly more complicated:

```
threshold(X,Y) :- thr(X,Y).

threshold(X) :- thr(X).

threshold :- thr.
```

Redefining Default Values for Arguments

It is a simple matter to redefine the default values for arguments of Prolog+ image processing arguments. Suppose that we wish to define new default values for the operator **thr**:

```
thr(X) :- thr(X, 200).     % Previous default value was 255

thr :- thr(75, 125).       % Not defined previously

thr :- thr(128, 255).      % Arbitrary but useful definition
```

Operators and Some Useful Control Facilities

A very useful operator "•" may be defined thus:

```
0•G.

N•G :-
      call(G),
      M is N –1,
      M•G.
```

The operator "•" may be used like a FOR-loop in a conventional programming language, since it permits the programmer to order the repetition of an operation G. In formal terms, N•G is a predicate which either succeeds or fails; if G fails on any of the N repetitions, then N•G will also fail. In order to understand the use of the "•" operator, notice that **process** may be redefined in the following way:

```
process :-
      frz,
      3•lpf,
      sca(3),
      rea(temp),
      swi,
      wri(temp),
      sub,
      abs,
      thr(1).
```

A very useful non-linear filter for detecting thin dark streaks and small spots may be defined thus:

```
filter :-
      sof,
      2•(3•lnb, neg),
      sub,
      son.
```

This is rather more easily read than the following equivalent version:

```
filter :-
      sof,
      lnb,
      lnb,
      lnb,
      neg,
      lnb,
      lnb,
      lnb,
      neg,
      sub,
      son.
```

The AND operator (&) may be defined in the following way:

```
op(900,xfy,"&"). % Precedence is 900 and "&" is left associative

A & B :- A , B.
```

The usefulness of this is simply that to a naive user "**&**" is easier to interpret than ","

The OR operator (**or**) may be defined in the following way:

```
op(900,xfy,or).

% Precedence is 900 and is left associative

A or B :- A; B.
```

The conditional statement **if_then_else** may be defined thus:

```
if_then_else(A,B,C) :- A, !, B.    % If A succeeds, test B
if_then_else (A,B,C) :- C.         % A has failed, so test C
```

The simpler **if_then** function could of course, be defined in the following way:

```
if_then(A,B) :- A, !, B.           % If A succeeds, test B
```

However, an operator, "->" may be defined as an alternative:

```
A -> B :- A, !, B.
```

A Prolog+ predicate, **case**, roughly equivalent to Pascal's CASE operator may be defined in the following way:

```
case(A, B) :-
      select_list_element(A, B, C), % Instantiate C to Ath element of list B
      call(C).

select_list_element(N,_,fail) :-
      N ≤ 0,
      !,
      fail.

select_list_element(1,[A],A).

select_list_element(A,[B|C],D) :-
      length(C,N),
      N > 0,
      X is A – 1,
      select_list_element(X,C,D).

select_list_element(_,_,_) :- fail.
```

Recognising Bakewell Tarts

Consider Figure 2.21, which shows the side and plan views of a small cake, popular in Britain and which is called a Bakewell tart (also see Figure 8.2). Now let us use Prolog+ to describe the image obtained by viewing such a cake from above.

```
bakewell_tart :-
    segment_image,    % Convert image to form shown in Figure 2.21(a)
    outer_edge,       % Check the outer edge
    cherry,           % Check the cherry
    icing.            % Check the icing
```

Programs written in Prolog+ are almost invariably written from the top level downwards. In this instance, **bakewell_tart** was the first predicate to be defined by the author when writing the program for inspecting these cakes. Notice that there are four obvious stages in verifying that the tart is a good one:

(a) Simplify the image (to a 4-level form), **segment_image**

(b) Check the integrity of the outer edge, **outer_edge**

(c) Check the presence, size and placing of the cherry, **cherry**

(d) Check the icing, **icing**

Even a novice programmer in Prolog+, or Prolog, can understand that **bakewell_tart** is only satisfied if all four of the subsidiary tests succeed. The secondary predicates, though not defined yet, are not necessary for an understanding of the process of recognising a Bakewell tart. Three of these are defined below. (**segment_image** is not given here, because it is problem specific and would distract us from the main point.)

```
outer_edge :-
    thr(1),                            % Select outer edge
    circular.                          % Standard test for circularity. Defined below

cherry :-
    thr(1),                            % Select outer edge
    cgr(X1,Y1),                        % Centroid of outer edge
    swi,                               % Switch images
    thr(200),                          % Select cherry
    swi,                               % Switch images – restore image for use later
    cgr(X2,Y2),                        % Centroid of the cherry
    distance([X1,Y1,],[X2,Y2],D),      % See Appendix II
    D < 20.                            % Are cherry and outer edge nearly concentric?

icing :-
    thr(128),                          % Select icing
    rea(mask),                         % Read annular mask image from disc
    xor,                               % Calculate differences between these images
    cwp(A),                            % Calculate area of white region
    A > 50.                            % Allow a few small defects in icing
```

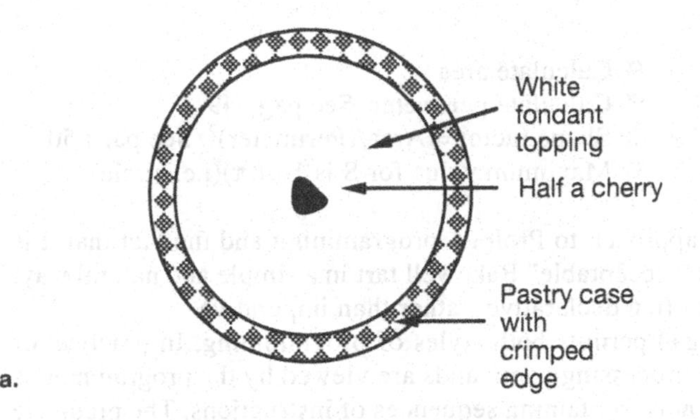

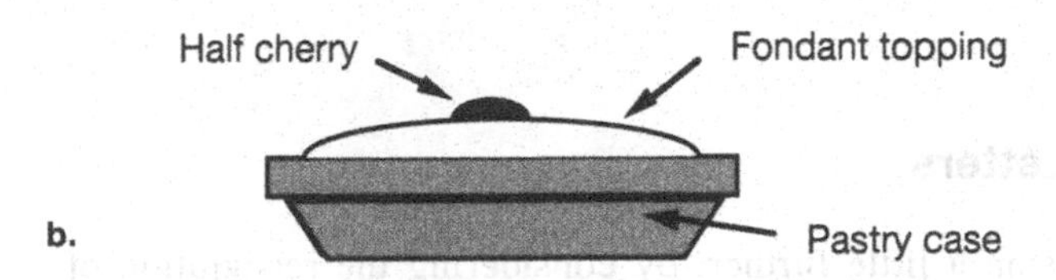

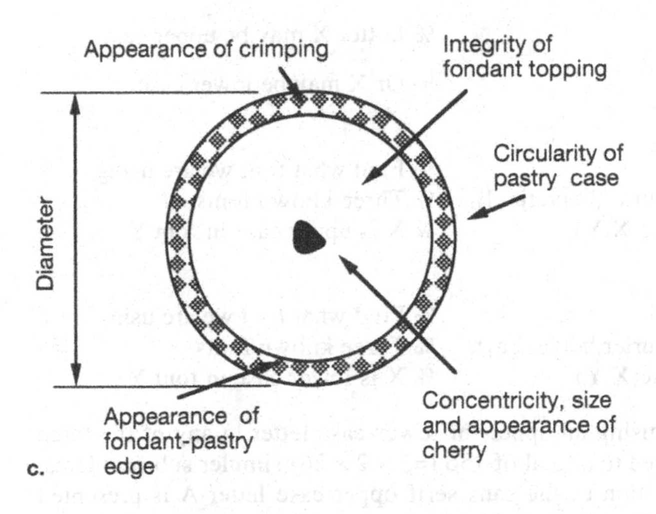

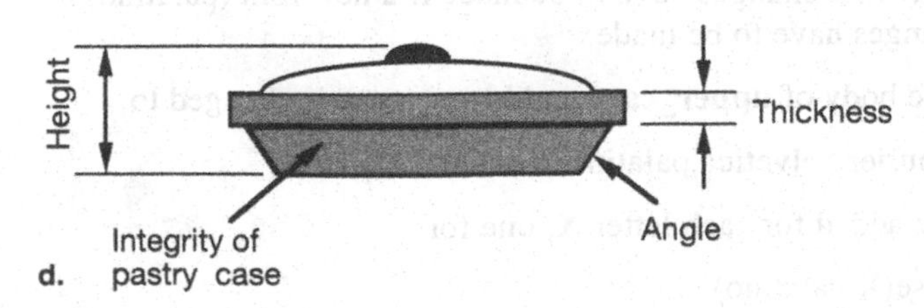

Figure 2.21. Bakewell tarts. (a) Plan view. (b) Side view. (c) Critical features, plan view. (d) Critical features, side view

```
circular :-
    cwp(A),                 % Calculate area
    perimeter(P),           % Calculate perimeter. See page 49
    S is A/(P*P),           % Shape factor = Area/(Perimeter)^2. See page 50
    S > 0.078.              % Maximum value for S is 1/(4.π)(i.e. circle)
```

Notice the highly modular approach to Prolog+ programming and the fact that it is possible to define what is an "acceptable" Bakewell tart in a simple and natural way. Programming in Prolog+ is often declarative, rather than imperative.

Prolog (and hence Prolog+) permits both styles of programming. In practice, of course, sequences of image processing commands are viewed by the programmer as ordinary (imperative) programs, containing sequences of instructions. The predicate **process**, defined on page 66, is effectively imperative, since none of its component goals ever fails.

Recognising Printed Letters

Now let us take the discussion a little further, by considering the recognition of printed letters. The top two layers of a Prolog+ program are given below:

```
letter(X) :- upper_case(X).                 % Letter X may be upper case

letter(X) :- lower_case(X).                 % Or X may be lower case

upper_case(X) :-
    font(Y),                                % Find what font we are using
    member(Y,[times,courier,helvetica]),    % Three known fonts
    recognise_upper_case(X,Y).              % X is upper case in font Y

lower_case(X) :-
    font(Y),                                % Find what font we are using
    member(Y,[times,courier,helvetica]),    % Three known fonts
    recognise_lower_case(X,Y).              % X is lower case in font Y
```

The complex task of recognising an upper- or lower-case letter in any of the three known fonts has been reduced to a total of 156 ($=3 \times 2 \times 26$) simpler sub-problems. (A simple declarative definition of the sans serif upper case letter A is presented later.) Now, let us consider what changes have to be made if a new font (palatino) is to be introduced. Two changes have to be made:

(i) the second line in the body of **upper_case** and **lower_case** is changed to

```
member(Y,[times,courier,helvetica,palatino])
```

(ii) Two new clauses are added for each letter X, one for

```
recognise_upper_case(X,palatino)
```

and another for

```
recognise_lower_case(X,palatino).
```

If we wanted to add recognition rules for the numeric characters, then 10 new clauses would be added, as in (ii). In other words, extending the scope of a Prolog+ program is conceptually simple, if rather tedious to accomplish.

Here, as promised, is a naive but quite effective declarative definition of the sans serif upper-case letter A:

```
recognise_upper_case(a,sans_serif) :-
    apex(A),                    % There is an apex called A.6
    tee(B),                     % There is a tee-joint called B
    tee(C),                     % There is a tee-joint called C
    line_end(D),                % There is a line_end called D
    line_end(E),                % There is a line_end called E
    above(A,B),                 % A is above B. See Appendix II and page 105
    above(A,C),
    about_same_height(B,C),     % See Appendix II
    about_same_height(D,E),
    above(B,D),
    above(C,E),
    connected(A,B),             % See Appendix II
    connected(A,C),
    connected(B,D),
    connected(C,E),
    connected(B,C),
    left(B,C),                  % See Appendix II
    left(D,E).
```

The reader should be able to understand the above program without detailed knowledge about how the predicates **apex, tee, line_end**, **about_same_height**, **above**, **connected** and **right** are defined. Figure 2.22 shows some of the objects which would be recognised by this program.

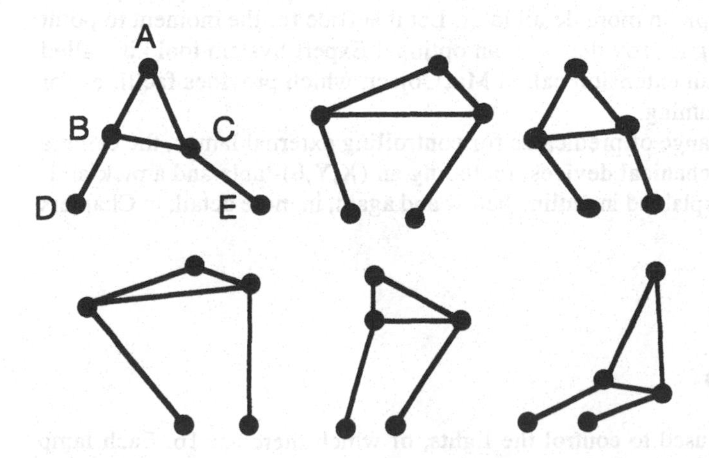

Figure 2.22. Objects which might be mistaken for an upper case, sans serif letter "A".

2.6 Controlling External Devices

Standard Edinburgh Prolog is unfortunately not well equipped for communicating with external devices. However, modern implementations, such as LPA MacProlog, often have better facilities in this regard. Prolog+ is very well suited to representing and manipulating symbolic knowledge-based rules for controlling external devices, for example, manufacturing machines, processing plant, etc.. Here is a simple example, which once again demonstrates how simple a program can be:

```
adjust_oven :-
        loaf_colour(normal),
        !.                          % Do not adjust the oven
adjust_oven :-
        loaf_colour(dark),
        reduce_temperature.         % This predicate reduces the temperature

adjust_oven :-
        loaf_colour(light),
        increase_temperature.       % This predicate increases the temperature
```

Further explanation seems to be unnecessary. Here is another example, this time for a hypothetical machine:

```
valve6(close) :-
        air_flow(above_normal),         % Test air flow rate
        temperature(water_supply,X),    % Measure water temperature
        X ≥ 56,                         % Test temperature
        solenoid15(operated),           % Test solenoid15
        close(valve6).                  % An action
```

This touches on another topic, namely the integration of Prolog+ and Expert Systems. We shall discuss this topic in more detail later. Let it suffice for the moment to point out that LPA MacProlog is provided with an optional Expert System tool kit, called *Flex*. It also possesses an extension called MacObject, which provides facilities for object oriented programming.

Prolog+ contains a range of predicates for controlling external lamps, the camera and various electro-mechanical devices, including an (X,Y,θ)-table and a pick-and-place arm.[7] These are explained in outline below and again, in more detail, in Chapters 3 and 6.

Controlling Lights

The predicate **light** is used to control the lights, of which there are 16. Each lamp can be set to any one of 16 brightness levels.

light(N,B). Switch lamp N to brightness level B. N must be instantiated to an integer in the range [0,15]. If B is not instantiated, it will be given a value equal to the present brightness value. If it is, then the lamp brightness will be adjusted accordingly.

light(0,B). Sets *all* lamps to the level specified by B. If B is not specified initially, it will be instantiated to the list of brightness levels of the 16 lamps.

The predicate **light** may be defined in the following way[8]:

```
light(0,B) :- all_lights(1,_,B).

light(N,B) :-
        var(B),                          % Check that B is variable
        get_prop(light,N,B).             % Instantiate B to brightness of lamp N

light(N,B) :-
        set_lamp(N,B),                   % Operate lamp controller
        set_prop(light,N,B).             % Remember brightness value

all_lights(17,_,_).                      % Terminate the recursion

all_lights(I,J,B) :-
        light(I,B),                      % Set lamp I to brightness level B
        J is I + 1,                      % Increment counter
        all_lights(J,B).                 % Recursion to set other lamps
```

Notice that this definition allows Prolog+ to keep a record of the lamp brightness values, without requiring feedback from the lamp controller; the predicate **set_lamp** is actually responsible for one-way communication with the lamp controller.

Operating an (X,Y,θ)-Table

Some of the primitives for operating the (X,Y,θ)-table are listed below:

origin Reset the table to its initial position.

calibrate Calibrate the coordinate axis transformation between the vision system and the robot

home Reset the table. Faster than **origin**, since it does not perform **calibrate**

move_x(X) Move the table to position X along the X axis

move_y(Y) Move the table to position Y along the Y axis

move(X,Y) Move the table to position (X,Y). This is defined below:

```
move(X,Y) :-
        move_x(X),
        move_y(Y).
```

move_t(T) Move the table by rotating it to position T
min_x Move the table to its minimum X position, keeping Y unchanged
max_x Move the table to its maximum X position, keeping Y unchanged
min_y Move the table to its minimum Y position, keeping X unchanged
max_y Move the table to its maximum Y position, keeping X unchanged
delta_x(X) Move the table by an amount X along the X axis. If the X limits would be violated, the goal will fail
delta_y(Y) Move the table by an amount Y along the Y axis. If the Y limits would be violated, the goal will fail
delta_t(T) Rotate the table by an amount T
where(X,Y,T) Where is the table in (X,Y,θ)-space?

The table has a finite range of movement, so any attempt to move it to a position outside this will result in failure. For example, the goal

move(100000,100000)

will fail, because the (X,Y)-limits of motion would be exceeded.

Pick-and-Place Arm

The primitive operators for the pick-and-place arm are:

ramin Move the arm inwards
ramout Move the arm outwards
up Move the gripper upwards
down Move the gripper downwards
grasp Operate the gripper
release Release the gripper

Additional functions may be defined in terms of these primitives:

```
pick :-
        down,
        grasp,
        up.

place :-
        down,
        release,
        up.

load_table :-
        ramout,
        pick,
        ramin,
        place.
```

```
clear_table :-
    ramin,
    pick,
    ramout,
    place.
```

Controlling the Camera and Lens

Several operators are reserved for controlling cameras and optics. Since there may be more than one camera, there is a command for selecting one of them. This is

csl(X) Select camera number X

(If X is outside the allowed range, this goal will fail.)
csl has two effects:

(a) The video multiplexer on the input to the frame store is set appropriately.
(b) Subsequent invocations of camera lens control operators will alter only the selected camera.

The following camera control facilities are provided:

aperture(X) Set the aperture to position X (X is an integer)

focus(X) Set the camera to focus at distance X

optics(X) Select an optical filter[9]

zoom(X) Set the zoom to X

Auxiliary I/O Functions

There are several unspecified I/O functions in Prolog+:

input_line(N) Succeeds if input line N is at logic level 1

output_line(N,X) Set output line N to logic level X (on or off). Always succeeds if N is within the appropriate limits

get_string(S) Instantiates S to the contents of the input buffer on the serial (RS 232) communications channel

send_string(S,D) Transmit string S, with a header identifying it as being intended for reception by device D. This provides a means of sending complex messages to external devices, via an additional serial data link

Of course, higher level predicates can be based upon these primitives, in order to achieve symbolic representation of external events.

Notes

1. It is of interest to note that the author was using operators of this type for several years before the current fashionable interest in morphological operators arose. New operators were invented simply by generating sequences of VCS commands, on a try-it-and-see basis.
2. The correct term is "application", since a program is, strictly speaking, a sequence of instructions. However, we shall continue to use the term "program", since this is more familiar to most readers.
3. **left** may be defined in terms of the more primitive relationship **location**:

```
left(A,B) :-
        location(A,Xa),
        location(B,Xb),
        Xa < Xb.
```

   Read this in English as:

```
left(A,B) "is proved to be true if"
        location(A,Xa) and
        location(B,Xb) and
        Xa < Xb.
```

   (Actually, this is perfectly valid in Prolog as well!)
4. The VCS image processor signals an error, which causes the failure of the Prolog goal, thr(186, 25).
5 The arity measures the number of arguments.
6. The name of a feature, such as an apex or tee-joint, may be the same as its address.
7 At the time of writing (December 1990), a multi-axis robot arm has just been interfaced to Prolog+.
8 The predicates **get_prop** and **set_prop** are specific to LPA MacProlog.
9. Imagine a wheel containing a number of optical filters, polarisers, etc. This can be rotated under the control of the Prolog+ program, thereby allowing the optics to be changed at will.

CHAPTER 3

OPERATING ENVIRONMENT

"Entia non sunt multiplicanda praeter necessitatem"
is normally paraphrased in English as:
"It is vain to do with more what can be done with less"
William of Occam

In addition to facilitating the writing of intelligent programs for image processing, Prolog+ is intended to operate a range of electro-mechanical equipment, forming what is known as a Flexible Inspection Cell (FIC). An FIC comprises a frame holding a set of lamps, several cameras, an (X,Y,θ)-table and a pick-and-place arm (see Figure 3.1(a)). Figure 3.1(b) shows another arrangement which incorporates a multi-axis robot. The control structure is shown in Figure 3.2. In order to make the operation of the cell as easy as possible, the author decided to provide a set of utilities around Prolog+. These will form the subject of this chapter.

3.1 Pull-Down Menus

The pull-down menus are provided in order to assist the user when developing programs and setting up a robot system (Figure 3.3). They are not intended to be a substitute for the more powerful keyboard commands, which they duplicate. The following menus have been made available:

Utility (see Figure 3.4)
Process (see Figure 3.5)
Analyse (see Figure 3.6)
Lights (see Figure 3.7)

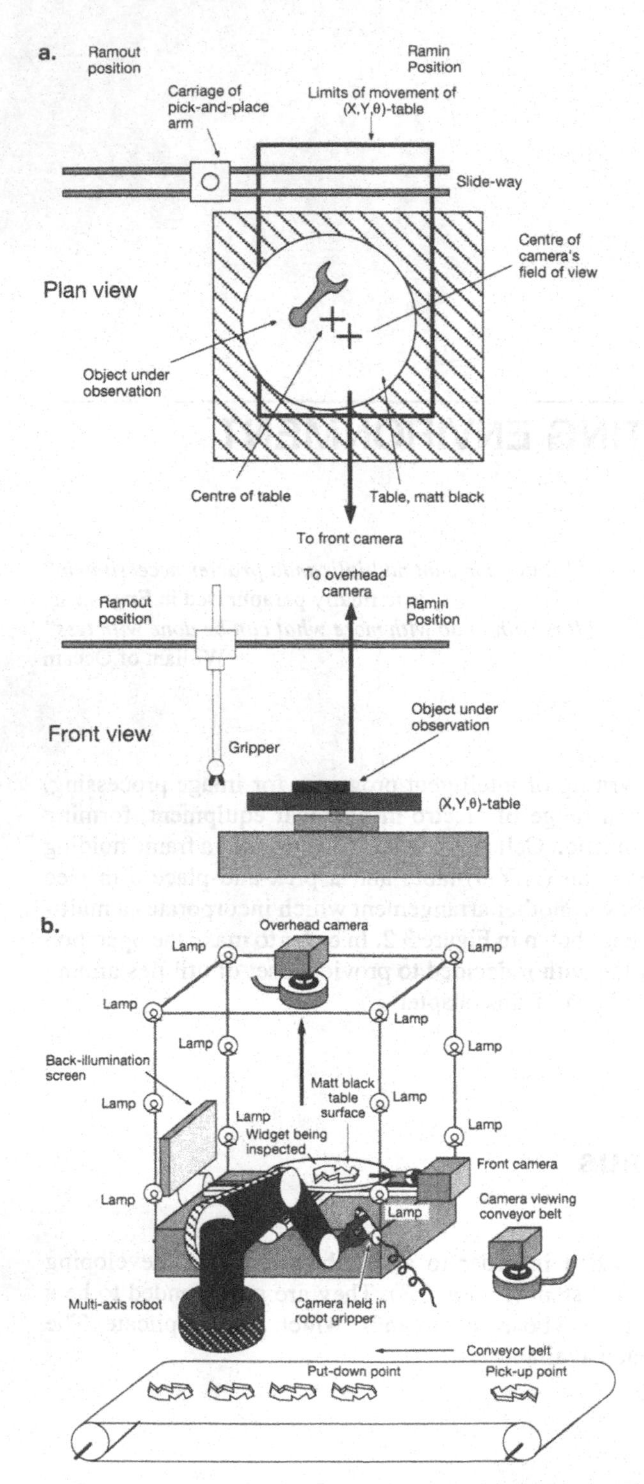

Figure 3.1. Flexible Inspection Cell. (a) Using an (X,Y,θ)-table. (b) Using a multi-axis robot.

Device (see Figure 3.8)
Applications (see Figure 3.9)

We shall now discuss each of these menus in turn.

Utility Menu

This menu provides a set of useful facilities for controlling the system configuration, although its exact contents depend upon the implementation. It is possible to:

(i) Initialise and reconfigure the system.
(ii) Switch to *Transparent mode* (explained below).
(iii) Control a video multiplexer on the input to the image processor.
(iv) Generate simple images, such as a wedge, cone, staircase, grid, etc.
(v) Read standard stored images from disc.
(vi) Extend the repertoire of Prolog+, to accommodate new image processing (i.e. VCS) commands. (An interactive dialogue allows the user to define the name and arity of a new image processing operator. This avoids the user having to worry about the details of how the image processing commands are declared within the Prolog+ software.)

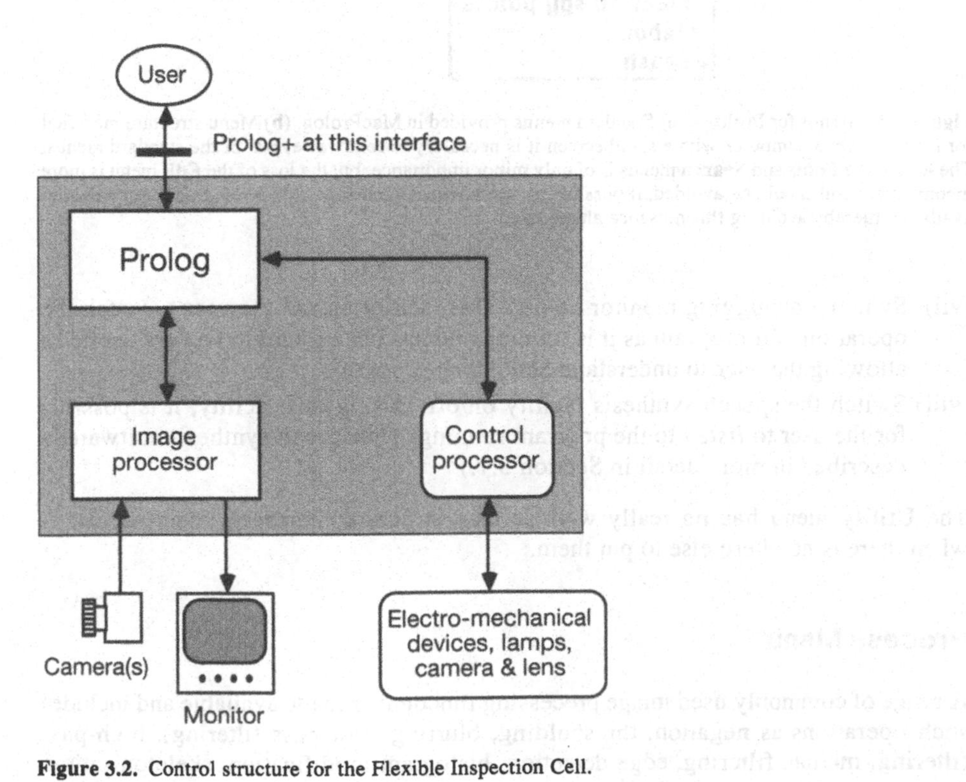

Figure 3.2. Control structure for the Flexible Inspection Cell.

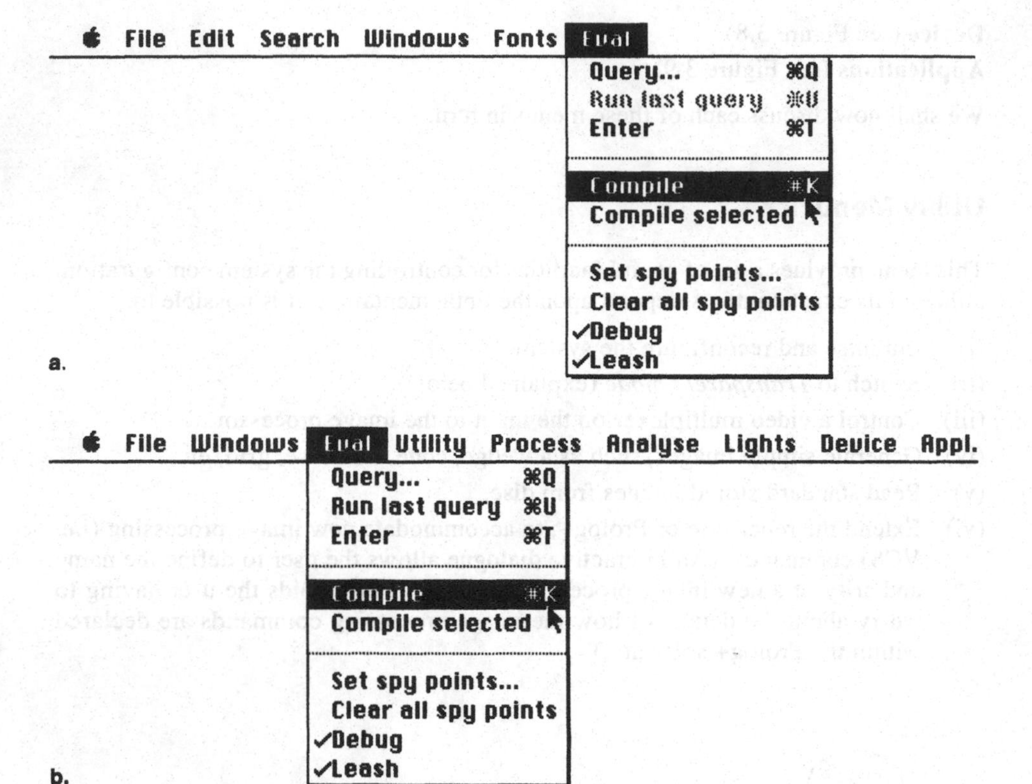

Figure 3.3. Menus for Prolog+.(**a**) Standard menus provided in MacProlog. (**b**) Menu structure modified for Prolog+. On a computer with a small screen it is necessary to remove several of the standard menus. The loss of the **Fonts** and **Search** menus is of only minor importance, but the loss of the **Edit** menu is more inconvenient and should be avoided, if possible. In more recent versions of MacProlog pop-up menus are available, thereby avoiding this nuisance altogether.

(vii) Switch a debugging monitor on/off. This facility allows the user to watch the operation of a program as it is running and has been found to be very useful in allowing the user to understand and debug programs.

(viii) Switch the speech synthesis facility on/off. (Using this facility, it is possible for the user to *listen* to the program running. The speech synthesis software is described in more detail in Section 3.7.)

The **Utility** menu has no really well defined structure; items are inserted into it when there is nowhere else to put them.

Process Menu

A range of commonly used image processing functions is made available and includes such operations as negation, thresholding, blurring (low-pass filtering), high-pass filtering, median filtering, edge detection, histogram modification, skeletonisation,

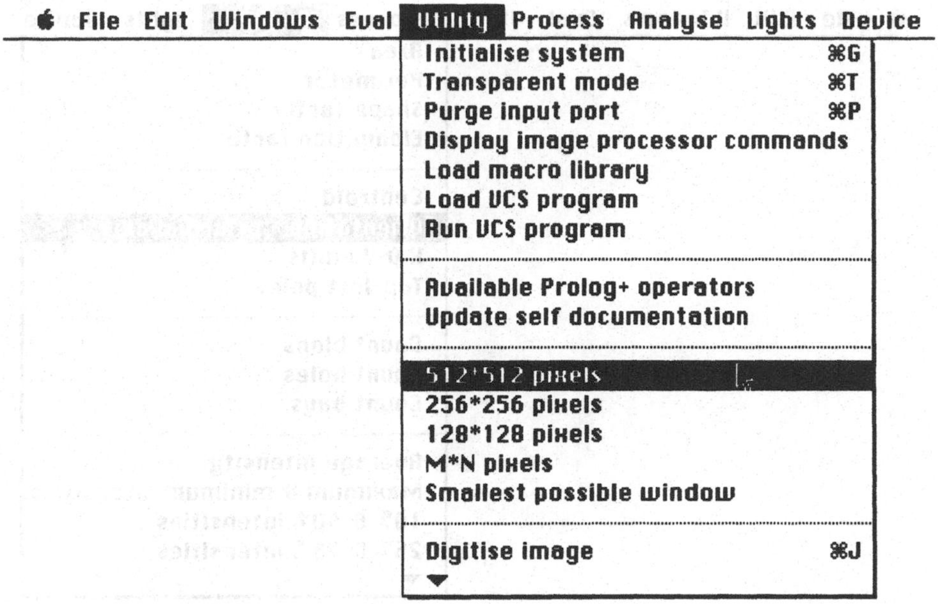

File Edit Windows Eval Utility Process Analyse Lights Device

Initialise system ⌘G
Transparent mode ⌘T
Purge input port ⌘P
Display image processor commands
Load macro library
Load UCS program
Run UCS program

Available Prolog+ operators
Update self documentation

512*512 pixels
256*256 pixels
128*128 pixels
M*N pixels
Smallest possible window

Digitise image ⌘J

Figure 3.4. The **Utility** menu (partial view). One important item not shown here is the **Extend menu**, by which the user can add new items to any existing menu, at will.

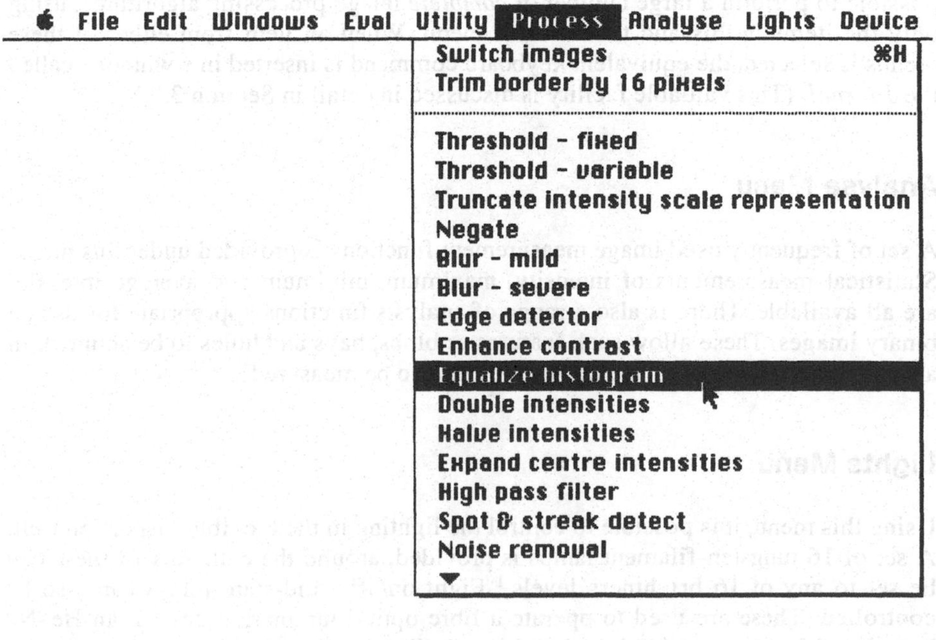

File Edit Windows Eval Utility Process Analyse Lights Device

Switch images ⌘H
Trim border by 16 pixels

Threshold - fixed
Threshold - variable
Truncate intensity scale representation
Negate
Blur - mild
Blur - severe
Edge detector
Enhance contrast
Equalize histogram
Double intensities
Halve intensities
Expand centre intensities
High pass filter
Spot & streak detect
Noise removal

Figure 3.5. The **Process** menu (partial view).

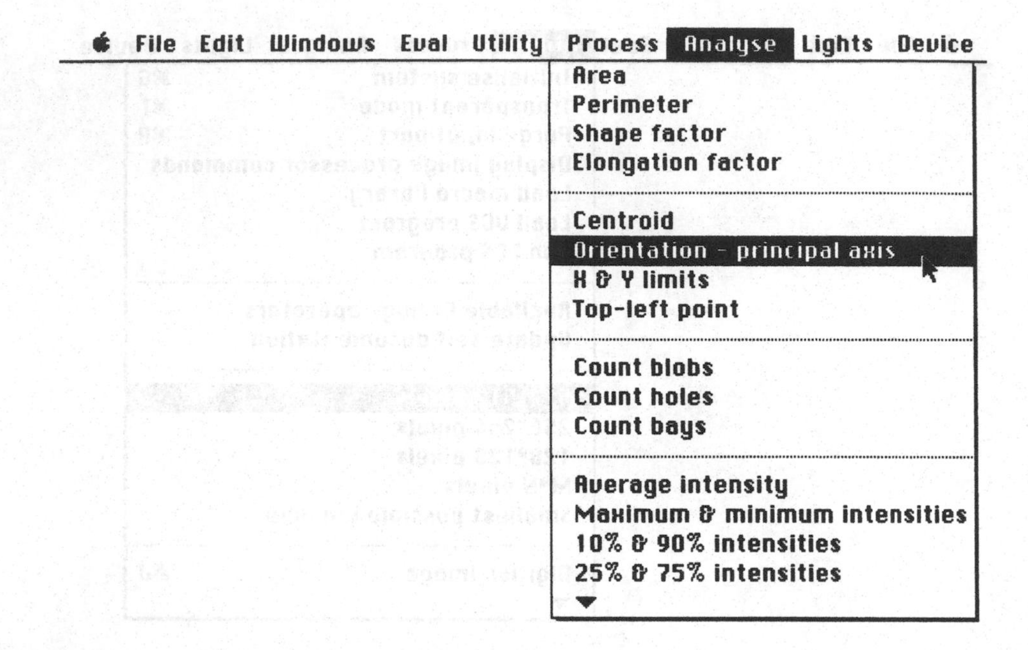

Figure 3.6. The **Analyse** menu (partial view).

drawing the convex hull of a blob, filling holes, isolating the biggest blob. It is possible to perform a large number of *complete* image processing algorithms, using only the items in this and the **Analyse** menu. When an item from either of these menus is selected, the equivalent keyboard command is inserted in a window, called the *Journal*. (This valuable facility is discussed in detail in Section 3.8.)

Analyse Menu

A set of frequently used image measurement functions is provided under this menu. Statistical measurements of intensity, maximum, minimum and average intensity are all available. There is also a range of analysis functions appropriate for use on binary images. These allow such features as blobs, bays and holes to be counted. In addition, the (white) area and perimeter can also be measured.

Lights Menu

Using this menu, it is possible to control the lighting in the Flexible Inspection Cell. A set of 16 tungsten-filament lamps is provided, around the cell. Any of these can be set to any of 16 brightness levels.[1] Eight on/off solid-state relays can also be controlled. These are used to operate a fibre-optic illumination device, an He–Ne laser (used for structured lighting) and a back-illumination device, as well as general-

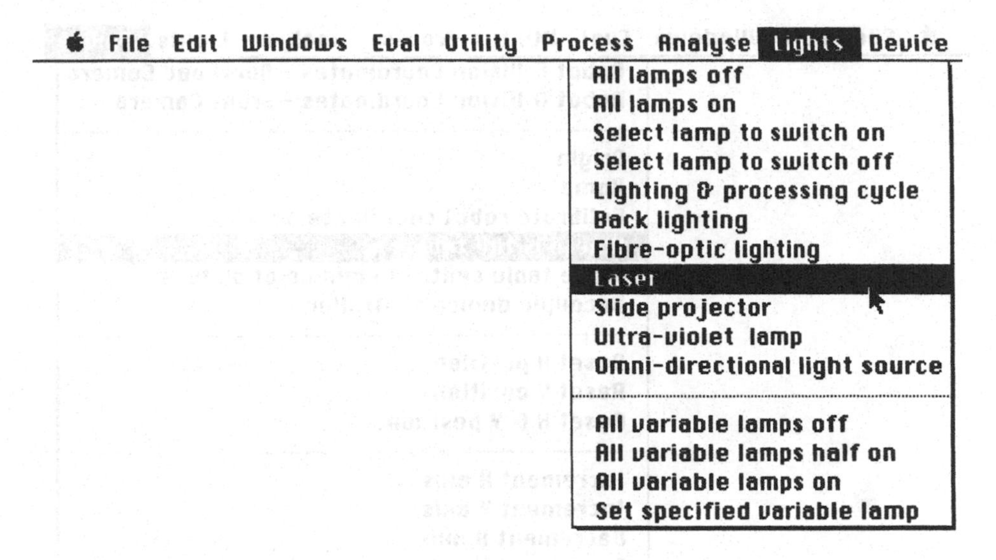

Figure 3.7. The **Lights** menu.

purpose flood-lamps. A slide projector could also be operated, allowing a wide variety of light patterns to be projected onto the scene being viewed.

Device Menu

The Flexible Inspection Cell also contains an (X,Y,θ)-table and a pick-and-place arm. Both of these can be controlled from the **Device** menu. The table coordinate system must somehow be related to the vision system coordinates and so the menu contains an item which initiates the calibration routine. It is also possible to order the table to move to its "home" position (i.e. beneath pick-and-place arm), or to the centre of the field of view of the (overhead) camera. In addition, the **Device** menu allows all three of the table axes to be incremented/decremented independently. The pick-and-place arm can also be operated using the **Device** menu.

It is in the operation of the cell that the importance of the pull-down menus becomes apparent. It is essential, when experimenting with the table and the lights, to be able to operate its various elements easily and without resorting to programming. It has been discovered by experience that the pull-down menus provide a very convenient method of experimenting with both the electro-mechanical equipment and lighting.

Applications Menu

This menu is provided for initiating demonstrations and other user-defined programs. Some interesting demonstrations are based upon the programs listed in Chapter 7 and include:

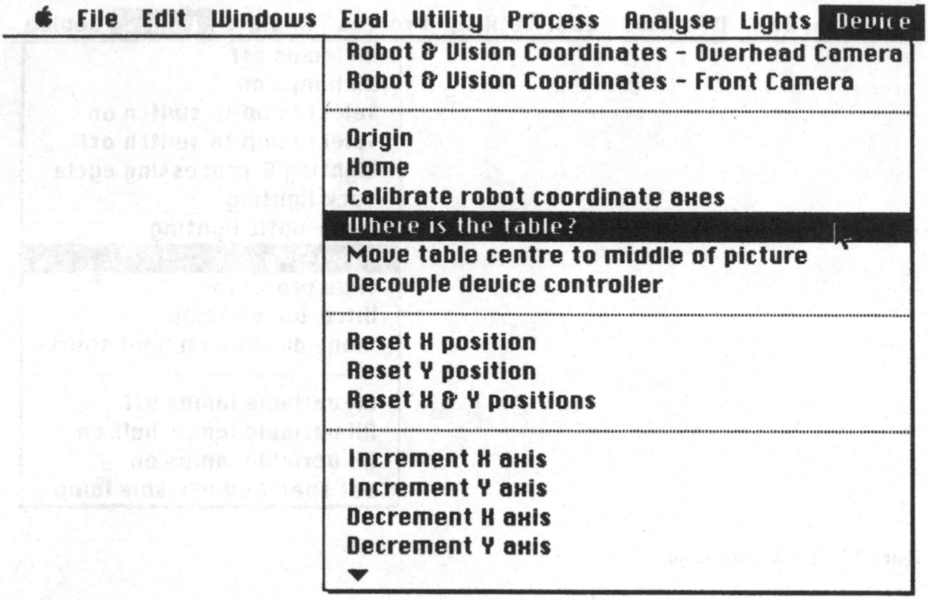

File Edit Windows Eval Utility Process Analyse Lights Device

Robot & Vision Coordinates - Overhead Camera
Robot & Vision Coordinates - Front Camera
Origin
Home
Calibrate robot coordinate axes
Where is the table?
Move table centre to middle of picture
Decouple device controller
Reset X position
Reset Y position
Reset X & Y positions
Increment X axis
Increment Y axis
Decrement X axis
Decrement Y axis

Figure 3.8. The **Device** menu.

Recognising the suit of a playing card
Picking up an automobile connecting rod (con-rod)
Dissecting a plant
Playing dice
Recognising cracks
Stacking blocks

In addition, the **Applications** menu can be used for other initiating utilities, such as an expert system that has been developed by the author for giving advice about lighting and viewing. The so-called Lighting Advisor is a program, written in MacProlog, which provides an interactive dialogue with the user and presents a series of questions about the type of object that is being viewed, what its important features are, as well as a variety of environmental and other factors. The Lighting Advisor then suggests one or more possible lighting and viewing configurations, giving its advice in the form of line drawings, references to the technical literature and short blocks of text. The Lighting Advisor is described elsewhere [BAT-89c]. The author is also developing a similar advisor, which will enable a vision systems engineer to select a suitable camera. A lens selection program is also needed. The latter is quite different from a conventional lens design program, which is principally concerned with creating multiple lenses from simple lenses. The type of lens selection procedure that is envisaged would concentrate upon choosing standard lens assemblies, close-up attachments, etc. Programs such as these three expert systems can be integrated with Prolog+ very simply, using the pull-down menus.

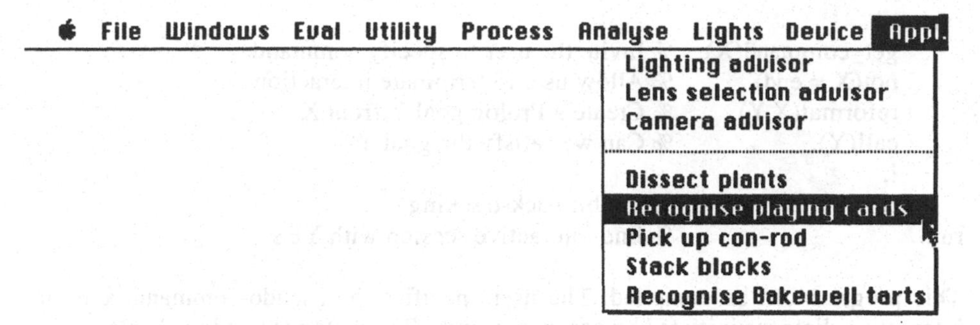

Figure 3.9. The **Applications** menu. This menu is intended to be extended by the user (see legend to Figure 3.4.).

3.2 Command Keys

The command keys provide the same functions as some of the menu items and allow even more rapid access to some of the frequently used system control operators. The following list shows the command keys currently in use on the author's Prolog+ system. (The character "" represents the Macintosh Command character.)

´G Initialise the system

´P Purge the I/O buffer, after an error, to make certain that the interprocessor communication continues correctly

´T Transparent mode – discussed in section 3.8

´J Digitise an image

´H Switch the current and alternative images

´1 Repeat previous X increment of (X,Y,θ)-table

´2 Repeat previous Y increment of (X,Y,θ)-table

´3 Repeat previous θ increment of (X,Y,θ)-table

3.3 Transparent (or Interactive) Mode

We stated earlier that an essential feature of Prolog+ is that it retains the interactive nature of its progenitors, the image processing languages such as Susie, Autoview and VCS. How can interaction be achieved via a compiled language? The answer will be immediately obvious to readers who are thoroughly familiar with Prolog. Consider the following program skeleton:

```
run :-
    get_command(X),  % Invite the user to specify command
    not(X = end),    % Allow user to terminate interaction
    reformat(X,Y),   % Créate a Prolog goal Y from X
    call(Y).         % Can we satisfy the goal Y?
    !,
    run.             % Inhibit back-tracking
run.                 % End interactive session with YES
```

This program can be compiled. The user specifies the pseudo-command X in an interactive dialogue initiated by **get_command**. (The dialogue window displayed by **get_command** is shown in Figure 3.10.) The predicate **reformat** is then used to create a properly formatted Prolog goal, Y, from X. Suppose that X is instantiated by **get_command** to **thr(125,193). reformat** then instantiates Y to

image_proc(thr(128,193),L)

and **call** tries to satisfy the goal.

call(image_proc(thr(128,193),L))

The effect of this is to order the image processor to perform the operation

thr(128,193).

The whole process is repeated, thereby achieving a semblance of interaction. The session terminates when the user orders the user pseudo-goal

end.

3.4 The @ Operator

There may be some situations in which the image processor requires a command which cannot be accommodated easily within Prolog+. The @ operator has been provided to assist here. The command

@'STRING'

simply sends the character string defined by STRING to the image processor, without trying to understand it. This is similar in concept to the # operator discussed in the following section, although their uses are quite different.

The principal use of the @ operator is to allow the user to program the image processor at a different level from normal. For example, Figure 2.20 shows a simple VCS program for blob analysis. To write a program like this from within a Prolog program, we would use the @ operator, to specify the program line by line. Here is

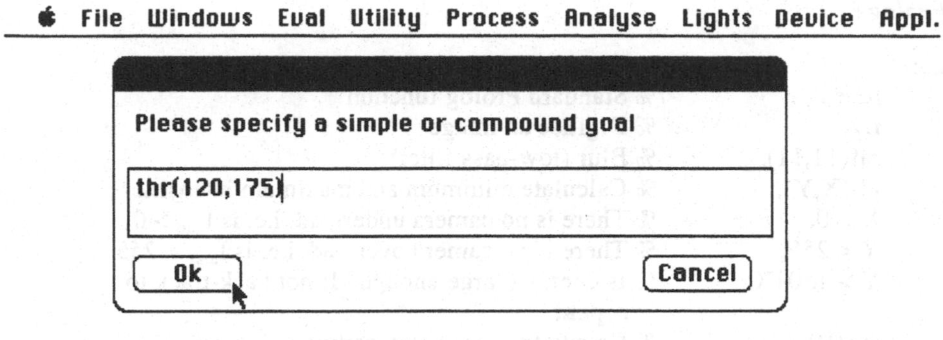

File Windows Eval Utility Process Analyse Lights Device Appl.

Please specify a simple or compound goal

thr(120,175)

Ok   Cancel

Figure 3.10. Transparent mode dialogue box.

(part of) a Prolog program which writes and then runs the VCS program shown in Figure 2.20:

```
write_vcs_prog :-
        @ '100          sof',
        @ '200          ndo 3 N',
        @ '300          S = 0',
        ......          % Lines 400 – 1300 go in here
        @ '1400         print "The total number of bays is", S'
        @ '1500         stop "Program finished"',
        @ run.
```

Although this seems to be a rather clumsy method of programming, it does allow the full range of facilities of the VCS image processing language to be employed. This has been very useful in the past, since it allows BASIC-type programs to be written which augment the VCS command repertoire.

3.5 Very Simple Prolog+

The objective of this section is to explain how a simple, but expandable "core" Prolog+ system may be implemented. This will be called VSP (Very Simple Prolog+). It should be noted that VSP is not committed to using any one specific image processor, although we shall continue to assume that VCS is being used. The reader is referred to Appendix III where the VSP software is listed in full, while Appendix IV describes another implementation using a different image processor, the Intelligent Camera.[2]

In order to understand the relationship between Prolog+ and VSP, compare the two programs below. These check that the minimum and maximum intensity limits and contrast are all within satisfactory limits.

Prolog+

```
go:-
    repeat,           % Standard Prolog function
    frz,              % Digitise an image
    raf(11,11),       % Blur (low-pass filter)
    gli(X,Y),         % Calculate minimum and maximum intensities
    X > 0,            % There is no camera underload, i.e. is Imin > 0?
    Y < 255,          % There is no camera overload. i.e. is Imax < 255
    Y > 150+X,        % Is contrast large enough? If not back-track to
                        repeat
    avr(R),           % Calculate average intensity
    nl,               % Print a message for the user
    write('[Minimum, Average, Maximum] = '),
    write([X,R,Y]).
```

VSP

```
go:-
    repeat,           % Standard Prolog function
    # frz,            % Digitise an image
    # raf(11,11),     % Blur (low-pass filter)
    # gli(X,Y),       % Calculate minimum and maximum intensities
    X > 0,            % There is no camera underload
    Y < 255,          % There is no camera overload
    Y > 150+X,        % Is contrast OK?
    # avr(R).         % Calculate average intensity
    nl,               % Message to the user
    write('[Minimum, Average, Maximum] = '),
    write([X,R,Y]).
```

In VSP, all commands to the image processor are prefixed by the # operator, which is defined in the following section. The # operator is the essential feature which makes the implementation of VSP simpler than that of Prolog+.

3.6 The # Operator

Prefixing all image processing commands with the # operator may, at first, seem a little clumsy, compared to ordinary Prolog+. Obviously, when typing programs which incorporate the # operator, the user must press one extra key for each command.

The advantage of using the # operator is that it makes the software interface between Prolog and the image processor much simpler and able to accommodate different types of image processor, without reprogramming. The # operator may be defined using LPA MacProlog in the following way:

```
# # A :- # A.              % Tolerate extra # in "Transparent mode"

# A :-
        A =.. [P|Q],        % Decompose the input term
        atom(P),            % Is the first part of input term an atom?
        constants(A,C),     % Extract list of constants from input term
        D =.. C,            % Construct command
        variables(A,E),     % Find the list of variables in input term
        !,
        image_proc(D,X),    % Send command D to image processor
        (append(E,_,X) ;    % Select results desired by the user
        append(X,_,E)).

# A:- message(['The command ',A,' was not recognised by the image processor').

constants([],[]).
constants([X|Y],[X|Z]) :- not(var(X)), constants(Y,Z).
constants([X|Y],Z) :- constants(Y,Z).

variables([],[]).
variables([X|Y],[X|Z]) :- var(X), variables(Y,Z).
variables([X|Y],Z) :- variables(Y,Z).
```

Notice that it is only necessary to rewrite **image_proc**, in order to accommodate a new image processor. In addition, this definition of # is such that it cannot be resatisfied on back-tracking. The command formats allowed by this definition are illustrated in Table 3.1.

Table 3.1

Type of arguments	Prolog+ command	String sent to the image processor	Instantiates variables
None	# neg	neg	None
Instantiated	# thr(12,34)	thr(12,34)	None
Uninstantiated	# avr(X)	avr	X
Mixture, case 1	# ndo(3,X)	ndo(3)	X
Mixture, case 2	# fun(1,A,2,B,3,C)	fun(1,2,3)	A,B,C[3]

It is a trivial matter to define new predicates which do not require the use of the # operator, as Table 3.2 shows.

Table 3.2

VSP	Function
thr(X,Y) :- # thr(X,Y).	Both arguments specified
thr(X) :- # thr(X,255).	Single argument option
thr :- # thr(128,255).	Default values for both arguments
threshold(X) :- # thr(X,255).	New name for an existing operator
thr(light_grey) :- # thr(192,255).	New argument type
thresh_neg(X) :- # thr(X), # neg.	New command

In this way, it is possible to define Prolog+ in terms of VSP. In the latter, error checking is performed by the image processor's command interpreter, rather than by Prolog.

3.7 Device Control in VSP

Device control may be achieved using another operator (¶), which is defined in a similar way to # (Table 3.3).

Table 3.3

VSP	Function
¶ home	Send the (X,Y-θ)-table to home position
¶ move_to(X,Y,T)	Move the table to (X,Y,T)
¶ rotate(D)	Rotate the table by D degrees
¶ pick	Lower the gripper, close it and then raise it
¶ ramout	Move the pick-and-place arm to out position
¶ light(7,9)	Set bulb number 7 to brightness level 9
¶ input(6,X)	Instantiate X to the input line number
¶ zoom(out)	Operate the camera's zoom lens

3.8 Pull-Down Menus in VSP

The VSP software is provided with four totally empty and one nearly empty pull-down menus. These are labelled as shown in Table 3.4.[4] (The items in brackets indicate the alternative keyboard commands.)

The *Extend menu* item under the **Utility** menu initiates an interactive dialogue which first asks the user to nominate which menu is to be extended. After this, the user is asked to specify the name of the new item and then what action it is expected to invoke. A fact of the following form is then placed in the database using the Prolog **assert** predicate.

user_menu(Menu,Item,Action).

Table 3.4

Menu name	Intended for	Initially contains
Utility	System commands	Extend menu (⌘M) Transparent mode (⌘T) Initialise system (⌘J)
Process	Processing grey-scale images	Empty
Analyse	Analysing images	Empty
Device	Operating electro-mechanical devices	Empty

3.9 Speech Synthesis

The VSP software supports a speech synthesiser, which makes it possible for a program to talk about what it is doing.[5] In order to control the speech synthesiser command from a VSP program, the user may use the **speak** predicate. Notice that **speak** does not need to be prefixed by either the # or ¶ operators, since it is implemented entirely within the Macintosh computer.[6]

The following goal

```
say_hello :-
        X = 'Hello dear friend',
        Y = 'This is a test of the speech synthesiser',
        concat(X,Y,Z),                  % Concatenate X and Y to create Z
        speak(Z).
```

will result in the utterance

"Hello dear friend. This is a test of the speech synthesiser"

In order to provide information about the flow of a program, three new predicates have been defined.

```
% _______________Use "repeats" instead of "repeat"_____________

repeats :-
        repeat,
        speak('Beginning of repeat loop').

% __________________Use "fails" instead of "fail"_______________

fails :-
        speak('Back-tracking'),
        fail.

% ____________________Use "!,cut" instead of "cut"______________

cut :-
        speak('Forwards through cut').
cut :-
        speak('Cut reached by back-tracking'),
        !,
        fail.
```

The following program demonstrates the use of speech and beautifully illustrates the concept of back-tracking to a novice programmer.

```
go :-
      member(X,[arm,leg]),          % Instantiate X to arm
      !, cut,                       % Modified cut; fix value of X
      member(Y,[cat,dog),           % Instantiate Y to cat, dog
      speak(X),                     % Speech synthesiser call
      speak(Y),                     % Speech synthesiser call
      fails.                        % Modified fail

go :- speak('Clause two. Goal go has succeeded.')% Forced success of goal "go"
```

The speech synthesiser output is then

"arm
Forwards through cut
cat
Back-tracking
dog
Back-tracking
Cut reached by back-tracking
Clause two. Goal go has succeeded."

Now, it may not be possible for the speech synthesiser to make any sensible utterance, given, say, a three-letter image processing or device control mnemonic command, such as **avr**, **thr**, **lpf**, etc.. One solution is to provide VSP with a dictionary, of the following form:

```
speech_dictionary(neg,negate).     % "# neg" will result in "negate" being said
speech_dictionary(thr,threshold).
speech_dictionary(con,'Three by three convolution operator').
.........
speech_dictionary(_,'No information available about this command').
```

This arrangement is not totally satisfactory, because it makes the VSP software specific to one particular image processor, which we have deliberately tried to avoid.

An alternative, which follows the spirit of VSP rather better, is to arrange for the functor of any term prefixed by # to be spelled. Assuming that there is no entry corresponding to **qwerty** in the dictionary, the following goal

```
# qwerty(1,2,X,3,4,Y)
```

would result in the "qwerty" being spelled letter by letter; the speech synthesiser would then say:

"queue
double ewe
ee

are
tee
wye"

In practice, a combination of these two approaches is likely to be favoured and is, in fact used in the author's VSP system.

It is important to be able to switch off the speech synthesis facility, because it becomes tedious after prolonged use and often seriously affects the operating speed.

3.10 The VSP Software and Other Implementations

Appendix III lists a Prolog program which combines the ideas described above. The **speak** and **prepare_to_speak** predicates are compatible with the speech synthesis software. Apart from this, and the unspecified driver for the image processor, the program is complete. In order to illustrate the programs, a driver for the VCS 512 image processor is presented.

Figure 3.11 shows the configurations for several Prolog+ systems which have been built in the past or are planned for construction in the near future. Of these, the one shown in Figure 3.11(f) is the most convenient, since it uses a single standard processor. The system organisation sketched in Figure 3.11(h) is the fastest yet envisaged. These two configurations will now be described in more detail. Appendix IV describes yet another implementation, which uses the Intelligent Camera (Fig 3.11(g)). The importance of this particular development is its small physical size and low cost. Plans to host the Prolog software within the body of the Intelligent Camera are also being considered.

Software

The system shown in Figure 3.11(f) uses a Macintosh II computer.[7] This has the following configuration:

Processor:	Motorola 68020/68030[8]
RAM:	≥ 5 Mbytes
Disc:	≥ 20 Mbytes
Display:	8-bit display option, monochrome monitor
Frame-store:	Data Translation
Image processor:	VCS, written in MPW C, Vision Dynamics Ltd, Hemel Hempstead, Herts, U.K.
Prolog:	MacProlog, version 2.5 or later, Logic Programming Associates Ltd, London, U.K.
Prolog+:	VSP software as listed in Appendix III, plus the Library of image processing operators described in Chapter 5

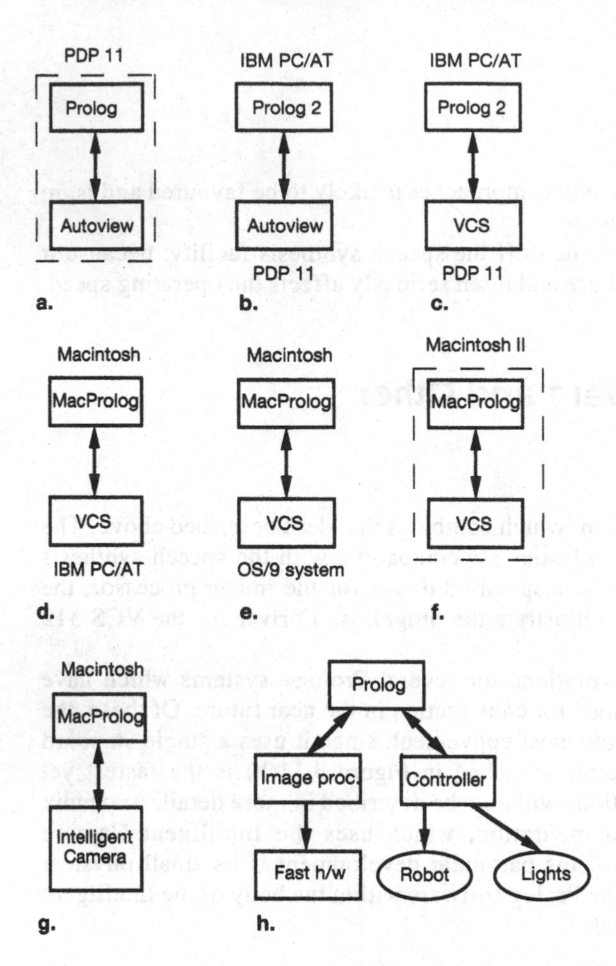

Figure 3.11. Possible ways of implementing Prolog+. Configurations (**a**)–(**e**) have been constructed previously, while (**f**) is under construction. (**g**) is the method of implementing VSP described in Appendix IV. Notice that a computer of modest size (e.g. Macintosh Plus or Macintosh SE) may be used. (**h**) Prolog+ may be used to control a high-speed image processor, as well as a robot, lights and other equipment (not shown). A system with this architecture is under construction.

Needless to say, a software implementation is slow in operation. However, it is relatively cheap and uses standard hardware. As a result, it is well suited as an educational and training tool and it is perfectly adequate as a prototyping tool for developing intelligent image processing techniques for inspection and robot vision.

Hardware

It is beyond the scope of this monograph to provide a comprehensive review of all of the numerous dedicated image processing systems that have been proposed in the past. Instead, we shall merely review the possibilities of using standard commercial plug-in boards for desk-top computers and then discuss just one dedicated image processor, which has a pipe-line organisation.

An accelerator is a *general-purpose* board which simply plugs into one of its host computer's expansion slots and permits essentially the same software to be used. This method of speeding up a desk-top computer is very quick and easy to accomplish, requiring only a cheque book, screw-driver and about an hour to install the new board (Figure 3.12). However, accelerators can provide only a relatively small speed improvement compared to their host.[9] Boards of this type cost roughly the same as a high-performance desk-top computer.

General-purpose, high-speed processing engines are available commercially. For example, array processors boasting processing speeds of roughly one order of magnitude greater than a desk-top computer are now available[10] and cost approximately 10 times as much (See Figure 3.13).

However, for much higher processing speeds, *dedicated* image processing equipment is needed. The organisation of one such system is sketched in Figure 3.14. It uses only standard, commercially available image processing boards and can achieve very high speed of operation (Table 3.5). Of course, such a dedicated image processor cannot be expanded nearly as easily as a software-based system; a very much higher level of programming skill is needed. For this reason, though it is capable of achieving a very high speed of operation, a dedicated image processor is essentially a closed system. What is remarkable is that it has been found to be possible to implement a comprehensive repertoire of image processing functions in a fixed architecture system of reasonable cost (see Appendix IV). Obviously, new operators have to be added from time to time, but this is a fairly infrequent event, since so much can be achieved by using combinations of commands; when the implementation is fast enough, deficiencies in the command repertoire can *almost invariably* be accommodated by programming at the Prolog+ level.

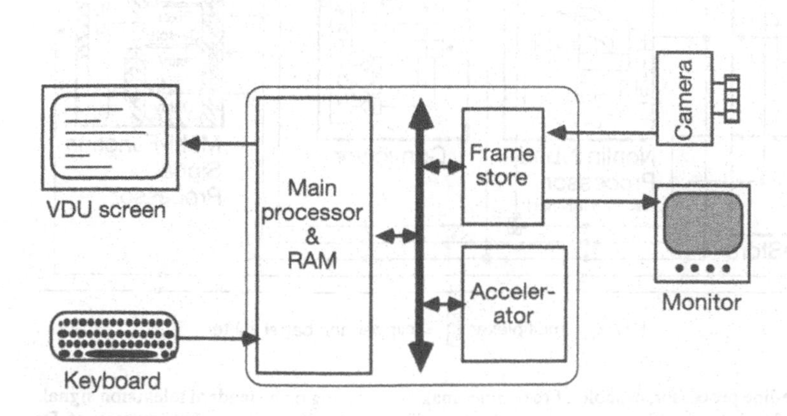

Figure 3.12. Hardware accelerator.

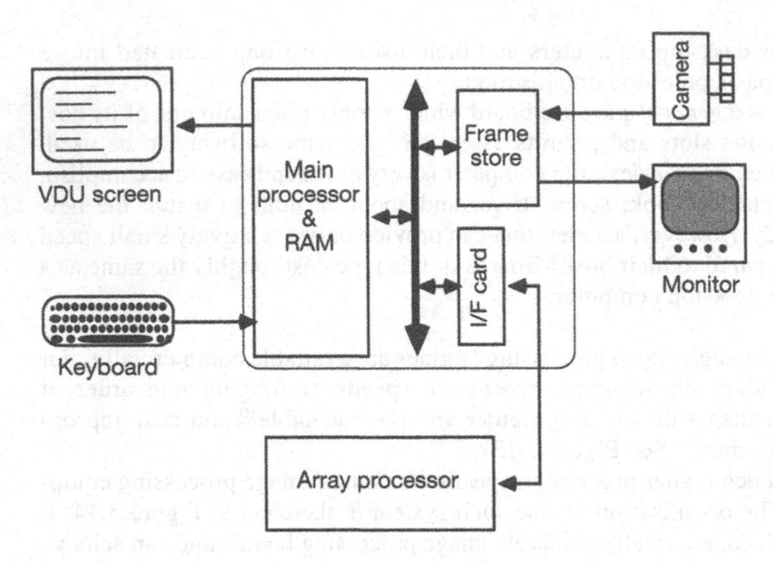

Figure 3.13. Array processor controlled from a desk-top computer.

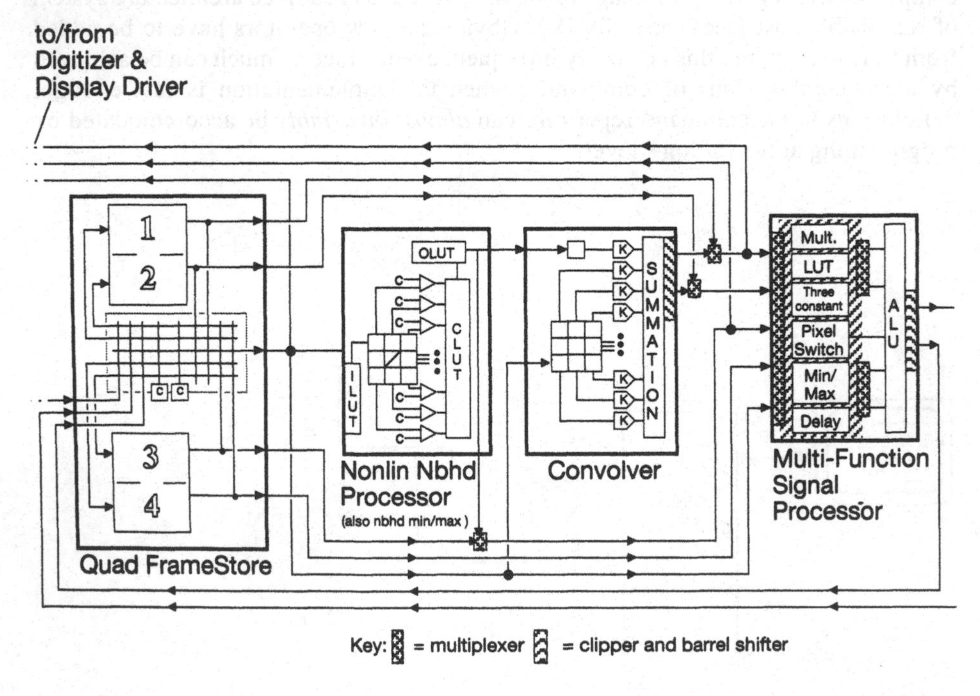

Figure 3.14. A pipe-line processor, capable of real-time image processing on a standard television signal. Computation times for this and other systems are given in Table 3.5. (Reproduced by courtesy of Dr Frederick Waltz.)

Table 3.5. Comparison of the execution times of various image processing functions implemented using different hardware systems. Notice the varying image resolution.

Function/mnemonic (Resolution)	**MAX** ($\sim 512^2$)	**VCS** (512^2)	**ICAM** (256^2)	**LAP** (256^2)	**VSP** (256^2)
Threshold (**thr**)	40	2600	37	7	403
Negate (**neg**)	40	2600	52	6	415
Add intensities (**add**)	40	3600	100	11	500
Max. intensities (**max**)	40	1300	105	11	522
3 × 3 low-pass filter (**lpf**)	40	24000	180	36	700
Sobel edge detector (**sed**)	120	8600	285	34	748
Average intensity (**avr**)	40	1500	165	90	732
Count white points (**cwp**)	40	1000	69	74	588

Key:

MAX Maxvideo image processing cards, Datacube, Inc., Peabody, MA, U.S.A. (The boards are operated in a fixed configuration which is capable of implementing a large number of functions. Times given here are based on the assumption that the system is operating on a CCIR video signal; 40 ms is the television frame period.)

VCS VCS 512 processor, Vision Dynamics, Hemel Hempstead, U.K. (M 68000 processor. Software written in C.)

ICAM Intelligent Camera, Image Inspection Ltd, Epsom, U.K. (Bit slice processor, micro-code software.)

LAP Linear Array Processor, National Physical Laboratory, Teddington, U.K.

VSP VSP using the Intelligent Camera. Comparison with the **ICAM** column indicates the computational overhead imposed by MacProlog, running on a Macintosh SE computer with a serial data link (RS423, 9600 baud).

Concurrent Processing

Another way to achieve greater speed in a software-based system is to use several image processors, running concurrently. This requires certain changes to the Prolog+ software, so a discussion of this topic will be deferred until Chapter 9.

3.11 Adding Other AI Facilities

LPA MacProlog is able to interface to a number of optional software tools, which greatly extend its usefulness. These are listed below:

(i) *Flex*, an expert systems tool kit

(ii) *Prolog++*, an object oriented extension to Prolog

(iii) *MacObject*, a graphical front end to Prolog++

(iv) KSL, a Knowledge Specification Language

(v) Graphics

The Prolog+ software may be regarded as providing a bridge between the low-level, (i.e. non-intelligent) image processing and the AI facilities available in Prolog and these extensions to it (see Figure 3.15). The potential offered by these extensions to MacProlog has yet to be realised.[11]

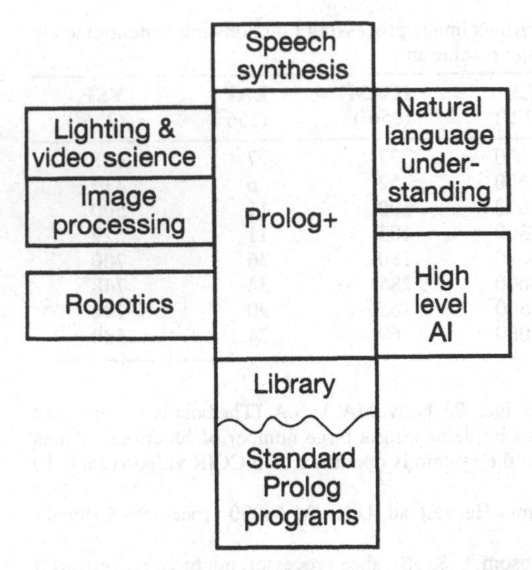

Figure 3.15. Prolog+ is a bridge between non-intelligent image processing and AI tools.

3.12 Summary

Prolog+ can be implemented using a standard software interface, called VSP, which is not committed to any one particular image processing engine. It is possible to define Prolog+ fully in terms of VSP. The VSP software makes use of the # operator, which prefixes all image processing commands. Similarly, the ¶ operator is used to control external devices, such as lights, the camera and an (X,Y,θ)-table. VSP is supplied with empty pull-down menus, but these can be extended easily by the user, without complicated programming. VSP also allows programs to gain access to the speech synthesis package. This has been found to be very useful in developing and debugging Prolog+/VSP programs. Its intended use, however, was to assist a user to follow the flow of a program as it analyses an image using back-tracking and recursion. Speech helps both the programmer and a casual onlooker to understand *why* a visually guided program is behaving in a certain way.

The implementation of Prolog+/VSP using software and dedicated hardware has also been discussed. We do not mean to suggest that the story is finished, but we are certainly at a point where a fast and flexible version of Prolog+ can be implemented in a machine of modest cost. *Prolog+ is neither too expensive, nor too slow for practical use!*

Notes

1. 0 (zero) represents off and 15 full on. There are, of course, 16^{16} different lighting patterns.
2. Image Inspection Ltd, Unit 7, First Quarter, Blenheim Road, Epsom, Surrey KT19 9QN, U.K. The command language is also derived from that of the Autoview image processor.

3. Depending upon how many values are calculated by the nonsense command **fun**, C, [B,C], or [A,B,C] will **share** with uninstantiated variables.
4. These menus apply to the Macintosh II version of the VSP software. The Macintosh Plus/SE version of the software uses only three special menus: "Utility", "Process" and "Analyse", because the screens of the these computers are narrower than that of the Macintosh II machine.
5. The speech synthesis software was developed for LPA MacProlog and was kindly made available to the author by Mr Terry Gritton, 65 Nunes Road, Watsonville, CA 95076, U.S.A.
6. The **speak** predicate, is of course, supported by a lot of "invisible" software, some of which is written in Prolog, while the remainder is in C. This controls the Macintalk speech synthesis package.
7. A word is perhaps in order about the choice of the Macintosh computer as the basis for VSP. Like most Macintosh users, the author has developed a very strong affection for this machine, on account of its very friendly user interface. He is well aware of the fact that the computer world is divided into those who love this computer and those who ridicule it as a toy. Having used this machine for several years for a wide variety of uses, including writing, drawing, spreadsheets, databases, expert systems, etc., he has gained great confidence in it and is now very firmly in the former category. The easy transportation of data between different applications was of particular value to the author while he was writing this book. Thus, it was possible to merge programs and diagrams into text and transport diagrams into Prolog. As a scientific computer, it lacks nothing. Indeed, it gains from being easy to use, so that the programmer can concentrate on the task in hand, not the operation of the machine. The most important single reason, however, is that MacProlog runs on this computer. This is an *excellent* implementation of the language and its use is made even more attractive by the WIMP operating environment.
8. The Macintosh II/C and II/CX computers were announced shortly before this section was written. These are both rather faster than the Macintosh II machine and would be suitable for this application.
9. In the February 1989 issue of MacUser magazine, the Day Star 030 accelerator was the only accelerator reviewed and provided a speed improvement over the Macintosh II computer of roughly 2.5:1. More recently, the much faster Macintosh IIfx computer has become available.
10. In the February 1989 issue of MacUser magazine, a 16-processor system, called Chorus, was reviewed. This machine boasts an operating speed of 80 Million Instructions Per Second (MIPS).
11. The role of Prolog+ as a bridge between image processing and AI is nicely summarised in the phrase "Pixels to Predicates", which the author recently discovered is the title of a book by another author. Pity!

CHAPTER 4

EXTENDING THE LANGUAGE

"When he's excited he uses language that would make your hair curl"
Ruddigore, W. S. Gilbert

4.1 The Library

A large number of predicates for image processing, robot control and other tasks are listed in Appendix II. Of course, such a Library is never "fully complete" and could be extended indefinitely. Several points should be made:

(a) All of the macros listed in the earlier publications of interactive image processing [BAT-79] can be rewritten in terms of Prolog+, although Appendix II mentions only a few of them. The author and his colleagues are adding further predicates to the Library, almost daily.

(b) The use of general data structures is well illustrated in the predicates **left** and **right**. We do not need to know what A and B are, in order to understand the statement

A is to the left of B.

Neither does Prolog+. Notice too that the predicate **right** can be defined conveniently in terms of **left** and that it is not necessary to understand how **left** works in order to understand **right**.

```
right(A,B) :- left(B,A).
```

This is also typical of Prolog-based programs. Think about the following English statement:

*The relationship **XYZ** between A and B exists if the relationship **ZYX** exists between B and A.*

We do not need to know what A and B are. Nor do we need to possess a definition of **ZYX**. The following examples emphasise this point, assuming that **XYZ** is the relationship **left** and **ZYX** is the relationship **right**.

A	**B**
john	mary
circle	square
letter_a	centre_of_image
3	5
biggest_hole	centroid_of_silhouette

Some of the possible definitions for the "reciprocal" relations **XYZ** and **ZYX** are as follows:

XYZ	**ZYX**
greater_area	smaller_area
inside	encloses
above	below
smoother	rougher
richer	poorer
>>	<<

Very few computer languages, apart from Prolog and its derivatives, can handle undefined data items and relationships in this way.

(c) Certain functions contained in the Library which are currently programmed using Prolog+ might be performed more efficiently in a conventional (i.e. imperative) language, such as C or Pascal. The point we wish to emphasise is that Prolog+ is versatile and flexible and does not place undue restrictions on the programmer. Computation speed is not our primary concern at this stage; user convenience and expressional power are.

(d) Extensive on-line documentation and *Help* facilities are provided to assist the programmer to make effective use of the Prolog+ Library. Some of these functions are implicit in MacProlog, whilst others have been added. The self documentation facility is described in more detail in a next section.

(e) Since the Library is continually being extended, keeping track of modifications could be troublesome. In order to reduce the difficulties, each window contains a small headless clause which places a record in the database indicating both the time and date when that window was last modified. This enables the user to keep track of the different versions of the software.

4.2 Self Documentation of the Library

The following is a short sample of code and comments, extracted from the Library. Each entry in the Library is preceded by a comment containing a character string enclosed within double triangular brackets: <<...>>.

```
/* << 'right(A,B)','image test','Is A to the right of B?' >> Succeeds if A is to right
of B. */

right(A,B) :- left(B,A).

/* << 'above(A,B)','image test','Is A above B?' >> Succeeds if A is above B. */

above(A,B) :-
        location(A,_,Ya),
        location(B,_,Yb),
        tolerance(P),
        Ya – Ya > P.
```

Notice that the brackets "<<" and ">>" enclose character strings which are invisible to the compiler. However, these text strings can be read by a predicate called **document_it**, in order to produce the documentation. **document_it** is defined as follows:

```
document_it :-
        retractall(descriptimember(A,B,C)),   % Clear database
        L = ['Library'],                      % Instantiate L to list of predicates
                                                in Library
        member(W,L),                          % Instantiate W to member of L[1]
        document_window(W,0),                 % Defined below
        fail.                                 % Force back-tracking

document_it :- message(['You MUST compile the program now']).

document_window(W,Z) :-
        wsearch(W, '<<',Z,A,B),      % Search window W for character string <<
        wsearch(W, '>>',Z,C,D),      % Search window W for character string >>
        P is A + 2,                  % Ignore first two characters (i.e. <<)
        Q is D – 2,                  % Ignore first two characters (i.e. >>)
        wsltxt(W,P,Q,X),             % Instantiate X to string between P and Q
        save_in_db(W),               % Save W in the database. Not defined here
        E is D + 1,                  % Move pointer past end of string just found
        document_window(W,E).        % Carry on to find next self-doc. char. string

document_window(_,_).
```

Once **document_it** has been run once, it is possible to use the database which it establishes to obtain "Help" information to display to the user.

The reason for using this mechanism for defining "Help" information is that it provides full compatibility with the program comments. Of course, not all comments are intended to be made available as "Help" information. Indeed, this is why the <<...>> construct was devised.

Two additional menu items might be made available under the **Utility** menu (see previous chapter) and to provide on-line access to the "Help" facility:

"Available Prolog+ Operators" calls **get_help** (defined below)

"Update self documentation" calls **document_it** (defined below)

Finding Prolog+ Operator Classes

The predicate **classes_available** lists all of the types of Library predicates available. This information is available in the second term between the <<...>> brackets.

```
classes_available :-
    nl,
    write('The following classes of Library operators are available:'),
    operator_classes([],_),
    nl.

operator_classes(L1,L2) :-
    description(A,B,C),
    not(member(B,L1)),
    nl,
    tab(10),
    write(B),
    append([B],L1,L3),
    operator_classes(L3,L2).

operator_classes(L,L).
```

What Does Each Class Contain?

The predicate **class_contains** lists all of Library predicates which have a given type, Z.

```
class_contains(Z) :-
    nl,
    write('The class'),
    write(Z),
    write('contains the following operators:'),
    description(A,Z,C),
    nl,
    tab(10),
    write(A),
    fail.

class_contains(_) :-
    nl.
```

List All Operators Available in the Library

```
operators_available(L1,L2) :-        % Append list of operators available to L1
    description(A,_,_),
    not(member(A,L1)),
    append([A],L1,L3),
    operators_available(L3,L2).

operators_available(L,L).
```

Finding "Help" Information about a Given Topic

The predicate **get_help** is used to obtain information about a given topic. Various calling options are supported.

```
get_help(all) :-                     % Get "Help" information on all predicates²
    description(A,B,C),
    nl,
    write('Goal:'),
    write(A),
    tab(10),
    write('Function type:'),
    write(B),
    nl,
    write('Answers the question:'),
    write(C),
    nl,
    fail.

get_help(all).

get_help(classes) :-                 % Find what operator classes are available
    nl,
    write('Operator classes available:'),
    operator_classes([],_),
    nl.

get_help(A) :-                       % Get help on a specific predicate
    description(A,B,C),
    nl,
    write('The goal"'),
    write(A),
    write('"performs a'),
    write(B),
    write('function and answers the question"'),
    write(C),
    write('"'),
```

```
        nl,
        nl.

get_help(_) :-
        message(['That operator is not available. – Perhaps you have got the name
                                                                    wrong']),
        get_help.

get_help :-                              % Use a scrolling menu to select the topic
        operators_available([],L1),
        append([all,classes],L1,L2),
        scroll_menu(['Select one topic to obtain information'],L2,[all],X),
        [Y] = X,
        get_help(Y).
```

4.3 Operators

A range of useful operators can be used to good effect to provide an alternative syntactic structure for Prolog+. Many of the Library predicates can be redefined, as the following examples show.

```
A left B :- left(A,B).
A left_of B :- left(A,B).
A bigger B :- bigger(A,B).
A inside B :- inside(A,B).
A touches B :- touches(A,B).
circular A :- circular(A).
line_end A :- line_end(A).
```

The major reason for defining operators like this is to make the syntax more natural to the programmer.

It is also possible, using MacProlog, to define the **if** operator thus:

```
:- op(1200,xfx,if).
term_expand((A if B),(A :- B)).
```

The logical connective operators **&** and **or** may be defined in the following way:

```
:- op(1100,xfy,';').
:- op(1000,xfy,',').
A & B :- A , B.
A or B :- A ; B.
```

Using operators in a fairly obvious way, it is possible to write a program for recognising a sans serif upper-case letter "A".

```
letter(A) if
        apex A &                        % Is A an apex?
        A above B &                     % Is A above B?
        B about_same_height C &         % Are B and C at about same height?
        A connected_to B &              % Are A and B connected?
        A connected_to C &              % Are A and C connected?
        B connected_to C &              % Are B and C connected?
        line_end D &                    % Is D a line end?
        line_end E &                    % Is E a line end?
        B above D &                     % Is B above D?
        D about_same_height E &         % Are D and E at about same height?
        B connected_to D &              % Are B and D connected?
        C connected_to E &              % Are C and E connected?
        B left C &                      % Is B to the left of C?
        D left E.                       % Is D to the left of E?
```

However, we can go somewhat further than this in creating a language resembling natural English, using the ability of Prolog to manipulate *Definite Clause Grammars* or *DCGs*.

4.4 Definite Clause Grammars

It has long been the ambition of computer scientists to use natural language for programming. However, the structure of a sophisticated modern language, such as English, German or Dutch, is very complex and representing human, as distinct from computer, languages in terms that a machine can understand is still the subject of much research. However, it is possible to represent a *small subset* of a natural language using various computer languages, including Prolog, Lisp and Pop11. For example, the sentences that might be needed to describe the positions of pieces on a chess board use both a restricted vocabulary and have a fairly simple structure. It is not difficult to define such a simple language using Prolog.

The conventional approach to linguistic analysis via Prolog relies upon what are called *Definite Clause Grammars* (DCGs). DCGs closely resemble the familiar Backus-Naur Form (BNF), which is used frequently to define computer language syntax.

A definite clause grammar consists of a set of *production rules* of the form

head ⟶ body

which may be interpreted as follows:

*To recognise **head** recognise **body**.*

It is no accident that this production rule closely resembles a normal Prolog rule and it is perhaps more natural for the Prolog+ programmer to read the production rule given above as

***head** is satisfied if **body** is satisfied*

Production rules may also contain the connective operators "|" and ",". The former allows options to be ORed together. For example,

head ⟶ test1 | test2 | test3 | ...... |testn

states that in order to satisfy **head**, it is sufficient to satisfy *any one* (or more) of the subsidiary tests: **test1**, **test2**, **test3**, ..., **testn**. An alternative is to use several separate rules. Thus:

head ⟶ test1

head ⟶ test2

head ⟶ test3

has the same meaning as:

head ⟶ test1 | test2 | test3

or

head ⟶ test1

head ⟶ test2 | test3

The operator "," (comma) is used to signify that two or more tests must *all* be satisfied in order that the head be satisfied. Thus:

head ⟶ test1,test2,test3

states that in order to satisfy **head**, all three of the subsidiary tests (**test1**, **test2**, **test3**) must be satisfied.

The body of a DCG is composed of terminal and non-terminal symbols and conditions, separated by commas. Here is a very simple DCG which describes ordinary British personal names

name ⟶ forename1, surname

forename1 ⟶ forename2 | forename2, forename1

forename2 ⟶ [angus] | [godfrey] | [mildred] | [nelly] | [david]

forename2 ⟶ [susie] | [george] | [mary] | [pat]

etc.

surname ⟶ [smith] | [higgins] | [davies] | [williams] | [jones]

Here are a few of the possible names which satisfy these grammar rules.

[angus,godfrey,smith]

[mildred,smith]

[charles,nelly,jones]

[godfrey, charles,godfrey, charles, godfrey, charles, godfrey, charles, jones]

Of course, some of these are quite acceptable, whilst others are just silly. This is a result of the very simple set of grammar rules that we have used to define a name.

We have already pointed out that there is a close resemblance between DCGs and Prolog rules, although it must be understood that they are not exactly the same. DCGs define the structure of a simple, formal language and can provide the basis for writing in a format approaching natural language. DCGs can be translated into Prolog clauses by adding extra arguments to each of the non-terminal symbols in the DCG. Indeed, most Prolog systems provide automatic translation, so that both notations can be used simultaneously within a single program. Further details are included in references.

Parsing is the process by which a sequence of symbols is analysed to determine whether it is a legitimate statement in the language defined by a set of production rules. Parsing of DCGs may be performed in Prolog using the built-in predicate **phrase**, although a simple parser will be defined later.

It should not be assumed that Prolog is limited to using DCGs when processing natural languages [GAZ-89].

Robotics

The following set of grammar rules defines a language for controlling a robot which moves pieces on a chess board:

```
move ⟶
    order,
    man,
    [to],
    position.

move ⟶
    order,
    man,
    [from],
    position,
    [to],
    position.

order ⟶
    movement |
    [please],
    movement.
```

```
movement ⟶
        [move] |
        [shift] |
        [translate] |
        [transfer] |
        [take].

man ⟶
        article,
        color,
        piece.

article ⟶
        [] |
        [my] |
        [your] |
        [a] |
        [the].

color ⟶
        [] |
        [white] |
        [black].

piece ⟶
        [pawn] |
        [rook] |
        [castle] |
        [knight] |
        [bishop] |
        [queen] |
        [king].

position ⟶
        [column],
        number,
        [row],
        number |
        [row],
        number,
        [column]
        number.

position ⟶
        number,
        number.

number ⟶
        [1] |
        [2] |
```

```
        [3] |
        [4] |
        [5] |
        [6] |
        [7] |
        [8] |
        [one] |
        [two] |
        [three] |
        [four] |
        [five] |
        [six] |
        [seven] |
        [eight].
```

Here are some of the sentences which are accepted by this grammar:

```
[please, move, my, white, pawn, to, column, 5, row, 6]
[shift, a, black, pawn, to, row, 4,column, 6]
[please, move, the, white, queen, from, column, 5, row, 6, to, column, 5, row, 6]
[transfer, my, bishop, from, 5, 6, to, 7, 3]
[take, the, black, rook, to, 7, 3]
[please, transfer, my, knight, from, row, 3, column, 2, to, 6, 8]
```

It is possible to parse sentences which are not complete. In the following example, notice there is an uninstantiated variable, Z. (The parser, **parse**, will be defined later.)

```
parse(move, [take, the, Z, to, row, 3, column, 2])
```

This instantiates Z to **pawn**, since this is the first terminal symbol listed for the rule piece. Finding all solutions of the same goal generates the following output:

```
Nº1  Z=pawn

Nº2  Z=rook

Nº3  Z=castle

Nº4  Z=knight

Nº5  Z=bishop

Nº6  Z=queen

Nº7  Z=king

No more solutions
```

This process can be taken further. For example, all sentences in the language can be generated and printed by finding all solutions to the call[3]

```
parse(move,X),
write(X),
nl.
```

All sentences which contain exactly eight words (i.e. in DCG notation, *symbols*) can be generated thus

```
parse(move,X),
X = [_,_,_,_,_,_,_,_],
write(X),
fail.
```

This presents one possible method of extracting "meaning" from a sentence. The procedure is as follows:

(a) Use the parser to check that the sentence is valid, according to the grammar rules.

(b) Use the unification process to search for specific symbols in certain places within the given sentence.

We shall return to this topic later, after we have defined a language for controlling a set of illumination devices.

Lighting Control

The following rules define a simple grammar describing operations for controlling a set of lamps. Notice that there is no attempt, at this stage, to switch lamps on/off; the rules merely define a pattern matching device, which can validate user input sentences. Notice that the comments indicate possible symbols that would be accepted by the parser

```
light -->
      light1 |
      light2 |
      light3 |
      light4 |
      light5 |
      light6 |
      light7 |
      light8.

light1 -->
      light_verb,          % set
      type1,               % every
      light_dev1,          % lamp
      light_par1.          % [to,level,5]
```

```
light2 --->
        light_verb,             % set
        [all],                  % all
        light_dev2,             % lamps
        light_par1.             % half_on

light3 --->
        light_verb,             % put
        light_dev1,             % lamp
        numbers,                % 5
        light_par2.             % on

light4 --->
        light_verb,             % [], empty string
        light_dev1,             % lamp
        [number],               % number, no option
        numbers,                % 5
        light_par2.             % half_on

light5 --->
        light_verb,             % switch
        light_dev3,             % front_light
        light_par2.             % off

light6 --->
        light_verb,             % switch
        light_dev1,             % lamp
        numbers,                % 3
        [to],                   % to, no option
        numbers.                % 7

light7 --->
        light_verb,             % put
        light_dev1,             % lamp
        numbers,                % 1
        [to],                   % to, no option
        lamp_intensity,         % [brightness, level]
        numbers.                % 8

light8 --->
        light_verb,             % switch
        light_dev1,             % lamp
        [number],               % number, no option
        numbers,                % 3
        [to],                   % to, no option
        lamp_intensity,         % level
        numbers.                % 5
```

```
light_verb ⟶
        [] |            % empty string
        [set] |
        [put] |
        [switch].

type1 ⟶
        [each] |
        [every].

light_dev1 ⟶
        [light] |
        [lamp].

light_dev2 ⟶
        [lights] |
        [lamps].

light_dev3 ⟶
        [laser] |
        [projector] |
        [back_light] |
        [front_light].

light_par1 ⟶
        [on] |
        [off] |
        [half_on] |
        [to],
        [level],
        numbers.

light_par2 ⟶
        [on] |
        [off] |
        [half_on].

lamp_intensity ⟶
        [] |
        [level] |
        [brightness] |
        [brightness, level] |
        [intensity] |
        [intensity,level].
```

Valid Sentences

Table 4.1 shows sentences which are valid sentences, according to this grammar.

Table 4.1

Clause	Accepted string
light1	[switch, each, lamp, off]
	[switch, every, light, on]
	[set, each, lamp, off]
	[put, each, lamp, half_on]
	[every, lamp, off]
	[each, lamp, half_on]
light2	[put, all, lamps, off]
	[set, all, lights, on]
	[all, lamps, half_on]
light3	[put,lamp, 6, off]
	[switch, light, 4, half_on]
light4	[put, lamp, number, 6, off]
	[light, number, 8, on]
light5	[switch,laser,off]
	[put, back_light, on]
	[projector, off]
light6	[set, lamp, 6, to, 7]
	[light, 5, to, 9]
light7	[put, lamp,6, to, level, 7]
	[light, 5, to, brightness,4]
light8	[put, lamp, number, 6, to, intensity, level, 7]
	[set, light, number, 6, to, level, 4]

4.5 Understanding Sentences

The following program defines a simple parser.

```
parse(X,Y) :- parse(X,Y,_).

parse(X,Y,A) :-
        Z =.. [X,Y,A],
        call(Z).
```

Its use is illustrated in the following program which invites the user to type a sentence in the form of a list, tests that it is valid according to the grammar rules defined for **sentence** and then says it, using the speech synthesiser.

```
parser :-
        prompt_read(['What do you want me to say?'],X),
        parse(sentence,X),              % Test that X is a valid sentence
        speakeasy(X),                   % It is, so try and say it
        message(['That sentence was parsed correctly']),
        !,
        parser.

parser :-
        speak('That sentence was not recognisable'),
        !,
        parser.
```

Used in this way, **parse** simply provides a means of testing that a given list defines a valid sentence; it does not provide any method of "understanding" the meaning of that sentence.

We shall develop several ideas, some very naive, for extracting "meaning" out of a sentence that has already been validated. Here is the first one, based upon the lighting DCG rule, **light7**. It simply discards all non-numbers in the sentence P, provided that P is valid according to rule **light7**. The output is an ordered pair, the first element of which is the lamp to be operated and the second is its new brightness value.

```
meaning(light7,P,Q) :-
      parse(light7,P),                % Is P a valid sentence?
      keep_numbers(P,Q),              % Discard all non-numbers in P
      Q = [_,_],                      % Are there just two elements left?
      write('The sentence'),
      write(P),
      write('has been reduced to the ordered pair'),
      writenl(Q).

keep_numbers([],[]).

keep_numbers([X|Y],[X|Z]) :-
      number(X),                      % Is X a number?
      !,
      keep_numbers(Y,Z).

keep_numbers([X|Y],Z) :-
      !,
      keep_numbers(Y,Z).
```

Here is another possibility, equally naive:

```
meaning(light7,P,[A,B]) :-
      parse(light7,P),
      P =[_,_,A,_,_,B].
```

Notice that this tries to match P to a list containing six elements, of which only the third and sixth are retained. However, the rule for **light_verb** permits both *empty* and *non-empty* strings, thereby causing some slight complication. Both of the following are valid sentences according to **light7**:

[put, lamp,6, to, level, 7]

and

[lamp,6, to, level, 7].

Hence, another clause, which matches P to a list containing only five elements is also needed:

```
% __________light_verb string non-empty, so P has six elements__________

meaning(light7,P,[A,B]) :-
      parse(light7,P),
      P =[_,_,A,_,_,B].

% __________light_verb string is empty, so P has only five elements__________

meaning(light7,P,[A,B]) :-
      parse(light7,P),
      P =[_,A,_,_,B].
```

In fact, **light7** requires *four* separate conditions, because there are four possible sentence structures accepted by this rule, as the following examples show:

[lamp, 6, to, level, 7]

[**put**, lamp,6, to, level, 7]

[lamp,6, to, **intensity**, level, 7]

[**put**, lamp,6, to, **intensity**, level, 7]

We are now in a position to define a complete program for "understanding" our lighting control language.

The goal **meaning(A,B,C)** tries to parse the given sentence A. If the parsing succeeds, then A is analysed, so that the lamp/lamps (B) and brightness level (C) can be found.

```
% Understanding sentences which are accepted by "light1"

meaning(P,all,Z) :-
      parse(light1,P),
      (P = [_,_,_,Z] ; P = [_,_,Z]).

% Understanding sentences which are accepted by "light2"

meaning(P,all,Z) :-
      parse(light2,P),
      (P = [_,_,_,Z] ; P = [_,_,Z]).

% Understanding sentences which are accepted by "light3"

meaning(P,X,Y) :-
      parse(light3,P),
      (P = [_,_,X,Y] ; P = [_,X,Y]).

% Understanding sentences which are accepted by "light4"

meaning(P,X,Y) :-
      parse(light4,P),
      (P = [_,_,_,X,Y] ; P = [_,_,X,Y]).
```

```
% Understanding sentences which are accepted by "light5"

meaning(P,X,Y) :-
        parse(light5,P),
        (P = [_,X,Y] ; P = [X,Y]).

% Understanding sentences which are accepted by "light6"

meaning(P,X,Y) :-
        parse(light6,P),
        (P = [_,_,X,_,Y] ; P = [_,X,_,Y]).

% Understanding sentences which are accepted by "light7"

meaning(P,X,Y) :-
        parse(light7,P),
        (P = [_,_,X,_,_,Y] ; P = [_,X,_,_,Y]),
        number(X).            % Resolve ambiguity with next clause

meaning(P,X,Y) :-
        parse(light7,P),
        (P = [_,_,X,_,_,_,Y] ; P = [_,X,_,_,_,Y]).
% Understanding sentences which are accepted by "light8"
meaning(P,X,Y) :-
        parse(light8,P),
        (P = [_,_,_,X,_,_,Y] ; P = [_,_,X,_,_,Y]),
        number(X).            % Resolve ambiguity with next clause

meaning(P,X,Y) :-
        parse(light8,P),
        (P = [_,_,_,X,_,_,_,Y] ; P = [_,_,X,_,_,_,Y]).

% Catch-all rule for use when the sentence is not recognised

meaning(_,unknown,unknown) :-
        message(['Sentence not recognised']).
```

Sample calls to **meaning(A,B,C)** are given in Table 4.2.

Table 4.2

A	B	C
[put,all,lamps,off]	all	off
[switch,each,lamp,half_on]	all	half_on
[set,laser,on]	laser	on
[lamp,3,to,level,2]	3	2
[switch,lamp,number,6,to,intensity,level, 5]	3	5
[switch,lamp,number,6,off]	6	off

Another Approach

The predicate **understanding**, which will be defined in a little while, uses a different mechanism for extracting information from sentences. A sequence of words (S) whose meaning is to be found is first presented as input to the parser, in order to verify that S is a valid sentence, according to the specified grammar rules. The trick we employ here to extract meaning from S is to include variables in the definitions of DCG rules. For example, the sentence

[the, circle, 'lies inside', a, rectangle]

is matched by the parser to Rule 4 in the definition of **sentence**. This unifies S to the structure

[the, **X**, **Y**, a, **Z**]

where, as expected in a Prolog program, **X, Y** and **Z** are variables. This instantiates X to **circle**, while Y is instantiated to **'lies inside'** and **Z** is instantiated to **rectangle**. The auxiliary predicate **save_info** then assembles the following clause from **X** and **Y** and then asserts it into the database.

'lies inside'(circle, rectangle).

Here is the program listing for the predicate **understanding**, which is the top level predicate in this particular program.

```
% ________________Try to extract the meaning from the sentence S__________

understanding(S) :- phrase(sentence,S).

% ________________________Save a new clause_________________________

save_info(X,Y,Z) :-
        Q =.. [Y,X,Z],         % Q will define a new fact for assertion into the DB
        not(Q),                % Check that the new clause does not already exist
        assert(Q).             % Insert Q as a fact in the DB

% Force this predicate to succeed even if the clause already exists

save_info(_,_,_).

% ______________DCG Rules for recognising sentences___________________

/* Rule 1 asserts the clause

iz(polygon, convex)

whilst satisfying the goal

parse([the, polygon, is, convex]).          */
```

```
sentence ⟶
        article, [X], is_verb, article, [Y],
        {save_info(X,iz,Y)}.

/* ________________________________________________

Rule 2 rule asserts the clause

named(circle, bonzo).

whilst satisfying the goal

parse([the, circle, is, called, bonzo]).

________________________________________________ */

sentence ⟶
        article, [X], is_verb, called, article, [Y],
        {save_info(X,named,Y)}.

/* ________________________________________________

Rule 3 asserts the clause

inside(cat, hexagon).

whilst satisfying the goal

parse([the, cat, is, inside, the, hexagon]).

________________________________________________ */

sentence ⟶
        article, [X], is_verb, [Y], article, [Z],
        {member(Y, [above,below, beside,inside,under]), save_info(X,Y,Z)}.

/* ________________________________________________

Rule 4 asserts the clause

contains(silhouette, polygon).

whilst satisfying the goal

parse([the, silhouette, contains, a, polygon]).

________________________________________________ */

sentence ⟶
        article, [X], [Y], article, [Z],
        {member(Y, [touches, contains, encloses, 'lies inside']), save_info(X,Y,Z)}.

/* ________________________________________________
```

Rule 5 asserts the clause

has(circle, texture).

whilst satisfying the goal

parse([the, circle, has, texture]).

```
_______________________________________________________________ */

sentence ——>
        article, [X], has_verb, article, [Y],
        {save_info(X,has,Y)}.

% ________________Definitions of secondary items________________

article ——>  [] | [a] | [the] | [some].

is_verb ——>  [is] | [are].

has_verb ——>  [has] | [have].

called ——>  [called] | [termed] | ['known as'].
```

Additions to the Database

Goal: parse([the, blob, is, above, the, spot])

New clause: above(blob, spot).

Goal: parse([the, circle, is, called, bonzo])

New clause: named(circle, bonzo).

Goal: parse([the, blob, touches, the, image_border])

New clause: touches(blob, image_border).

Goal: parse([the, silhouette, contains, a, polygon])

New clause: contains(silhouette, polygon).

Goal: parse([the, circle, 'lies inside', a, rectangle])

New clause: 'lies inside'(circle, rectangle).

Goal: parse([the, polygon, is, convex])

New clause: iz(polygon, convex).[4]

4.6 Simple Language for Describing Pictures

The following rules define the grammar of a simple language for describing pictures. It is very far from complete and there is no program for extracting meaning from the sentences, which the grammar accepts.

```
/* ____________________Top level – Sentences____________________ */

sentence -->
        noun_phrase,
        verb_phrase.

sentence -->
        sentence,
        [and],
        sentence.

/* ________________________Articles________________________ */

article -->
        [] |            % Empty set
        [a] |           % Indefinite article
        [some] |        % Indefinite article, plural
        [the].          % Definite article

/* ____________________Adjectives____________________ */

intensity -->
        [black] |
        ['very dark grey'] |
        ['dark grey'] |
        ['mid grey'] |
        ['light grey'] |
        ['very light grey'] |
        [white].

adj_misc -->
        [fuzzy] |
        [sharp].

size -->
        ['very small'] |
        [small] |
        ['medium size'] |
        [large] |
        ['very large'].
```

```
numbers --->
        [none] |
        [zero] |
        [one] |
        [two] |
        [three] |
        [four] |
        [five] |
        [six] |
        [seven] |
        [eight] |
        [nine] |
        [zero] |
        [several] |
        [many] |
        ['very many'].

position --->
        [left] |
        [right] |
        [top] |
        [bottom] |
        [center] |
        [inside] |
        [outside].

location --->
        position |
        position, position.

blob --->
        [solid] |
        ['has some holes'] |
        [hollow] |
        [match_stick].

histogram --->
        [unimodal] |
        [bimodal] |
        ['trailing tail at top end'] |
        ['trailing tail at bottom end'] |
        [multimodal].

adjective --->
        intensity |
        size |
        numbers |
        blob |
        location |
        histogram |
        adj_misc.
```

```
/* ____________________Adjective Phrases________________________ */

adjective_phrase ⟶
      [] |
      adjective |
      adjective, adjective.

/* ________________________Nouns____________________________ */

noun ⟶
      [] |
      [histogram] |
      [edges] |
      [spots] |
      [streaks] |
      [corners] |
      [shading] |
      [circle] |
      [square] |
      [ellipse] |
      [ovoid] |
      [square] |
      [object] |
      [background] |
      [feature].

/* ____________________Noun Phrases__________________________ */

noun_phrase ⟶
      article,
      adjective_phrase,
      noun.

noun_phrase ⟶
      article,
      noun,
      adjective_phrase.

noun_phrase ⟶
      noun_phrase,
      [and],
      noun_phrase.

/* ________________________Verbs____________________________ */

verb ⟶
      [is] |
      [are] |
      [has] |
```

```
    [contains] |
    [encloses] |
    ['lies inside'].

/* ____________________Verb Phrase________________________ */

verb_phrase -->
    verb,
    noun_phrase.
```

The following are valid sentences in this language and make sense:

[the, black, spots, are, fuzzy]

[the, fuzzy, black, spots, are, inside, the, square]

[the, hollow, 'mid grey', circle, has, 'dark grey', streaks]

[the, background, is, 'dark grey', and, has, black, streaks]

[the, background, is, 'dark grey', and, has, a, solid, white, square]

[the, histogram, is, bimodal, and, the, object, is, a, white, square]

[a, square, has, four, corners]

While the following sentences are also grammatically valid, they do not convey a significant meaning.

[the, black, spots, contains, fuzzy]

[the, black, background, is, white]

[a, square, is]

[the, histogram, has, spots]

[the, fuzzy, black, spots,are, sharp,white, spots]

4.7 Summary

The language described in the previous chapter forms the starting point for the development of a much larger and more useful software system than is obtained by merely embedding image processing operations within Prolog. In this chapter, we have begun to see how the new language can be extended in three ways.

The Library holds a large collection of Prolog+ programs for a wide range of tasks and is constantly growing. It is essential, in such a situation, that proper attention be paid to documentation and providing good "Help" facilities for the user. This is made easier by the self-documentation process, which extracts data from within program comments. In this way, only one set of notes is needed for each predicate. Since these comments are located adjacent to the predicates to which they relate, keeping the documentation up to date is quite straightforward.

The use of (Prolog) operators to simplify the syntax of Prolog+ programs has been discussed and some fairly obvious examples have been presented. However, the main route to achieve this objective is through the use of Definite Clause Grammars. A DCG for controlling the lighting in a Flexible Inspection Cell and a program for understanding sentences which conform to that grammar have also been provided. Several programs for extracting meaning from DCG sentences have also been given. Finally, a grammar for describing pictures has been listed.

It is not suggested that the lighting control and picture description languages are anywhere near to being complete; our objective in this chapter has been merely to show *how* DCGs may be used to analyse and "understand" a restricted subset of English. Clearly, very much more programming is needed before we can claim to have achieved a proper, fully developed means of programming vision systems using natural language. However, we can, at least, see one possible route ahead. Representing a more comprehensive subset of English in DCGs is mechanical and straightforward, if a little tedious. For a more complete discussion of the use of Prolog to understand natural language, the reader is referred to the book by Gazdar & Mellish [GAZ-89]. Also see [CLO-81, STE-86]. The role of Prolog+ as a bridge between image processing and other areas of work under the umbrella term Artificial Intelligence has been demonstrated, even though the full potential of the new language has not yet been realised.

We shall return to the topic of natural language understanding again in Chapter 8, where we shall use the ideas discussed in this chapter when we attempt to describe pictures in English, as a prelude to generating recognition programs automatically.

Notes

1. The list membership operator on MacProlog is **on**, although we shall continue to use **member** to ensure greater clarity.
2. This might be used to prepare a list such as that in Appendix II.
3. Warning: this would take a very long time.
4. We cannot use "is" for the name of this relation, since this is reserved in Prolog for testing the results of arithmetic operations.

CHAPTER 5

ANALYSING VECTOR REPRESENTATIONS OF IMAGES

"Yes, I have a pair of eyes ... but bein' only eyes, you see my wision's limited."
Pickwick Papers, Charles Dickens

5.1 Introduction

Innumerable applications give rise to situations in which there are sharp edges in the images to be analysed. A favourite form of processing in this type of situation is represented by the following sequence:

```
process :-
    edges,                        % Edge enhancement operator – see below
    threshold,                    % Could be fixed or adaptive
    noise_reduction,              % Numerous option exist – no details given here
    skeleton,                     % E.g. mdl
    analyse.                      % Will be discussed later in this section
```

Here are just a few of the numerous options for **edges:**

```
edges :- sed.                         % Sobel edge detector
edges :- red.                         % Roberts edge detector
edges :- gra.                         % Cheap and cheerful edge detector
edges :- raf(5,5), sub.               % Simple (linear) high-pass filter, 5 × 5 window
edges :- raf(11,11), sub.             % High-pass filter, 11 × 11 window
edges :- wri,2•(5•lnb,neg),rea,sub.   % Non-linear "high-pass" filter
```

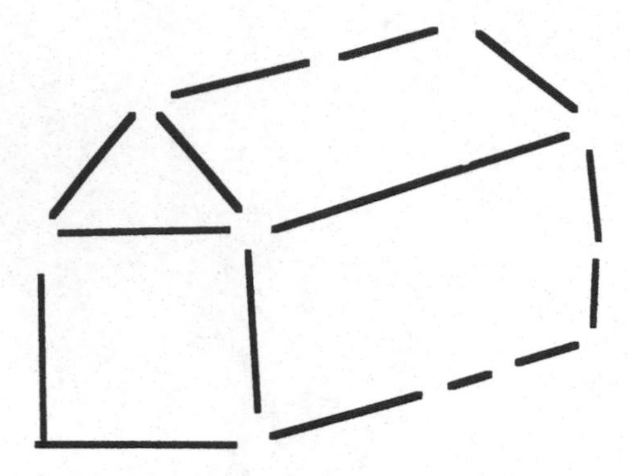

Figure 5.1. Images consisting of disconnected linear segments form the subject of this chapter.

This processing scheme generates a binary image in which there is a set of digital arcs, forming a set of disconnected graph-like structures. It is not difficult to represent a digital arc by a collection of straight-line segments (Figure 5.1). Analysing and understanding images of this type is the topic of this chapter.

Imagine a set of lines which fall within the camera's field of view. Suppose that we want to recognise precisely or sloppily drawn rectangles, simple letters, parallelograms, trees, etc. formed from such lines. These are just some of the tasks that we shall discuss in this chapter.

There are several ways of representing lines of finite length. Each has its own advantages and disadvantages. Here are *some* of the possibilities:

(i) Specify line ends in the form of database facts:

line([100,200],[250,350]). % Line joins [100,200] and [250,350]

line([150,220],[270,230]). % Line joins [150,220] and [270,230]

(ii) Specify line ends in a list

L = [... [100,200],[250,350], [150,220],[270,230], ...].

(iii) Specify a "starting point" for each line, its orientation and length:

[15,45,67,23], % Length 23 units, inclination 67°, starting point [15,45]

Clearly, lines coded in this way may be represented individually in a set of database facts (as in (i)), or grouped together in a list (as in (ii)).

(iv) As (iii), but omit the "starting point", except possibly for the first point.

Initially, we shall assume that either method (i) or (ii) is being used.

5.2 Acquiring Data from a Graph

When it is applied to a binary image, the skeletonisation operator, **mdl**, yields a graph-like structure that is well suited for further analysis, using Prolog-level decision-making. In order to do so, it is necessary to acquire coordinate data about the positions and interconnections of particular skeleton features, i.e. limb ends and junctions.[1] We shall define a predicate, called **acquire_data**, which generates a list whose elements are of the form:

[[X1,Y1], [X2,Y2], N]

from an image containing a skeleton. [X1,Y1] and [X2,Y2] are the ends of an arc containing N pixels. Here is the listing of **acquire_data**:

```
acquire_data(L) :-
    mdl,                                  % Skeleton
    wri(temp),                            % Save image on disc
    cnw,                                  % Count white neighbours
    min,                                  % Ignore background points
    hil(3,3,0),                           % Eliminate everything but joints
                                            and limb ends
    thr(1),                               % White spots on black background
    rea(temp),                            % Read skeleton image from disc
    xor,                                  % Break skeleton at joints
    ndo,                                  % Shade skeleton limbs
    wri(temp1),                           % Save image for further analysis
    get_data([],L).                       % Build list – see below for example

get_data(L,L) :-
    rea(temp1),                           % Read shaded, broken skeleton
                                            image
    gli(_,0).                             % Is maximum intensity = 0?

get_data(L1,L2) :-
    rea(temp1),                           % Read remains of broken skeleton
                                            image
    gli(_,Z),                             % Find maximum intensity
    hil(Z,Z,0),                           % Remove brightest limb
    wri(temp1),                           % Save other bits of skeleton
    swi,                                  % Switch images
    thr(Z),                               % Now analyse limb just removed
    cwp(N),                               % Count number of white pixels
    cnw,                                  % Count white neighbours
    min,                                  % Ignore background points
    thr(2,2),                             % Get end points
    top_left(X1,Y1),                      % Top-left end point at [X1,Y1]
    pfx(X1,Y1,0),                         % Set that pixel to black
    top_left(X2,Y2),                      % Find other end point [X2.Y2]
    get_data([[[X1,Y1],[X2,Y2],N]|L1],L2). % Go and do it again

top_left1(X,Y) :-
    dim(_,_,Y,_),                         % Find top-most white pixel
    bve(1,Y,511,Y,X,_,_,_).               % Find its X-coordinate
```

Format for Output List

The elements of the output list all have an identical structure of the following form:

[[X1,Y1],[X2,Y2],D]

where D is the distance between the points [X1,Y1] and [X2,Y2]. Here is a sample of the output list:

[[[205, 117], [203, 232], 116], [[323, 226], [205, 235], 119], [[202, 235], [85, 381], 156]]

Plotting a Tree-Like Structure

The following predicate plots a tree-like structure, whose branches are all straight lines, given a list in the format calculated by **acquire_data**.

```
tree_plot([]).

tree_plot([[[X1,Y1],[X2,Y2],N]|L]) :-
    vpl(X1,Y1,X2,Y2,N),    % Line [[X1,Y1], [X2,Y2]],
                           %   intensity N
    tree_plot(L).          % Do the same for rest of the list
```

Notice that **acquire_data** or **tree_plot** could be enhanced by adding a facility for detecting points of high curvature on the skeleton limbs. This is left as an exercise for the reader. (*Hint:* Compare the directions of two consecutive straight-line segments.)

5.3 Low-Level Predicates

Now, let us briefly review some of the Library predicates which are available for analysing line drawings. Table 5.1 lists those predicates which are of particular interest to us here and which we shall now discuss in more detail.

Table 5.1. Low-level predicates for analysing line drawings.

Predicate	Operation
distance(P,Q,R)	Match R to the distance between P and Q
angle(P,Q,R)	Match R to the angle formed by lines P and Q
near(P,Q)	Is P near to Q?
far_away(P,Q)	Is P far_away from Q ?
collinear(P,Q,R)	Are the points P, Q and R collinear?
right_angle(P,Q,R)	Do the points P,Q and R form a right angle?
parallel([P,Q],[R,S])	Are lines [P,Q] and [R,S] parallel? (P, Q, R and S are points)
perpendicular([P,Q],[R,S])	Are the lines [P,Q] and [R,S] perpendicular?
lines_cross([P,Q],[R,S])	Do the lines [P,Q] and [R,S] cross?
y_junction(A,[B,C,D])	Is A a Y-junction? (i.e. connected to three other points)
crossover(L,A,[P,Q,R,S])	Is A a "cross-over point", defined in the list L?

Here is a perfectly general definition of **distance**, expressed in terms of an unspecified predicate called **locate**:

```
distance(A,B,D) :-
    locate(A,[Xa,Ya]),          % Not specified but examples are given below
    locate(B,[Xb,Yb]),
    Dx is Xa – Xb,
    Dy is Ya – Yb,
    D2 is Dx*Dx + Dy*Dy,        % D2 is the distance squared
    sqrt(D2,D).                 % Square root function
```

Notice that the predicate **locate** is, as yet, undefined. One possibility is given by the following naive definition:

```
locate([X,Y],[X,Y]).
```

This states that the *name* of the object is [X,Y] and its *address* within the image is also [X,Y]. Another possibility is given by:

```
locate(A,[X,Y]) :-
    isolate(A),                 % Isolate the named feature (this leaves a
                                  single blob)
    cgr(X,Y).                   % Find the centroid [X,Y] of the blob
```

The predicate **angle**, may be defined in a similar way to **distance**, and again is based upon the undefined predicate **locate**:

```
angle(A,B,C) :-
    locate(A,Xa,Ya),
    locate(B,Xb,Yb),
    X1 is Xb – Xa,
    abs(X1,X),
    Y1 is Yb – Ya,
    abs(Y1,Y),
    ((not(X = 0), Z is Y/X,tan(D,Z)); D = 1.571),
    adjust_for_quadrant(X1,Y1,D,E),
    deg_rad(F,E),
    mod(F,360,C).

% Quadrant one

adjust_for_quadrant(A,B,C,C) :-
    A ≥ 0,
    B ≥ 0.

% Quadrant two

adjust_for_quadrant(A,B,C,D) :-
    A ≤ 0,
    B ≥ 0,
    D is 3.142 – C.
```

```
% Quadrant three

adjust_for_quadrant(A,B,C,D) :-
        A ≤ 0,
        B ≤0,
        D is C + 3.142.

% Quadrant four

adjust_for_quadrant(_,_,C,D) :-
        D is 6.284 – C.
```

The predicates **near** and **far_away** can be defined in a fairly obvious way, using **distance**:

```
near(A,B) :-
        tolerance1(V),          % Refer to DB for tolerance parameter
        distance(A,B,D),
        D < V.

far_away(A,B) :- not(near(A,B)).
```

The predicate **collinear** is defined in the following way:

```
collinear(A,B,C) :-
        tolerance2(P),          % Tolerance parameter, held in DB
        angle(A,B,Sab),
        angle(B,C,Sbc),
        Q is Sab – Sbc,
        abs(Q,R),
        R < P.

collinear(A,B,C) :-
        tolerance2(P),          % Tolerance parameter
        angle(A,B,Sab),
        angle(A,C,Sac),
        Q is Sab – Sac,
        abs(Q,R),
        R < P.
```

Three points, A, B and C, form a right angle if the predicate **right_angle** can be satisfied:

```
right_angle(A,B,C) :-
        angle(A,B,C,Z),
        tolerance3(P),
        Q is Z – 90.0,
        abs(Q,R),
        R < P.
```

```
right_angle(A,B,C) :-
    angle(A,B,C,Z),
    tolerance3(P),
    Q is Z – 270.0,
    abs(Q,R),
    R < P.
```

Supposedly parallel lines may be tested using **parallel**:

```
parallel(A,B) :-
    angle(P,Q,R),          % Calculate angle between P and Q
    tolerance4(V),         % Refer to DB for tolerance value
    R ≤ V.
```

and, of course, perpendicular lines may be identified using a similar rule[2]

```
perpendicular(A,B) :-
    angle(A,B,P),
    tolerance4(V),         % Refer to DB for tolerance value
    Z is P – 90,
    abs(Q,R),
    R ≤ V.
```

Testing whether two lines, A and B, intersect can be accomplished using **lines_cross**, which is defined thus:

```
lines_cross([[X1,Y1],[X2,Y2]],[[X3,Y3],[X4,Y4]]) :-
    A is X1 – X2,
    B is X3 – X4,
    C is X4 – X2,
    D is Y1 – Y2,
    E is Y3 – Y4,
    F is Y4 – Y2,
    D1 is A*B – B*D,
    N1 is C*E – B*F,
    N2 is C*D – A*F,
    test_lines(D1,N1,N2).

test_lines(D,N1,N2) :-
    D > 0,
    N1 > 0,
    N2 > 0,
    N1 ≤ D,
    N2 < D.

test_lines(D,N1,N2) :-
    D < 0,
    N1 < 0,
    N2 < 0,
    N1 ≥ D,
    N2 ≥ D.
```

A Y-junction is a point (A) which is connected to just *three* other points (B,C, D) and may be recognised using **y_junction**, which is defined as follows:

```
y_junction(A,[B,C,D]) :-
        line_connected(A,B),
        line_connected(A,C),
        not(B = C),
        line_connected(A,D),
        not(B = D),
        not(C = D).

line_connected(A,B) :- line(A,B).

line_connected(A,B) :- line(B,A).
```

A point (A) is a *crossover point*, if it is connected to four other points, X1, X2, X3 and X4, in the following way:

The rotational ordering of X1, X2, R and S is important, in that X1 is "opposite" X2 and X3 is "opposite" X4. (Relative, not absolute, values of angles are important.) The predicate **detect_crossovers** is true if A is a crossover point, connected to X1, X2, X3 and X4. This predicate is defined thus:

```
detect_crossovers([[X1,A,X3],[X2,A,X4]]) :-
        cross_over(A,[B,C,D,E]),
        angle(A,B,Xb),
        angle(A,C,Xc),
        angle(A,D,Xd),
        angle(A,E,Xe),
        sort([[Xb,B],[Xc,C],[Xd,D],[Xe,E]],[[_,X1],[_,X2],[_,X3],[_,X4]]).

cross_over(A,[B,C,D,E]) :-
        line_connected(A,B),
        line_connected(A,C),
        not(B = C),
        line_connected(A,D),
        not(B = D),
        not(C = D),
        line_connected(A,E),
        not(B = E),
        not(C = E),
        not(D = E).
```

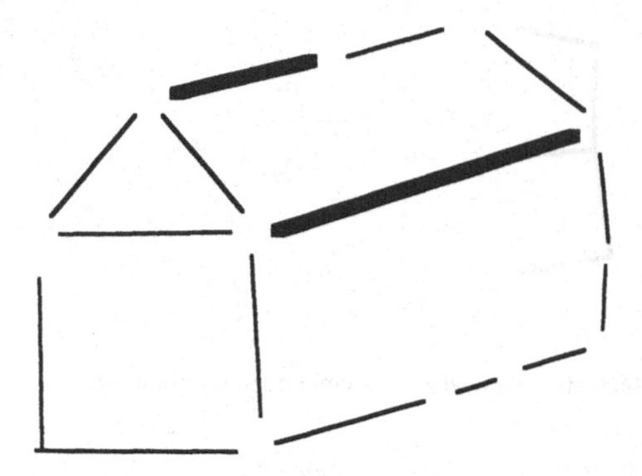

Figure 5.2. Pairs of parallel straight lines are detected by the predicate **parallel_lines**.

5.4 Identifying Simple Geometric Figures

The predicate **parallel_lines** finds one pair of parallel lines from a list of lines, L.

```
parallel_lines(L,X,Y) :-
    member(X,L),
    member(Y,L),
    not(X = Y),
    parallel(X,Y).
```

Notice that the list L may contain any number of lines; all lines, in addition to those two detected by **parallel_lines**, will simply be ignored (see Figure 5.2). If back-tracking occurs, **parallel_lines** tries to find another pair of parallel lines. It should also be noted that the predicate **parallel_lines** does not require that we specify exactly how the lines are represented.

parallelogram is a predicate for finding *precisely drawn* parallelograms, of any size and at any orientation:

```
parallelogram(A,B,C,D) :-
    parallel_lines([[X1,Y1],[X2,Y2]], [[X3,Y3],[X4,Y4]]),
    parallel_lines([[X1,Y1],[X3,Y3]], [[X2,Y2],[X4,Y4]]).
```

whereas **rectangle** finds precisely drawn rectangles

```
rectangle(A,B,C,D) :-
    parallel_lines([[X1,Y1],[X2,Y2]], [[X3,Y3],[X4,Y4]]),
    parallel_lines([[X1,Y1],[X3,Y3]], [[X2,Y2],[X4,Y4]]),
    right_angle([X1,Y1],[X2,Y2], [X3,Y3]).
```

Detecting rectangular figures which may have open corners is just a little more complicated but can be accomplished using **near_rectangle**:

Figure 5.3. The E-shaped figure can be detected by **letter(e)**, even when it is embedded within an image containing other objects.

```
near_rectangle(A,B,C,D) :-
    parallel_lines([[X1,Y1],[X2,Y2]], [[X3,Y3],[X4,Y4]]),
    parallel_lines([[X5,Y5],[X6,Y6]], [[X7,Y7],[X8,Y8]]),
    perpendicular([[X1,Y1],[X2,Y2]],[[X5,Y5],[X6,Y6]]),
    near([X1,Y1],[X5,Y5]),
    near([X2,Y2],[X6,Y6]),
    near([X3,Y3],[X7,Y7]),
    near([X4,Y4],[X8,Y8]).
```

A skeletal sans serif upper case letter **E**, may be recognised using the following predicate

```
letter(e) :-
    parallel_lines([[X1,Y1],[X2,Y2]], [[X3,Y3],[X4,Y4]]),
    parallel_lines([[X1,Y1],[X2,Y2]], [[X5,Y5],[X6,Y6]]),
    perpendicular([[X1,Y1],[X2,Y2]], [[X1,Y1],[X3,Y3]]),
    perpendicular([[X3,Y3],[X4,Y4]], [[X3,Y3],[X5,Y5]]),
    distance([X1,Y1],[X2,Y2],D12),
    distance([X3,Y3],[X4,Y4],D34),
    distance([X5,Y5],[X6,Y6],D56),
    D12 = D34,
    D12 = D56,
    distance([X1,Y1],[X3,Y3],D13),
    distance([X3,Y3],[X5,Y5],D35),
    D13 = D35.
```

Notice that while **letter(e)** does recognise **E** correctly, it also succeeds when it is applied to a more complex image, provided that **E** is present somewhere (see Figure 5.3).

5.5 Trees and Forests

A tree-like structure may be detected within a set of straight-line segments, using the following predicate

```
tree([],T,T).

tree([[A,B]|C], T1,T2) :-
    not(member([A,B],T1)),
        (
        member([A,C],T1);
        member([C,A],T1);
        member([B,C],T1);
        member([C,B],T1)
        ),
    tree(C,[[A,B]|T1],T2).

tree([[A|B], T1,T2) :- tree(B,T1,T2)
```

Now, consider two or more trees which are superimposed (Figure 5.4(a)). Such a figure will be called a *forest*. Notice that this is slightly different from the graph-theoretic terminology, since the trees in our forests have fixed (but unknown) physical dimensions; they do not merely define (graph-theoretic) connectivity. This type of structure could occur, for example, when analysing (botanical) plants, viewed in silhouette and which are packed close together. Finding all of the nodes on a tree in a forest can be accomplished using the predicate **tree_in_forest**:

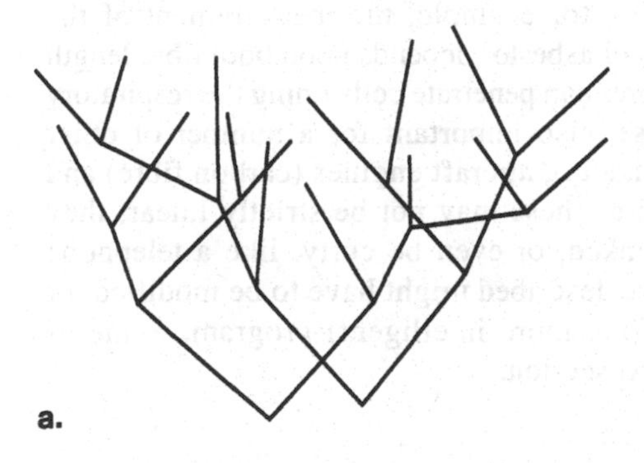

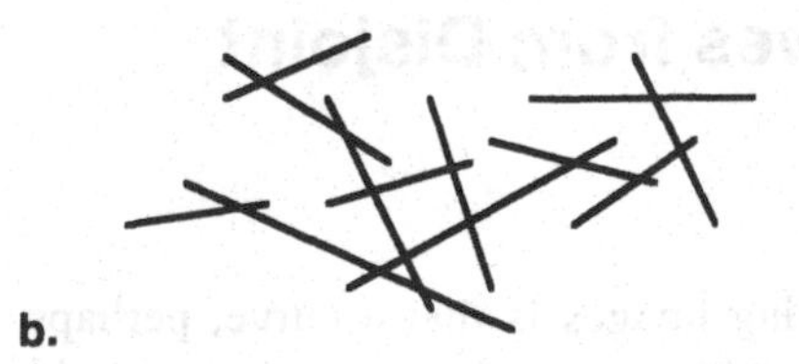

Figure 5.4. Collections of linear objects. (**a**) There are two overlapping tree-like structures. The predicate **tree_in_forest** may be used to separate them. (**b**) A number of important, industrial vision problems can be reduced to analysing pictures of this form. Calculate the histogram of lengths of the sticks. Find the longest/shortest one. What is the mean length? What is the average aspect ratio, length:width?

```
% New node is a crossover point

tree_in_forest(L1,L2) :-
        member(A,L1),
        detect_crossovers([[A,B,C],[_,B,_]]),
        not(member(B,L1)),
        tree_in_forest([B,A|L1],L2).    % Repeat until all nodes have been found

% New node is not a crossover point

tree_in_forest(L1,L2) :-
        member(A,L1),                   % Find a node (A) known to be on the tree
        line_connected(A,B),            % Find a node (B) connected to A
        not(member(B,L1)),              % Is B already known?
        tree_in_forest([B,A|L1],L2).    % Repeat until all nodes have been found

% Terminate recursion

tree_in_forest(L,L).
```

Notice that the predicate **tree_in_forest** is not algorithmic. When a branch of one tree intersects a node (i.e. a bifurcation or limb end) of another, the procedure will not work properly. Hence, **tree_in_forest** is a heuristic procedure, since it is a rule-of-thumb which usually works satisfactorily but does not always does so.

A variant of **tree_in_forest** is able to resolve images like that shown in Figure 5.4(b), where several linear objects ("sticks") are dropped haphazardly onto a table. The ability to measure the lengths of "sticks" in images like this is important for a number of practical applications, including for example, the measurement of the lengths of fibres. The carcinogenic property of asbestos depends upon both fibre length and straightness, since only long straight fibres can penetrate cells lining the respiratory tract. Measuring fibre length is, of course, also important for a number of other industries, including textiles (cotton, wool, etc.), aircraft engines (carbon fibre) and paper. Of course, real-world objects such as these may not be strictly linear; they may be gently curved (like a banana), kinked, or even be curly, like a telephone cable. In this event, the techniques we have described might have to be modified, or more likely, used within a higher level (i.e. more intelligent) program. Some of these ideas are discussed further in the next section.

5.6 Creating Continuous Curves from Disjoint Lines

A common problem that occurs when processing images is that a curve, perhaps tracing the edge of an object, appears broken, when it was hoped that we would generate a continuous arc. This happens quite often, for example, when we apply an edge detector, followed by fixed-parameter thresholding. The following command sequence might well create a discontinuous curve:

```
edge_detector,          % It does not matter which edge detector is used
thr(?)                  % Threshold parameter is chosen for optimal results
```

Choosing a low value for the threshold parameter usually results in a lot of background noise (evident as "snow"), while the edge remains intact. Raising the threshold eliminates the snow but often breaks the edge contour. As we have emphasised, this is a very common difficulty when we process images from a real-world scene in which the objects and background are both imperfect and unclean, while the camera is noisy and the optics are dusty. There are two ways out of this dilemma:

(a) Use a low value for the threshold parameter and process the image to eliminate the effects of snow.

(b) Use a higher value for the threshold and then join together the arc segments so generated.

In the past, method (a) has usually been favoured for industrial inspection machines but we shall now consider the other alternative in a little more detail.

For the sake of simplicity, we shall assume that an arc is composed of linear segments. This is easy to achieve, by simply "breaking" an arc at points of high curvature, or decomposing the arc into linear segments of limited length.

In this section, we shall discuss how we might reconstruct a continuous curve, from a set of discontinuous linear segments. (For example, consider how a *continuous outline* of the object in Figure 5.1 might be drawn.) Suppose that we are given a database containing a set of facts of the form

```
line_segment(A,B).
```

Each fact specifies the ends of a line, which may or may not join together, to form a continuous arc. The algorithm used by the predicate **join_arc** operates on the principle that if two lines, [A,B] and [C,D] are such that B and C lie within a specified distance from each other, they may be connected together by adding a third line [B,C]. The maximum inter-line gap is determined by **tolerance5**.

```
join_arc(Z) :-
    line_segment(A,B),              % Find one line segment
    retract(line_segment(A,B)),     % Delete it from the database
    remember(arc, [A,B]),           % Store it as a property
    joining,                        % Find all lines which form an arc
    recall(arc,Z).                  % Recover list from property database

joining :-
    line_segment(A,B),              % Find next line segment
    join(A,B),                      % Try to add on one more element
    joining.                        % Do it again to add more lines to arc

joining.                            % Force this predicate to succeed

tail_end(A,B) :- reverse(A,[B|_]).  % Find last element (B) in list A

tolerance5(V).                      /* Defines maximum gap size for
                                       joining line ends together */
```

```
% Add line [B,A] to beginning of list

join(A,B) :-
    recall(arc,[C|D]),               % Recall property for object "arc"
    distance(A,C,E),                 % Find distance between A and end of arc
    tolerance5(V),                   % Find gap tolerance value
    E ≤ V,                           % Is gap small enough?
    remember(arc,[B,A,C|D]),         % Remember new arc
    retract(line_segment(A,B)),      % Delete line just added to the arc
    !.

% Add line [A,B] to beginning of list

join(A,B) :-
    recall(arc,[C|D]),
    distance(B,C,E),
    tolerance5(V),
    E ≤ V,
    remember(arc,[A,B,C|D]),
    retract(line_segment(A,B)),
    !.

% Add line [A,B] to end of list

join(A,B) :-
    recall(arc,C),
    tail_end(C,D),
    distance(A,D,E),
    tolerance5(V),
    E ≤ V,
    append(C,[A,B],F),
    remember(arc,F),
    retract(line_segment(A,B)),
    !.

% Add line [B,A] to end of list

join(A,B) :-
    recall(arc,C),
    tail_end(C,D),
    distance(B,D,E),
    tolerance5(V),
    E ≤ V,
    append(C,[B,A],F),
    remember(arc,F),
    retract(line_segment(A,B)),
    !.

join.
```

Sample "Input" Database

```
line_segment([100,100],[120,122]).
line_segment([179,183],[201,202]).
line_segment([500,500],[900,900]).  % This line does not connect with others
line_segment([123,125],[140,141]).
line_segment([0,0],[20,22]).
line_segment([23,25],[40,41]).
line_segment([79,83],[101,102]).
line_segment([144,142],[162,159]).
line_segment([44,42],[62,59]).
line_segment([60,60],[79,81]).
line_segment([160,160],[179,181]).

% Dummy - included to ensure that Prolog knows what to do when DB is empty3

line_segment.
```

Output

When it is run on the above "input" database, the goal **join_arc** instantiates Z to the list

[[0, 0], [20, 22], [23, 25], [40, 41], [44, 42], [62, 59], [60, 60], [79, 81], [79, 83], [101, 102], [100, 100], [120, 122], [123, 125], [140, 141], [144, 142], [162, 159], [160, 160], [179, 181], [179, 183], [201, 202]]

This is, of course, a list of the coordinate pairs of the vertices of a polygonal arc. Only one fact

line_segment([500,500],[900,900]).

remains in the "input" database, all of the others having been removed by **retract**.

5.7 An Alternative Representation for Edges

So far, we have represented the edges in a scene by linear segments described by giving their end points. As we explained earlier, other data formats are possible and, in this section, we shall describe one of them.

The **chain code** (also called the **Freeman code**) is a popular method of representing the edge of a blob-like figure. A higher level edge-coding technique, which may be derived from it, is called the **polar vector representation** (**PVR**). In the PVR, an edge is represented by a list of vectors of the form

$\{(R_1,\emptyset_1), (R_2,\emptyset_2), (R_3,\emptyset_3),\ldots,(R_n,\emptyset_n)\}$

where $(R_i, Ø_i)$ denotes a line of length R_i and inclined at an angle $Ø_i$ to the horizontal axis. The Library predicate **pvr** can be used to generate the polar vector representation of a figure in a binary image. Using VSP coupled to a VCS image processor, the chain code is calculated first (by VCS), is input into the VSP software and then the PVR is derived from it. Here is a definition for **pvr** on this basis.

```
pvr(Y) :-
        chain_code(L),
        find_runs(L,[],[_|Z]),          % Find runs of chain code elements. Ignore
                                           the first element of the chain code list.
        pvr_convert(Z,[],Y).            % Convert to pvr format

% Look for runs of chain code elements in the input list

find_runs([],L,L).                      % Terminate recursion

% Next chain code element same as previous one

find_runs([X|Y],[[X,A]|B],Q) :-
        C is A + 1,                     % Increment run-length counter by one
        !,                              % Inhibit back-tracking
        find_runs(Y,[[X,C]|B],Q).       % Do it again on the tail of the input list

% Chain code element is not same, so re-initialise the run length counter

find_runs([X|Y],A,Q) :-
        !,
        find_runs(Y,[[X,1]|A],Q).

% End of the list – ignore the last two elements (peculiar to VCS software)

pvr_convert([_,_],L,L).

% Convert runs of chain code elements to chain code

pvr_convert([[C,N]|Y],L1,L2) :-
        pvr_table(C,A,B),               % C = chain code, A = vector length,
                                           B = orientation
        R is N*A,                       % Find total length of the vector
        append([[R,B]],L1,L3),          % Add new PVR element to list
        !,
        pvr_convert(Y,L3,L2).

% Conversion table. Chain code to PVR

pvr_table(0,1,0).
pvr_table(1,1.414,45).
pvr_table(2,1,90).
pvr_table(3,1.414,135).
pvr_table(4,1,180).
pvr_table(5,1.414,225).
pvr_table(6,1,270).
pvr_table(7,1.414,315).
```

We shall now show how the PVR of the edge of a blob can be used to good effect.

5.8 Constructing Piece-Wise Linear Models of Edges

When it is first computed, the PVR list contains a large number of elements, representing short vectors and which are approximately aligned in groups. The predicate **smooth_pvr** may be applied to reduce the length of the PVR list, so that it contains a small number of long vectors. Here is the listing of **smooth_pvr**:

```
% Terminating recursion
smooth_pvr([],[],_,_).

% Smooth if either A < U, or C < U.

smooth_pvr([[A,B],[C,D]|E],F,U,V) :-
      (A < U; C < U),
      new_pvr([A,B],[C,D],[G,H]),
      !,
      smooth_pvr([[G,H]|E],F,U,V).

% Smooth if the difference of angles < V

smooth_pvr([[A,B],[C,D]|E],F,U,V) :-
      P is B – D,
      abs(P,V),
      Q < V,
      new_pvr([A,B],[C,D],[G,H]),
      !,
      smooth_pvr([[G,H]|E],F,U,V).

% Smooth if the difference of angles < V and B > 360 -V and D ≥ 0

smooth_pvr([[A,B],[C,D]|E],F,U,V) :-
      B > 360 – V,
      D ≥ 0,
      P is 360 – B + D,
      abs(P,Q),
      Q < V,
      new_pvr([A,B],[C,D],[G,H],U,V),
      !,
      smooth_pvr([[G,H]|E],F).
% Smooth if the difference of angles < V and D > 360 – V and B ≥ 0
```

```
smooth_pvr([[A,B],[C,D]|E],F,U,V) :-
        D > 360 – V,
        B ≥ 0,
        P is B + 360 – D,
        abs(P,Q),
        Q < V,
        new_pvr([A,B],[C,D],[G,H]),
        !,
        smooth_pvr([[G,H]|E],F,U,V).

% No smoothing, so go on one step

smooth_pvr([A|B],[A|C],U,V) :-
        smooth_pvr(B,C,U,V).

/* Calculating one new PVR element from two old ones. This process is equivalent
to vector addition. */

new_pvr([A,B],[C,D],[E,F]) :-
        cosine(B,X1),
        sine(B,Y1),
        cosine(D,X2),
        sine(D,Y2),
        X is X1*A + X2*C,
        Y is Y1*A + Y2*C,
        Z is X*X + Y*Y,
        sqrt(Z,E),
        angle([0,0],[X,Y],F).
```

Notice the use of the two smoothing control parameters; the third and fourth arguments of **smooth_pvr**. In order to achieve effective smoothing effect, it may be necessary to apply **smooth_pvr** several times, as indicated in the following call sequence:

```
pvr_smooth(L1,L2) :-
        smooth_pvr(L1,Z1,3,20),
        smooth_pvr(Z1,Z2,6,20),
        smooth_pvr(Z2,Z3,6,20),
        smooth_pvr(Z3,L2,6,20).
```

Notice the changing values of the first parameter. **smooth_pvr** is not the only edge smoothing operator we possess. For example, the Library predicate **smooth_arc** achieves a similar effect, but uses a data format of the following form:

[..., [18,21], [31,28], [42,39], [48,52], [53,54], [80,90], [56,58], [69,72], ...]

Of course, it is a trivial matter to write a program which converts between these two methods of representing edges. However, we digress from our main theme.

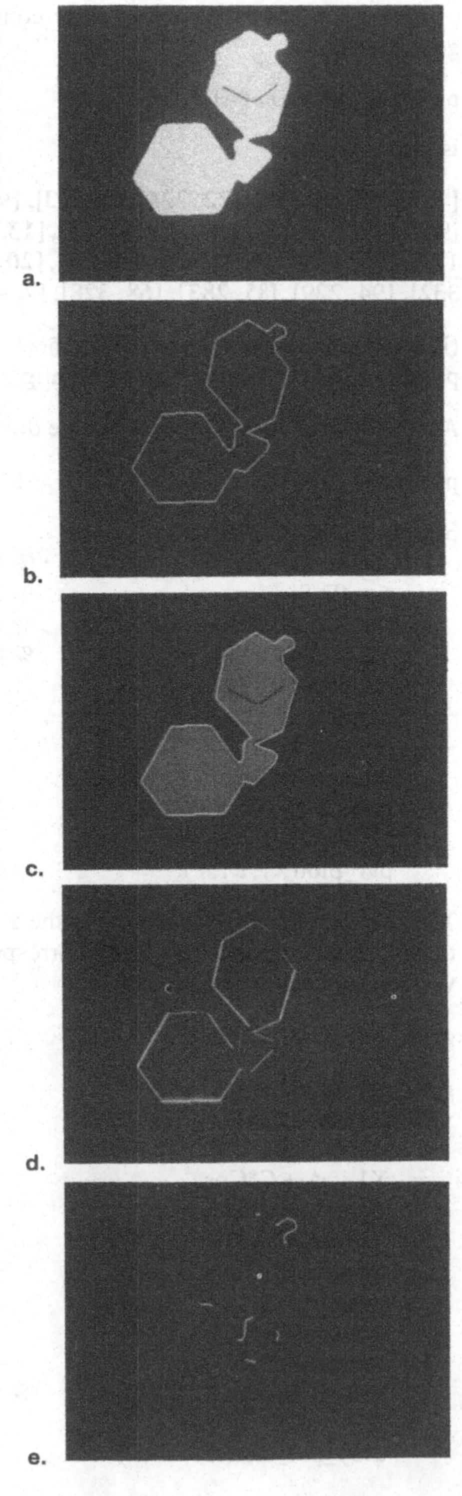

Figure 5.5. Edge smoothing, using **grey_pvr_plot**. (**a**) Binary image. The internal edge will be ignored. (This picture was obtained by processing a grey-scale image. Details of this processing are not necessary to understand the edge smoothing procedure. The application details are confidential.) (**b**) The smoothed edge. (**c**) The smoothed edge fitted to the image shown in (**a**). (**d**) Edge segments in (**b**) shaded according to their lengths. (**e**) Short edge segments highlighted.

Figure 5.5 shows a binary image containing a single blob, while the PVR list, generated by

pvr(L1), pvr_smooth(L1,L2)

is given below:

[[38, 30], **[6, 351],** [22, 324], **[15, 23],** **[9, 69],** **[8, 97],** [27, 154], **[17, 87],** [42, 52], [96, 109], [61, 158], [45, 20], **[6, 51],** **[13, 113],** [60, 139], **[20, 195],** [69, 122], [102, 179], [75, 236], [111, 300], [51, 357], **[20, 6],** [77, 57], **[6, 9],** **[8, 310],** [31, 278], **[15, 332],** [94, 229], [85, 283], [68, 328], **[7, 0]**]

(Bold type indicates those PVR elements whose lengths are less than or equal to 20 pixels – this is a simple way of highlighting some of the corners.)

Any VPR list can be plotted as a line drawing, using the predicate **pvr_plot**:

```
pvr_plot(_,_,[]).                     % Terminate recursion

pvr(plot(A,B,[[C,D]|T) :-
      cosine(D,Cos),                  % Cosine uses degrees
      sine(D,Sin),                    % Sine does too
      X1 is A + C*Cos,                % Calculate coordinates of end of PVR element
      Y1 is B + C*Sin,                % Calculate coordinates of end of PVR element
      integer(X1,X2),
      integer(Y1,Y2),
      integer(A,A1),
      integer(B,B1),
      vpl(A1,B1,X2,Y2,255),           % Plot a white line
      !,
      pvr_plot(X1,Y1,T).              % Do it for the rest of PVR list as well
```

The corners of a blob, coded using the PVR, can be highlighted, simply by shading those elements in the list which correspond to short vectors. This requires only a very minor change to **pvr_plot**:

```
grey_pvr_plot(_,_,[]).

grey_pvr_plot(A,B,[[C,D]|T) :-
      cosine(D,Cos),
      sine(D,Sin),
      X1 is A + C*Cos,
      Y1 is B + C*Sin,
      integer(X1,X2),
      integer(Y1,Y2),
      integer(A,A1),
      integer(B,B1),
      integer(C,C1),
      vpl(A1,B1,X2,Y2,C1),            % Shade vector to indicate its length
      !,
      grey_pvr_plot(X1,Y1,T).
```

In Figure 5.5(d), the long linear segments in the silhouette of the test object have been highlighted by **grey_pvr_plot**. Once such an image has been created,[4] it is a trivial matter to threshold it, so that either the long (nearly straight) edge elements, or the corners can be isolated, as desired (see Figure 5.5(e)).

5.9 Curve Parsing

Consider a digital arc, which is specified as illustrated below, by listing its nodal points:

[[0,0], [50,0], [60,5], [70,12], [80,40], [90,60], [100,100], [80,90], [70,70], [60,50], [50,30],[40,10],[30,50],[20,70],[10,90],[0,100]].

The predicate **parse_curve** simply annotates such an arc in the following way:

[[0, 0], [50, 0], [60, 5], [70, 12], [80, 40], **left**, [90, 60], **right,** [100, 100], [80, 90], [70, 70], **left**, [60, 50], [50, 30], [40, 10], **straight**, [30, 50], **right**, [20, 70], **left**, [10, 90], **straight**, [0, 100], left].

The words **left**, **right** and **straight** indicate the direction of curvature of that section of the arc listed to the left. Here is the listing of **parse_curve**.

```
parse_curve(L) :-
        curve([A,B|C]),                          % Refer to database for two points
        curve_parser([A,B|C],[B,A],L1),          % Curve parser
        reverse(L1,L2).                          % Reverse list

curve_parser([A,B,C|D],[E|F],G) :-
        direction(A,B,C,Z),                      % In what direction is curve moving?
        build_list(Z,E,C,[E|F],L),
        curve_parser([B,C|D],L,G).               % Do it all again

curve_parser(_,L,L).                             % Catch-all clause to end recursion

% Is the curve changing direction?

build_list(Z,Z,A,[P|Q],[P,A|Q]).                 % First two arguments are same

build_list(Z,_,A,X,[Z,A|X]).                     % First two arguments are not same

% Find direction that curve is turning

direction([X0,Y0],[X1,Y1],[X2,Y2],left) :-
        (X1 - X0)*(Y2 – Y0) > (Y1 – Y0)*(X2 – X0).   % Going left
```

```
direction([X0,Y0],[X1,Y1],[X2,Y2],right) :-
    (X1 – X0)*(Y2 – Y0) < (Y1 – Y0)*(X2 – X0).   % Going right

direction([X0,Y0],[X1,Y1],[X2,Y2],straight).        % Not turning

% Illustrating format for the "input" data

curve([[0,0], [50,0], [60,5], [70,12], ...]).
```

5.10 Summary

Images formed by collections of straight line segments ("line drawings") have been the subject of this chapter. The ability to "understand" them has been our main concern. The construction of a connected figure from a set of disconnected lines has been considered. Relationships between pairs of straight-line segments, such as crossing, parallel, perpendicular etc. can be detected and simple geometric figures can be identified. The smoothing of a polygonal arc has also been discussed, as has the "parsing" of an arc into segments of monotonic curvature.

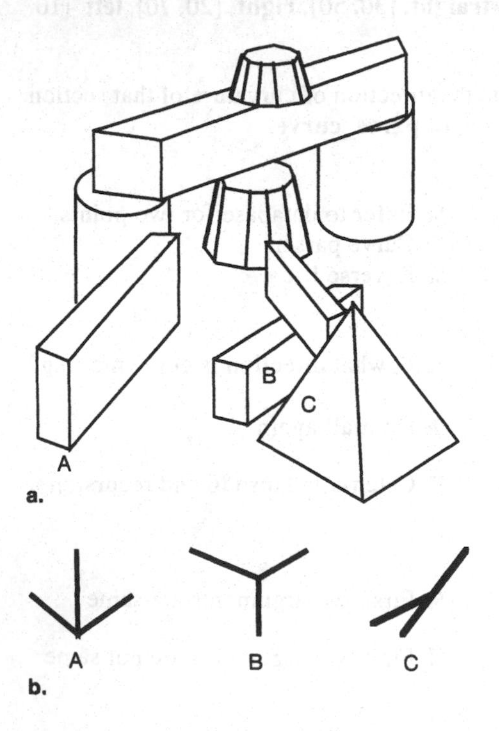

Figure 5.6. The Blocks World. (**a**) A scene composed of blocks stacked upon each other. (**b**) Detail of the intersections of the edges. Analysing patterns such as these has been one of the major successes of AI research. Almost every student of AI learns about this.

We are now in a position to be able to provide visual input into some popular, well established AI programs. It is frequent practice in AI textbooks to regard "line drawings" as representations of the real world, without any regard as to how the data might be acquired by an AI program. For example, many textbooks analyse scenes that can occur in the so-called *Blocks World*, where all of the objects are solid, convex, polyhedral bodies. Figure 5.6 represents a type of diagram that has been published numerous times, in many technical articles on AI. Indeed, analysing such "drawings" is one of the standard lessons that every student of AI learns. However, the question as to how the data from a real image get into a Prolog program is never discussed; that question is always conveniently ignored. Prolog+ provides one possible route for acquiring such data and analysing them.

As an illustration of this, we may point to the Warplan program, discussed in Chapter 7. This is an abstract task planner and may be used, for example, in formulating a sequence of moves by which a set of blocks may be stacked in some pre-defined order. Sensing the initial stacking of the blocks is not part of the Warplan software, which has been written in Prolog and is commonly regarded as representing one of the significant achievements of AI research. Providing a visual input to Warplan is easy for Prolog+ and of course, allows a real robot to move real blocks about. There are many more examples and we may cite recent work by Brian Garner and his colleagues at Deakin University, Victoria, Australia [GAR-89]. This article describes some sophisticated methods, again using Prolog, for recognising a *chair*. Of course, this is a very variable type of object and Garner describes how stools, office chairs, armchairs, etc. can be regarded as sub-classes of the more general concept, *chair*. A program is described which can recognise a chair from a "line drawing". Once again, the program has no input, which Prolog+ could provide.[5]

Notes

1. Corners of skeletons are also important. We shall see later how to detect these features.
2. It may be preferable to define **perpendicular** in such a way that the angle between A and B is either 90° or 270°: similarly for **parallel**.
3. This has no effect at all on the running of the program, except that when the "input" database is otherwise empty, the goal

   ```
   line_segment(A,B).
   ```

   will fail, rather than flag a system error.
4. In fact, the thresholding can be performed directly on the list. However, it is conceptually easier to consider it as being performed on an image like that shown in Figure 5.5.
5. Professor Garner and the author recently discussed this point. As a result, the VSP software is now being modified to provide an interface between the real world (i.e. a video camera) and the software written by Garner and his colleagues. This is exactly the role that the present author hoped Prolog+ would find and more examples like this would be most welcome!

CHAPTER 6

ROBOT AND LIGHTING CONTROL

"Slaves obey your masters"
St Paul's Letter to the Ephesians

6.1 Introduction

As we have already seen, Prolog+ is able to control electro-mechanical devices, such as an (X,Y,θ)-table, a pick-and-place arm, cameras and lighting. In this chapter, we shall discuss some higher level control facilities than we have considered so far. Figures 3.1 and 3.2 show, in diagrammatic form, the structure and organisation of a so-called Flexible Inspection Cell. This should be borne in mind as we discuss the calibration and use of a visually guided robot. We shall concentrate initially upon the use of the overhead camera to acquire images.

6.2 Calibration of a Robot Vision System

In nearly all visually guided robots, the image processor and manipulator arm rely upon totally different coordinate systems. It is important therefore that, given the address of a point in vision system space the system should be transformed into the different set of coordinates used by the robot. In the same way, the robot may have placed, or discovered, an object in a position which is known in terms of its own coordinates. In this case, the vision system needs to be able to interpret that data so that it can properly interpret images.

When a visually guided robot is first set up, it is necessary, therefore, to calibrate it, so that the two parts of the system can communicate together effectively. In other words it is necessary to provide some mechanism so that vision system coordinates

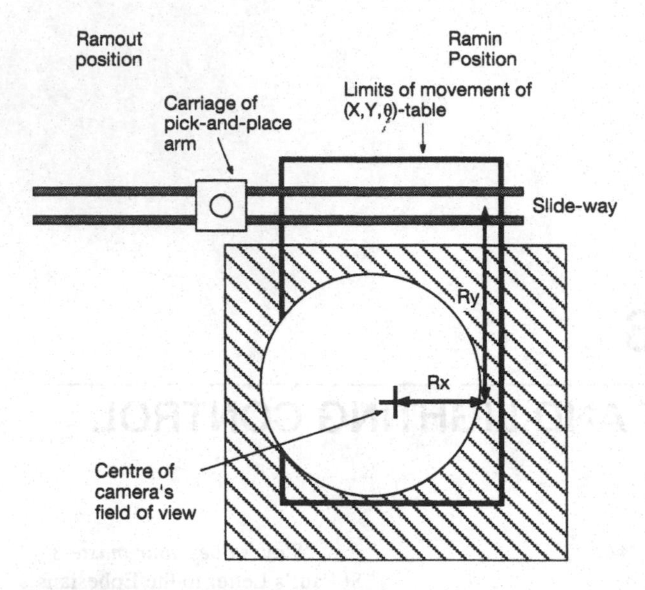

Figure 6.1. Flexible Inspection Cell calibration parameters. (Also see Figure 3.1.)

can be mapped into robot system coordinates and vice versa. To begin, we need to calibrate the system and the predicate **calibrate_robot** is a procedure for calibrating an (X,Y,θ)-table, which is viewed using an overhead camera (see Figures 3.1 and 6.1).

```
calibrate_robot :-
    message(['Place a white disc at, or near, the centre of the (X,Y,Theta)-table']),
                                /* The user must respond to this message
                                before the program continues. This disc is
                                called the target. */
    home,                       % Put table in the "home" position
    ctm,                        % Connect the camera directly to the monitor1
    offset(Rx1,Ry1),            % Consult db for initial offset parameters
    move_table(Rx1,Ry1),        % Move table to [Rx1,Ry1]
    locate_target(X11,Y11),     % Use vision system to locate the target
    turntable(180),             % Rotate table by 180°
    locate_target(X12,Y12),     % Locate the target again, using vision
    turntable(0),               % Reset the table to 0° position
    X1 is 0.5*(X11 + X12),      % Now find centre of the table – X coordinate
    Y1 is 0.5*(Y11 + Y12),      % Now find centre of the table – Y coordinate
    get_scales(Sx,Sy),          % Calculate scale factors
    Rx2 is Rx1 + Sx*(256 – X1), % Calculate movement of table
    Ry2 is Ry1 + Sy*(256 – Y1),
    integer(Rx2,Rx3),
    integer(Ry2,Ry3),
    move_table(Rx3,Ry3),        % Shift table by known amount
    locate_target(X2,Y2),       % Locate the target again
```

```
    nl,
    write('New offset parameters are:'),
    write([Rx3,Ry3]),
    nl,
    write('The centroid of the target is now at'),
    write([X2,Y2]),
    nl,
    assert(offset(Rx3,Ry3)),        % Revise the offset parameters ...
    retract(offset(Rx1,Ry1)),       % ... and forget the old values
    frz,                            % Digitise an image
    vpl(256,1,256,511,255),         % Plot vertical white line
    vpl(1,256,511,256,255),         % Plot horizontal white line
    define_robot_limits.            % Draw box to show limits of travel of the table

get_scales(Sx,Sy) :-
    calibrate_incs(Xinc,Yinc),      % Consult database
    locate_target(X1,Y1),           % Find where target is in terms of vision coords.
    delta_x(Xinc),                  % Move table ...
    delta_y(Yinc),                  % ... by fixed amount
    locate_target(X2,Y2),           % Locate target again
    Sx is Xinc/(X2 - X1),           % Calculate the ...
    Sy is Yinc/(Y2- Y1),            % ... scale factors
    nl,                             % Information for the user
    write('Scale parameters:'),
    write([Sx,Sy]),
    scale_multipliers(S1,S2),
    assert(scale_multipliers(Sx,Sy)),
    retract(scale_multipliers(S1,S2)),
    nl.

% This simple predicate is based on the assumption that the target is a white disc

locate_target(X,Y) :-
    grab,                           % Digitise an image. Not specified here
    thr(64),                        % Arbitrary threshold parameter
    ndo,                            % Shade objects by size
    thr(255,255),                   % Select the biggest to isolate target
    blb,                            % Make sure target appears as a solid figure
    cgr(X,Y).                       % Calculate centroid of the blob

% Initial values of calibration constants

calibrate_incs(500,500).            % Constants for the additive term of
                                    conversion formula

scale_multipliers(22.727, -20).     % Scale factors for axis conversion

offset(5134, -1223).                % Offset constants (i.e. The movement needed
                                    to shift the table from pick_up point to
                                    centre of camera's field of view.
```

```
/* The first clause is merely a convenience for the user. It draws a rectangle on the
image display so that the range of possible movements of the centre of the table can
be visualised */

define_robot_limits :-
        define_robot_limits(_,_,_,_).

define_robot_limits(X3,Y3,X4,Y4) :-
        axis_transformation(-7500,-5000,X1,Y1),    % Limits of table motion
        axis_transformation(7500,5000,X2,Y2),      % Limits of table motion
        image_limits(X1,Y1,X3,Y3),                 % Check valid vision system
                                                     coordinates
        image_limits(X2,Y2,X4,Y4),                 % Check valid vision system
                                                     coordinates
        nl,                                        % Information for the user
        write('Robot movement limits:'),
        write([[X3,Y3],[X4,Y4]]),
        nl,
        vpl(X3,Y3,X4,Y3,255),                      % Draw a rectangle
        vpl(X3,Y3,X3,Y4,255),
        vpl(X4,Y3,X4,Y4,255),
        vpl(X3,Y4,X4,Y4,255).

% What to do if the point is outside the picture

image_limits(X1,Y1,X2,Y2) :-
        size_limits(X1,X2),                        % Check X variable
        size_limits(Y1,Y2).                        % Check Y variable

size_limits(A,0) :- A < 0.                         % Hard limiting at edge of
                                                     the image
size_limits(A,511) :- A > 511.                     % Hard limiting at edge of
                                                     the image

size_limits(A,A).                                  % Address variable is OK so
                                                     keep it
```

6.3 Converting Vision System Coordinates to Robot Coordinates

The predicate **axis_transformation** uses the parameters calculated by **calibrate_robot** to transform vision system coordinates to robot coordinates, or vice versa.

```
% Convert vision system coordinates to robot coordinates

axis_transformation(Xr,Yr,Xv,Yv) :-
    not(var(Xv)),                   % Vision system coordinates are known
    not(var(Yv)),                   % Vision system coordinates are known
    var(Xr),                        % Robot coordinates are not known
    var(Yr),                        % Robot coordinates are not known
    offset(Dx,Dy),                  % Consult database
    scale_multipliers(Sx,Sy),       % Consult database
    Xr1 is Dx + Sx*(Xv – 256),      % Calculate robot's X coordinate, floating point
    Yr1 is Dy + Sy*(Yv – 256),      % Calculate robot's Y coordinate, floating point
    integer(Xr1,Xr),                % Convert to integer
    integer(Yr1,Yr),                % Convert to integer
    Xr ≤ 7500,                      % Check range of answer; can robot reach point?
    Xr ≥ – 7500,                    % Check range of answer; can robot reach point?
    Yr ≤ 5000,                      % Check range of answer; can robot reach point?
    Yr ≥ – 5000.                    % Check range of answer; can robot reach point?

% The robot cannot reach the point specified

axis_transformation(Xr,Yr,Xv,Yv) :-
    not(var(Xv)),                   % Vision system coordinates are known
    not(var(Yv)),                   % Vision system coordinates are known
    var(Xr),                        % Robot coordinates are not known
    var(Yr),                        % Robot coordinates are not known
    message(['Out of range of the robot']),    % Tell user about the problem
    !,
    fail.                           % Force failure if we get this far

% Convert robot coordinates to vision system coordinates

axis_transformation(Xr,Yr,Xv,Yv) :-
    not(var(Xr)),                   % Robot coordinates are known
    not(var(Yr)),                   % Robot coordinates are known
    var(Xv),                        % Vision system coordinates are not known
    var(Yv),                        % Vision system coordinates are not known
    offset(Dx,Dy),                  % Consult database
    scale_multipliers(Sx,Sy),       % Consult database
    Xv1 is 256.5 + (Xr – Dx)/Sx,    % Calculate vision system X coordinate
    Yv1 is 256.5 + (Yr – Dy)/Sy,    % Calculate vision system Y coordinate
    integer(Xv1,Xv),                % Convert to integer
    integer(Yv1,Yv).                % Convert to integer

axis_transformation(_,_,_,_) :-
    message(['Error in the way you have specified the parameters']),
    !,
    fail.                           % Force failure if we get this far
```

6.4 Moving the Robot

Let us consider now how **axis_transformation** can be used. Normally, of course, the vision system will calculate the position of some feature in an image before the robot is moved to that point, in order to pick up an object, drill a hole, insert a screw, weld a work-piece, or perform some similar task. This is how it can be coded in Prolog+:

```
locate_feature(X,Y),                      % Use vision system to locate a feature
axis_transformation(Xrobot,Yrobot,X,Y)    % Find where feature is in robot space
```

The predicate **move_robot_to** allows the programmer to work with the vision system coordinates without ever bothering about the fact that the robot uses a different set of coordinates.

```
move_robot_to(X,Y) :-
    axis_transformation(Xr,Yr,X,Y),       % Convert to robot system coordinates
    move_table(Xr,Yr).                    % Move the table
```

Notice that we have defined the software in such a way that

```
move_robot_to(256,256)
```

locates the table so that its centre is at the centre of the field of view of the camera.

We are now in a position to be able to control the robot in a very simple way:

```
locate_feature(X,Y),                      % Use vision system to locate a feature
move_robot_to(X,Y),                       % Move robot to feature found above
pick_up,                                  % Grasp object, pick it up. Not defined
```

In some situations, the robot coordinates may be known. For example, the robot may encounter an obstacle so that it cannot move further in a given direction. Assuming that the robot knows where it is in space, it would then be helpful if the vision system coordinates could be calculated, so that the obstacle might be identified in an image. This task could be achieved in the following way:

```
obstacle(X,Y),                            % Robot locates an obstacle
axis_transformation(X,Y,Xvision,Yvision), % Transform the axes
interpet_obstacle(Xvision,Yvision)        % Interpret the scene visually
```

We have ignored an important point here: the robot may have no way of knowing where it is in space. However, Prolog+ can keep a track of its movements. The use of *properties*[2] is generally to be preferred to that of **assert** and **retract**, since they can be adjusted more easily and are faster. Here is a revised definition of **move_robot_to**:

```
move_robot_to(X,Y) :-
    axis_transformation(Xr,Yr,X,Y),       % Convert to robot system coordinates
    move_table(Xr,Yr),                    % Move the table
    set_prop(robot,position,[Xr,Yr]).     % Remember robot position
```

In order to find where the robot is, in terms of its own coordinate system, we may use the following predicate:

```
locate_robot(X,Y) :-
        get_prop(robot,position,[X,Y]).    % Recall robot position
```

whereas, the predicate **where_is_robot** locates the robot in terms of the vision system coordinates

```
where_is_robot(X,Y) :-
        get_prop(robot,position,[Xr,Yr]),
        axis_transformation(Xr,Yr,X,Y).
```

The final topic in this section is that of moving the robot in such a way that an object which has been dropped onto the (X,Y,θ)-table in random position is normalised. Suppose that the vision system, using an overhead camera, has been able to determine that the centroid of an object is at [X,Y] (using vision system coordinates) and at orientation A. The normalisation process moves the table so that the centroid is located at the middle of the camera's field of view and rotates the object to an orientation of 0°.

```
normalise(X,Y,A) :-
        B is –A,                          % Calculate negative of A
        turntable(B),                     % Rotate table by angle B
        cosine(A,Cos),                    % Calculate cosine, A is in degrees
        sine(A,Sin),                      % Calculate sine, A is in degrees
        scale_multipliers(Sx,Sy),         % Consult database for scale multipliers
        X1 is –Sx*((X – 256)*Cos + (Y – 256)*Sin),
        Y1 is –Sy*(–(X – 256)*Sin + (Y – 256)*Cos),
        integer(X1,X2),
        integer(Y1,Y2),
        delta_x(X2),                      % Increment X axis of table
        delta_y(Y2),                      % Increment Y axis of table
        grab,                             % Digitise another image
        vpl(1,256,511,256,255),           % Draw cross-lines ...
        vpl(256,1,256,511,255),           % ... at centre of the image
        draw_robot_jaws.                  % See next section for details
```

Later, we shall illustrate the use of this predicate.

6.5 Additional Robot Facilities

Of course, it would be helpful to know where the gripper will be placed when the robot tries to pick up an object. No gripper is of insignificant size, so a misaimed move might put the object in such a position that it and/or the gripper and robot could be damaged. In fact this is quite a simple task, as we shall show. Let us assume first that the robot gripper has two parallel jaws, which open and shut in a simple pincer movement. It is possible to draw lines on the image which represent the gripper jaws in the open and/or shut state. Here is a simple example. The jaws lie parallel with the X-axis and are 30 pixels long. When open, the jaws are 24 pixels apart and are 12 pixels apart when shut.

```
draw_robot_jaws(open) :-
    vpl(236,234,276,234,255),
    vpl(236,258,276,258,255),

draw_robot_jaws(shut) :-
    vpl(236,250,276,250,255),
    vpl(276,262,276,262,255).
```

Each of these two predicates simply draws a pair of parallel horizontal lines on the image. The user can, of course, verify visually that the gripper will not collide with the object that is to be picked up. However, a little more processing allows this decision to be made automatically.

```
safe_to_lift(X,Y) :-
    move_robot_to(X,Y),        % Move the robot as described above
    grab,                      % Digitise another image
    convert_to_binary,         % Convert image to binary. Object is white and
                               %   background is black. Not specified
    swi,                       % Switch images
    draw_robot_jaws(open),     % Draw the jaws
    min,                       % AND the two images together
    cwp(0),                    % There is no overlap. So it is safe to lift object
    lift_object.               % Operate pick and place arm. Not specified

safe_to_lift(_,_) :-
    message(['Warning: The robot cannot safely lift the object']),
    !,
    fail.                      % Force failure
```

Another possibility is to count the blobs, using **ndo** or **eul**. There will be one blob for the silhouette of the object that is to be picked up and one each for the two gripper jaws (see Figure 6.3).

Of course, a different type of gripper (e.g. magnetic or vacuum) could be modelled using other graphical forms, such as a circle:

```
draw_suction_gripper :-
    hic(256,256),              % Intensity cone
    thr(200,255).              % Threshold creates a circle in the image
```

When using a multi-camera vision system in conjunction with an (X,Y,θ)-table, it is difficult for the user to remember how the different coordinate axes are related to each other. For this reason, a simple program like that shown below is most helpful and saves a lot of unnecessary frustration.

```
axis_directions :-
        axis_directions(overhead_camera),
        axis_directions(front_camera).

axis_directions(overhead_camera) :-
        nl,
        write('O/H Camera = RIGHT: Robot = Y+ Vision = X+'),
        nl,
        write('O/H Camera = DOWN: Robot = X+ Vision = Y+'),
        nl,
        write('O/H Camera = ANTICLOCK: Robot = T+ Vision = ANTICLOCK'),
        nl.

axis_directions(front_camera) :-
        nl,
        write('O/H Camera = RIGHT: Robot = Y– Vision = X–'),
        nl,
        write('O/H Camera = DOWN: Robot = X+ Vision = Focus'),
        nl,
        write('O/H Camera = ANTICLOCK: Robot = T+ Vision = X+'),
        nl.
```

It is also helpful for the user to be able to see the images directly from each of the cameras, as well as those being displayed by the vision system. For this reason, the video signal from each camera should be displayed on a separate video monitor. These two seemingly minor points will avoid much distress, while the user is trying to write software!

6.6 Two Demonstrations

White Disc with Black Spot

In the following program, the overhead camera is used to view a white disc which has an eccentric black spot. The table surface is assumed to be nominally matt black (see Figure 6.2).

```
demon(disc) :-
        move_robot_to(256,256),      % Move table to centre of field of view
        turntable(0),                % Set table orientation to 0°
        disc_centroid(X1,Y1),        % Calculate the disc centroid, [X,Y]
        xor,                         % Highlight the hole
        ndo,                         % Shade blobs in order of their sizes
        thr(255,255),                % Select brightest (i.e. biggest) blob
        blb,                         % Make sure spot is solid white blob
        xor,                         % Show the spot
```

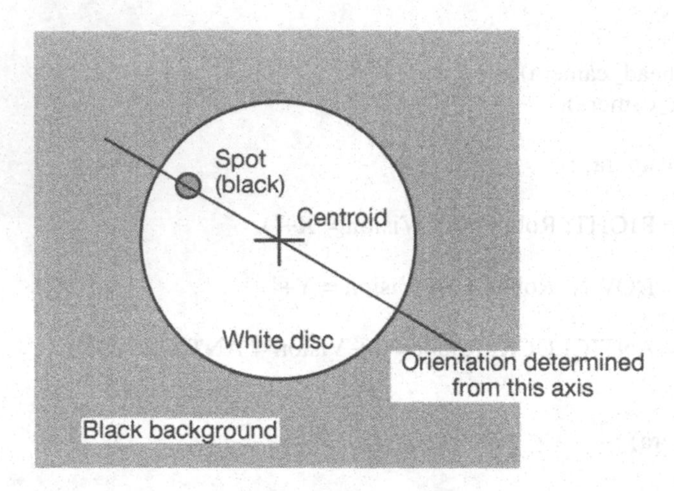

Figure 6.2. First demonstration. Picking up a white disc with an eccentric black spot.

```
        cgr(X2,Y2),                     % Find centroid of the spot
        angle([X1,Y1],[X2,Y2],A),       % Find orientation of the disc
        normalise(X1,Y1,A).             % Normalise position and orientation

disc_centroid(X,Y) :-
        frz,                            % Digitise an image
        yxt.                            % Rotate image by 90°
        thr(100),                       % Poor method but it works here
        ndo,                            % Shade blobs in order of their sizes
        thr(255,255),                   % Select brightest (i.e. biggest)
        blb,                            % Remove the spot, for now
        cgr(X,Y).                       % Calculate centroid of the disc
```

The result of satisfying demon(disc) is to position the disc so that it is in the middle of the camera's field of view and the spot lies along the X, axis to the right of the disc centre.

Automobile Connecting Rod (Con-rod)

Here is the second of our two demonstration programs (see Figure 6.3). This one directs the robot to pick up a con-rod.

```
demon(conrod) :-
        move_robot_to(256,256),   % Put table in centre of the camera's field of view
        turntable(0),             % Put table to 0° position
        frz,                      % Digitise image
```

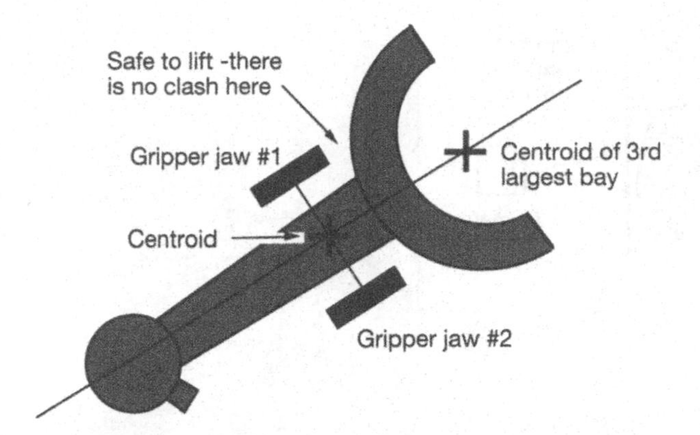

Figure 6.3. Second demonstration. Picking up a con-rod. The gripper jaws are placed with reference to the centroid of the blob silhouette and the line joining this point to the centroid of the third largest "bay". We can project the gripper axes onto the image of the con-rod and determine whether it is safe for the robot to be lowered over it. Two small rectangles are drawn representing the tips of the jaws of the robot gripper. The number of blobs in the resulting image is then counted, using either **eul** or **ndo**. If there are exactly three blobs (one for each of the two jaw tips and one for the con-rod), there is no possibility that the robot will collide with the con-rod. An extension of this idea can be used to decide whether the gripper will be able to hold the con-rod without slipping. (Size, not weight, is the criterion by which this will be judged.)

```
yxt,                             % Rotate image by 90°. Camera and robot X-
                                   axes are at 90°
thr,                             % Naive thresholding – poor technique but it
                                   works!
cgr(X1,Y1),                      % Find centroid of the silhouette of the conrod
chu,                             % Convex hull
max,                             % Superimpose convex hull on the silhouette
blb,                             % Fill holes (i.e. "bays")
xor,                             % Show the bays
ndo,                             % Shade bays according to their sizes
thr(253,253),                    % Select third largest bay
cgr(X2,Y2),                      % Centroid of third largest bay
angle([X1,Y1],[X2,Y2],A),        % Calculate the orientation
normalise(X1,Y1,A),              % Normalise the position and orientation
offset(U,V),                     % Consult database for offset parameters
U1 is –U,
V1 is –V,
delta_x(U1),                     % Move table to sit beneath …
delta_y(V1),                     % … the pick-and-place arm
ramin,                           % Move arm over table
pick,                            % Lower gripper, close it, then lift
ramout,                          % Move arm out
place.                           % Put con-rod down
```

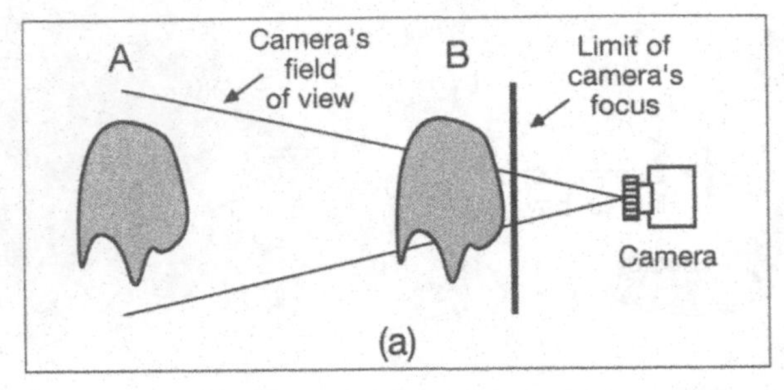

a.

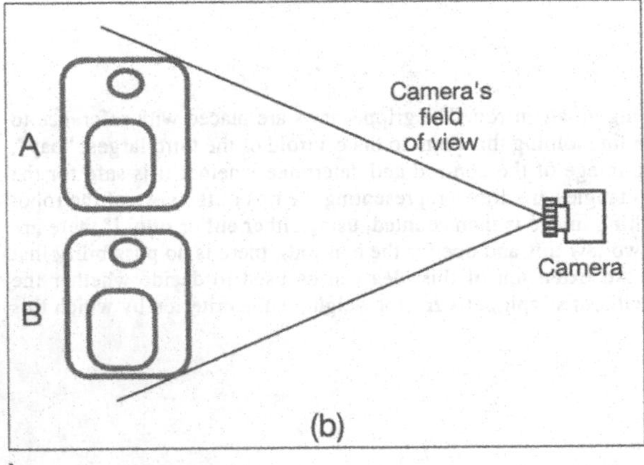

b.

Figure 6.4. The front camera can be used in a variety of interesting ways. **(a)** Zooming, using object motion. The camera has a wide-angle lens which can remain in focus over a broad range of distances. In position A, the whole object may be viewed, whereas in position B, only part of it may be seen. The object may be moved along the vertical axis and rotated in order to obtain a number of high-resolution views covering its complete surface. **(b)** Stereoscopy based upon motion of the object. It is assumed that lighting variations, caused by the object motion, can be ignored. The number of views is not limited to any particular number, as it would be if different cameras were used. Stereoscopic viewing, by *rotating* the object, is possible but has no direct counterpart in human/animal vision.

6.7 The Front Camera

So far we have only considered the overhead camera. Refer again to Figure 3.1, where it will be seen that the front camera looks horizontally towards the workpiece. The overhead camera can fix a point in (X,Y)-space and the front camera a point in (X,Z)-space. The front camera can be used in the normal way for image acquisition, prior to processing along the same lines as discussed elsewhere in this book. However, there are several ways in which the front camera can be useful in other, distinctive and interesting ways (see Figure 6.4).

(i) *Automatic zooming*. Suppose that the front camera is fitted with a short focal-length lens, capable of retaining sharp focus over a very wide range of camera–

object distances. (Some lenses for example can provide a sharp, well focused image over distances ranging from about 10 cm to infinity.) If the object under examination is moved along the camera's optical path, zooming is achieved.

(ii) *Stereoscopy*. The effect of stereoscopic vision can be achieved using a single camera and moving it, or the object under examination.

However, it should be noted that it may be necessary to employ *automatic focusing* when using the front camera,[3] since the object under examination can be moved both laterally (X-axis) and along the camera's optical path (Y-axis). The task of automatically focusing the camera might be assigned to specialised hardware and might use optical, infra-red, ultra-sonic, or some other form of range sensing. Range (measured along the Y-axis) might be sensed using the overhead camera. The measurement so obtained would then be used to control the focus of the front camera. Alternatively, the point of optimal focus may be determined by maximising the high-frequency content in the images derived from the front camera, as the object–camera distance is varied. The high-frequency content in an image may be estimated[4] using a simple command sequence:

```
hf_content1(A) :-
        raf(11,11),          % Blur the image
        swi,                 % Switch current and alternative images
        sub,                 % Subtract the original from the blurred image
        pct(95,A).           % Find intensity (A) to threshold image so
                               that 95% is black
```

The parameter A is then maximised as the object is moved, along the Y-axis. This involves a simple search, in which the object is moved progressively, using **move_y**. Writing a Prolog+ program which will perform a one-dimensional high-climber is left as an exercise for the reader. (You may safely assume that the function A(Y), where Y is the horizontal position, has only one maximum value.)

An alternative measurement of the high-frequency content of an image may be based upon the magnitude of the maximum gradient.

```
hf_content2(A) :-
        sed,                 % Blur the image
        edg(0,1),            % Mask edge effects
        gli(_,A).            % Find maximum intensity
```

So far, we have not made any serious attempt to fuse data from the overhead and front cameras. One obvious way to do this is to use one camera to identify a certain feature on the object being examined and then use that knowledge to assist in the interpretation of the images from the other camera. For example, suppose that a die were to be viewed, using the overhead camera first and that a "6" (i.e. six spots) were seen on its top surface. Then, we could deduce that the front camera could see a "2", "3", "4" or "5", but not "1" or "6". Furthermore, the orientation of a die can be semi-determined[5] from the orientation of the spots on the top surface. Hence, we can make a slightly more precise prediction: when the overhead camera sees a "6", the front camera will see either "2" or "5" (if the orientation is 0°), or "3" or "4" (if the orientation is 90°.) Here is a simple Prolog+ program to make such a prediction.

```
front_camera_sees(A) :-
    select(overhead_camera),   % Use csl to do this
    count(spots,N),            % Instantiate N to the number of spots on top
                               %   surface
    member(N,[2,3,6]),         % Check we can measure orientation properly
    orientation(P),            % P = 0° or 90°. No other values are allowed
    prediction(N,P,A).         % Refer to the database

front_camera_sees(A) :-
    count(spots,N),            % Repeat processing – for simplicity
    prediction(N,_,A).         % Refer to the database

prediction(1,_,[2,3,4,5]).     % Cannot determine orientation from one spot
prediction(2,0,[1,6]).         % Two spots. Orientation is 0°
prediction(2,_,[3,4]).         % Two spots. Orientation is 90°
prediction(3,0,[1,6]).         % Three spots. Orientation is 0°
prediction(3,_,[2,5]).         % Three spots. Orientation is 90°
prediction(4,_,[1,2,5,6]).     % Cannot determine orientation from four spots
prediction(5,_,[1,3,4,6]).     % Cannot determine orientation from five spots
prediction(6,0,[2,5]).         % Six spots. Orientation is 0°
prediction(6,_,[3,4]).         % Six spots. Orientation is 90°
```

Clearly, this particular type of cross-reference between cameras is particularly well suited to Prolog+ and could be applied to other objects of greater commercial importance, such as printed containers and a variety of machined components.

The fusing of data from the overhead and front cameras is often problem specific and does not demonstrate any clearly discernible general principles. For this reason, we shall turn our attention now to the use of yet another camera, but this time not shown in Figure 3.1.

6.8 Range Maps

A popular method of obtaining range information in robot vision systems is illustrated diagrammatically in Figure 6.5. Triangulation by structured lighting, as this technique is called, could easily be incorporated into a Flexible Inspection Cell.[6] Simple geometric optics, based upon the assumption that the camera is a pin-hole device, yields a formula which relates the *apparent* deviation of the laser line to the surface height. However, the method is such that a single image yields information about only one vertical section through the object being viewed. For this reason, the object has to be moved past the camera, in a series of steps, each of which yields height information about only one slice. Thus, a 512×512 range map can be acquired in a minimum of 512 television frame periods (20.5 seconds using the U.K. television standard). Although this method is slow, it is widely used in industrial vision systems. It is a simple matter to write a Prolog+ program to construct a range map.

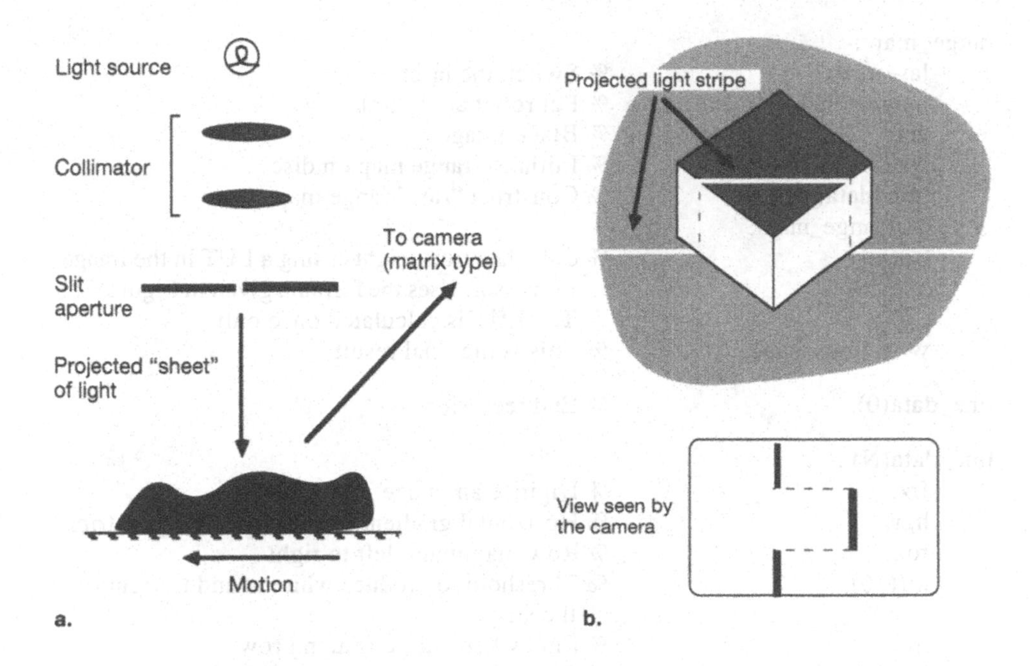

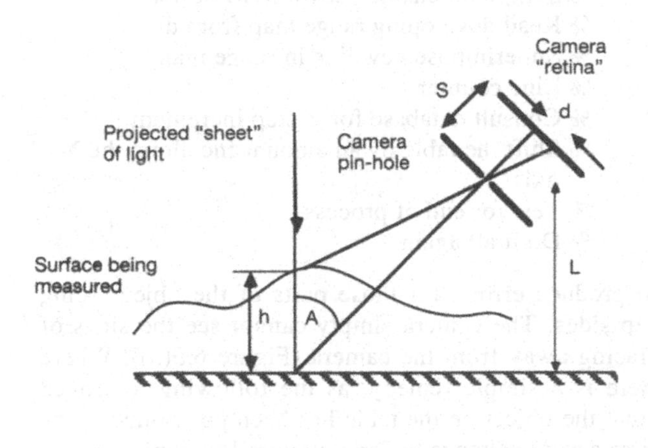

$$h = \frac{d.L.(1 + \tan^2 A)}{S.\tan A + d}$$

Figure 6.5. Generating a range map. The ideas implicit in this diagram can be incorporated into the Flexible Inspection Cell (Figure 3.1). (**a**) Physical layout (cross section of the optical system). In practice a light stripe is projected onto the object being observed; light from a laser is expanded into a fan-shaped beam, using a cylindrical lens. A matrix camera is used. Thus, one vector of height data (i.e. one row in a range map) can be derived during every television frame period. Most publications show the light source at the side and the camera overhead. This arrangement makes the decoding of the range data rather easier and is therefore to be preferred. (**b**) Light stripe projected onto a rectangular block on a plane table. (*Top*: What a person would see. *Bottom*: What a camera sees. The camera is rotated by 90°, to make decoding of the image easier.) (**c**) Geometrical optics may be applied to derive the height h from the observed deviation, d, of the light stripe.

```
range_map :-
    laser(on),                  % Switch the laser on
    move-robot_to(1,256),       % Put robot at [1,256]
    zer,                        % Black image
    wri(range_map),             % Initialise range map on disc
    line_data(512),             % Construct "raw" range map
    rea(range_map),
    tra(lut),                   % Calculate true heights using a LUT in the image
                                  processor. Uses the formula given in Figure 6.5.
                                  The LUT is calculated once only
    wri(range_map).             % This is the final result

line_data(0).                   % End recursion

line_data(N) :-
    frz,                        % Digitise an image
    hgr,                        % Horizontal gradient to enhance the light stripe
    rox,                        % Row maximum, left to right
    thr(16),                    % Threshold to produce white at and to right of
                                  the stripe
    int,                        % Find where stripe is along row
    lgt(512,X),                 % Sample intensities down RHS of image
    wri(range_map),             % Read developing range map from disc
    lfx(N,X),                   % Superimpose new line in range map
    M is N –1,                  % Line counter
    step(Inc),                  % Consult database for Y step increment
    delta_y(Inc),               % Shift the table by an amount Inc along the Y
                                  axis[7]
    M ≥ 1,                      % Test for end of process
    line_data(M).               % Do it all again
```

Now, **range_map** is prone to produce errors, for those parts of the object being scanned which have very steep sides. The camera simply cannot see the sides of very steep cliffs, if they are facing away from the camera (Figure 6.5(a)). Where occlusion occurs like this, there is a simple remedy, as the following improved version shows. It is assumed that the object on the table has been positioned at the centre of the table. (This requires a minor change to demon(conrod); simply remove **ramout**.)

```
improved_range_map :-
    range_map,                  % Create first range map
    wri(temp),                  % Save result in safe place on disc
    step(I1),                   % Consult database for step size
    I2 is –I1,                  % Negate increment
    asserta(step(I2)),          % Put negated step increment at top of database
    turntable(180),             % Rotate table and object by 180°
    range_map,                  % Create second range map
    lrt,                        % Flip along horizontal axis
    retract(step(I2)),          % Restore step increment to previous value
```

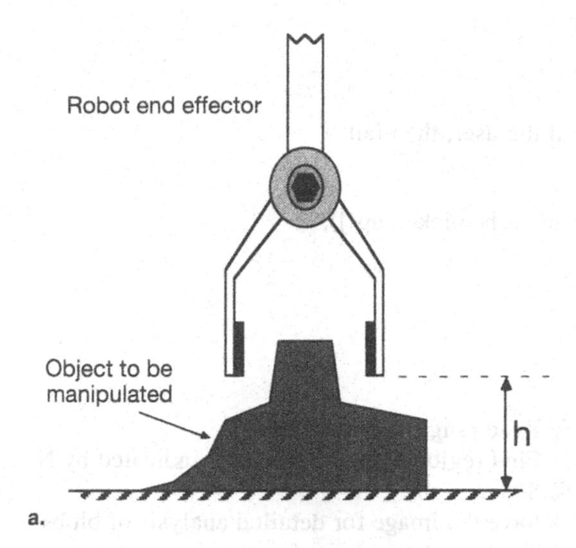

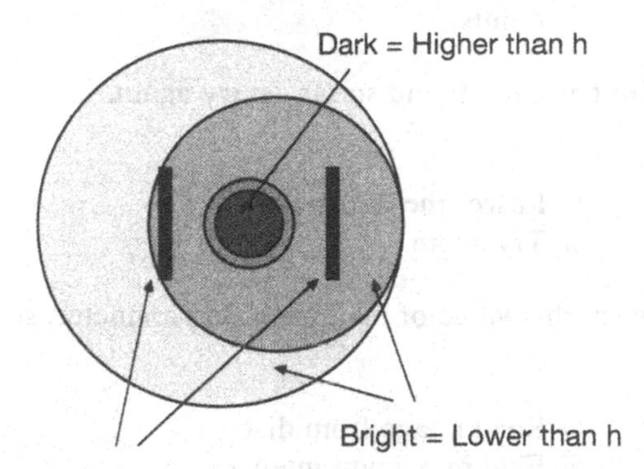

Figure 6.6. Using a range map to decide where to place the jaws of a gripper. (**a**) Side view showing the gripper above the work-piece (stippled). (**b**) Range map and gripper-jaw silhouette projected onto it.

```
rea(temp),            % Read first range map image from disc
max,                  % Fill in "holes" where occlusion occurs
wri(range_map).       % This is the final result
```

We next consider how Prolog+ can be used to calculate gripping points on an object which has been scanned using **improved_range_map**. We shall assume that the object lies on a table beneath a robot that is capable of X, Y and Z movements, but for the sake of clarity, we shall ignore rotation (θ). The gripper is assumed to be a simple two-finger pincer type (see Figure 6.6).

```
gripping_points :-
    search_grip_points(255).

% Object cannot be picked up so tell the user, then fail

search_grip_points(0) :-
    message(['Sorry this object cannot be picked up']),
    !,
    fail.

% Try to pick up the object

search_grip_points(N) :-
    rea(range_map),              % Read range map from disc
    thr(N,255),                  % Find regions higher than level indicated by N
    ndo,                         % Shade blobs
    wri(temp1),                  % Save the image for detailed analysis of blobs
    analyse_blobs(N).            % Try to pick up object for possible gripping
                                   points

% No satisfactory gripping point has been found so far, so try again.

search_grip_points(N) :-
    M is N - 1,                  % Lower the threshold
    search_grip_points(M).       % Try again

% No more blobs left to analyse at this value of the threshold parameter, so fail

analyse_blobs(_) :-
    rea(temp1),                  % Read image from disc
    gli(_,X),                    % Find maximum intensity
    X = 0,                       % Is maximum intensity = 0?
    !,
    fail.                        % Forced failure

% Try to pick up the object

analyse_blobs(N) :-
    rea(temp1),                  % Read image from disc
    gli(_,W),                    % Find maximum intensity
    thr(W,W),                    % Isolate that blob
    cgr(X,Y),                    % Find centroid of the blob
    safe_to_lift(X,Y),           % Can the robot safely lift the object?[8]
    normalise(X,Y,0)             % Move robot to (X,Y). Theta is zero
    convert(N,Z),                % Consult database; convert range map
                                   intensity to height
    lift_object(Z).              % Lift the object, placing gripper at height Z

% Try another part of the object as a possible place to put gripper.
```

```
analyse_blobs(A) :-
    rea(temp1),          % Read image from disc
    gli(_,X),            % Select blob
    hil(X,X,0),          % Eliminate this blob
    wr(temp1),           % Save the image
    analyse blobs(A).    % Try again
```

6.9 Lighting in a Flexible Inspection Cell

Figure 3.1 shows in diagrammatic form the lighting arrangement in a Flexible Inspection Cell. The question that naturally arises is how the lights can be adjusted automatically. Perhaps we should first question *whether* it is possible to adjust the lights automatically, so that the features of interest can be clearly seen.

The author is of the opinion that it is will not be possible to write a program which can optimise the lighting for a given inspection task, for a long time to come. This comment reflects some experience working in this area; the author studied lighting for inspection and robot vision for several years. Recently, he has spent a considerable amount of time developing an expert system which advises an engineer about appropriate lighting methods for Automated Visual Inspection. The *Lighting Advisor*, as this program is called, is written in MacProlog [BAT-89c]. This means that the software is available for consultation, at any time, by the Prolog+ programmer. The advisor uses simple rules for deciding what advice to present. Many of the rules could be modified for direct control of the lamps in a Flexible Inspection Cell. For example, one of the rules is of the form:

IF	*The object is opaque.*
AND	*The object has a matt surface.*
AND	*The silhouette of the object is to be viewed.*
THEN	*Use back illumination.*

Clearly, the advice given could easily be changed to, or supplemented by, a series of Prolog+ commands:

```
all_lights(off),        % Switch all lights off
back_lighting(on),      % Switch the back-lighting unit on
csl(4).                 % Select the camera which faces the back-lighting unit
```

and, of course, the conditional terms in such a rule can be expressed easily in Prolog+. However, not all of the advice terms in the Lighting Advisor rules can realistically be represented in this way. Moreover, the program is intended to interact with an engineer, so that it cannot, of its own accord, decide how to illuminate a given scene; it always needs human input. For this reason, the Lighting Advisor does not offer any real opportunity for development into an automatic illumination controller; we need to adopt a different approach to illumination control.

Remember that Prolog+, is a declarative language and hence is well suited for defining tests for recognising "acceptable" lighting patterns. This follows the spirit of Prolog, where the nature of an acceptable or satisfactory solution can be specified, rather than an algorithm for finding such a solution. Prolog+ is a good language for finding *satisfactory* solutions, whereas imperative languages, like C and Pascal are excellent for finding *optimal* solutions. Of course, the method of search is quite different in declarative and imperative languages but so too is the type of solution found. The words "satisfactory" and "optimal" were set in italics to emphasise this point. With these thoughts in mind, we can approach the control of lighting. The rules are *permissive,* in that they allow certain lighting patterns but inhibit others.

Here are some extremely simple heuristic requirements for good lighting. The rules are expressed in English, but will later be rewritten in Prolog+.

1. A well lit scene has no points so dark that the images obtained from it are black.
2. A well lit scene has no points so bright that the images obtained from it are white.
3. A well lit scene produces a reasonably high *local* contrast in all parts of an image. (This assumes that there is some variation in reflectivity over the whole scene.)
4. If a (large) region in an image is very dark, compared to the remainder of the picture, then the lamp which illuminates the corresponding part of the scene should be made brighter.
5. If a (large) region in an image is very bright, compared to the remainder of the picture, then the lamp which illuminates the corresponding part of the scene should be made darker.
6. Suppose that image A (obtained using lighting pattern La) shows features which are not visible in another image, B (obtained using lighting pattern Lb), but that all features visible in B are also evident in A. Clearly, lighting pattern La is better than Lb, and is likely to be a better basis for further experimentation.
7. If a certain feature is visible in one lighting pattern, then it must also be visible in another, which purports to be "better".
8. A good lighting pattern will highlight features that are visible in a number of other, less satisfactory, lighting arrangements.

Here is a procedure which makes use of the principle implicit in Rule 8 to discover what features must be visible in any supposedly "good" lighting pattern. Switch all lights off. Then, turn on one of the lamps. Digitise an image and analyse it for features which are manifest by creating large intensity gradients. (A useful heuristic is to use a high-pass filter or an edge detector, followed by thresholding. This will successfully detect spots, streaks, and edges.) The result will be a binary image, B_i, which will contain a number of real features of interest, as well as artifacts around the edges of shadows. The latter can be identified since they are invariably associated with the edges of regions which are very dark. It is possible therefore to process B_i, in order to obtain a simplified image, B_i', in which the edges of shadows have been eliminated. Now, a feature visible in B_i' may also be manifest in a slightly different form in another image, B_j'. Some way must be found, therefore, of associating these features together. If we simply **OR** together the features in the set of images $\{B_1', B_2',$

B_3', ... } using **max**, some of the blobs will merge, even though there may not be a perfect overlap between them. The resulting image obviously contains a set of features which must *all* be visible in any lighting pattern which purports to be optimal, or nearly so. In this way, we can generate a criterion by which new lighting patterns can be judged.

Now, this procedure, though not expressed formally in terms of Prolog+, presents no real problems. Since the subject of this book is Prolog+ and not lighting, we shall not pursue the discussion much further. One final remark is in order though. What we have done is to express some initial ideas about the automatic control of lighting informally, in terms of English and this is a situation that the author has encountered many times during the development of Prolog+. Of course, to make further progress, we need to formalise such a procedure using some suitable computing language. The ability of Prolog+ to provide a convenient vehicle with which to formulate and experiment with such ideas is, of course, the theme of this book. The reader is invited to investigate the formulation of the above lighting control procedure in terms of Prolog+. (Warning: this as a topic of such magnitude that it could well form the basis of a Ph.D. research programme.) The direction of the path ahead is clear, even though its end is not yet discernible.

6.10 Additional Remarks

The present chapter has merely introduced the idea of controlling a robot using an intelligent vision system, based on Prolog+. There is a great deal of very exciting work to be done in this area and, no doubt, the reader will see numerous possibilities for future developments. We conclude this chapter by briefly discussing the role and suitability of Prolog+ as a robot control language and the problems of interfacing vision systems to a robot.

Is Prolog+ a suitable language for controlling a robot? The answer is affirmative but the role of Prolog needs to be defined very carefully. Present-day robot control languages are imperative, often resembling Basic or Pascal, whereas Prolog+ is declarative and requires a totally different approach to writing programs. Any person, with prior programming experience and trying to use Prolog+ for the first time, has to surmount a major intellectual barrier. Doing so is not easy and it requires considerable confidence to believe that the investment of time and effort will be worthwhile. A novice Prolog+ programmer attempting to use the language to control a robot would probably find it helpful to rely upon the use of predicates such as **case, if_then, if_then_else** and operators akin to "•". These are, of course, included in the language to assist novice Prolog+ programmers, and do nothing to promote good programming style. The reader should remember that, while it is possible to write perfectly good procedural programs in Prolog+, this means that the most useful features of the language (i.e. recursion, back-tracking, data matching and unification) are likely to be ignored.

The way that we have used the robot throughout this book is as a slave that moves to any position defined by the vision system and operates its gripper on command. This is not the way that commercial robot manufacturers have traditionally seen their products being used; traditional robot languages are principally concerned with the task of *how* to reach a defined state in $(X,Y,Z,\theta_X,\theta_Y,\theta_Z)$-space.[9] In Prolog+, abstract concepts are being manipulated and complex movement patterns which contain variable position parameters for some of the moves will be constructed. These will, in turn, be integrated with vision, computational and other (non-robot) operations. Prolog+ should not be regarded as representing an alternative to specialised robot control languages, such as AMT or VAL. Indeed, it would not compete with them at doing those numeric computations that are needed for performing, say, the rapid transformation of coordinate axes used while operating an articulated-arm robot. Prolog+ should be seen instead as providing the foundation upon which a pyramid of high-level symbolic computation, incorporating both vision and (robot) control, can be built. A specialised robot control language may well operate beneath Prolog+, just as an image processing language like VCS does.

Both robot manufacturers and industrial robot users are unused to thinking about robots in the way that we have been discussing. In a "traditional" robot movement sequence, all of the moves are fixed in advance. However, a visually guided robot executes a movement sequence in which certain moves are unspecified until the moment just prior to their being performed.

Most industrial robots are not designed to be used in conjunction with a vision system; getting the robot to incorporate position parameters, derived from a vision system or elsewhere, into its movement sequence, while the robot is actually moving, is not easy. Many robots simply do not provide a data port and appropriate software by which position parameters may be inserted into a movement sequence while it is being executed. It is to be hoped that robot manufacturers will soon provide the hardware and software facilities needed to make use of the Prolog+ style of using a robot.

Notes

1. This is done to allow the user to see the table as the vision system will later. It has been found to be very useful in providing a check that the camera is in focus, zoom is set correctly, etc.
2. Properties are a feature of LPA MacProlog but are not supported in Edinburgh Prolog. They closely resemble Lisp's properties in function and use.
3. The reason that we have not encountered focusing problems when working with the overhead camera is that it has been assumed that we are using a robotic device capable of moving only in the X, Y and θ directions. Object height variations are also assumed to be small compared to the height of the camera above the (X,Y,θ)-table. When these assumptions are valid, the overhead camera will always remain in focus.
4. This simple parameter is one of many possible which could be used in this way.
5. We cannot resolve rotations of P° and (P + 180)°. Hence, the orientation can only be partially determined, leaving a choice of one out of two options.
6. The author has incorporated equipment for structured lighting into the FIC which he has built in his laboratory at Cardiff. This uses a laser, plus a beam expander overhead, and a video camera, mounted on its side and looking downwards at about 45°. As is explained later, this arrangement reduces the processing to a minimum.
7. Recall that earlier we stated that the X and Y axes of the robot and vision system are interchanged; see the definition of **axis_directions**.

8. The version of **safe_lift** given earlier needs to be modified slightly. However, the revised version follows the same general principles.
9. Prolog+ contains no facility for defining paths for robot movement. Nor is there any means of controlling the acceleration/deceleration of the arm. There is no facility for choosing which arm attitude is best when trying to reach a defined position. Remember that there may be many different arm positions which can achieve the goal of placing the end effector at a given point in $(X,Y,Z,\theta_X,\theta_Y,\theta_Z)$-space. Thus, there are good and bad ways to try and lift an object.

CHAPTER 7

FURTHER APPLICATIONS

"What impossible matter will he make easy next?"
The Tempest, William Shakespeare

7.1 Introduction

The objective of the present chapter is to demonstrate that Prolog+ can be used to good effect in a wide variety of industrial and other applications. The examples discussed here were selected carefully to reflect the ability of the language to represent a wide range of useful inspection and control procedures. The important issue at this stage is proving that Prolog+ has the expressional power that will make it useful in a range of practical situations. We shall not consider processing speed, nor compare Prolog+ with other languages. However, it should be noted that the four essential features of Prolog, namely recursion, back-tracking, unification and declarative programming, are all put to good use in the programs listed in this chapter; the move away from conventional imperative programming languages does not signal a loss of functionality, nor ease of use.

7.2 Recognising Playing Cards

There are several sub-problems:

(i) Identifying the suit. This can be applied to both picture and non-picture cards.
(ii) Identifying the value of non-picture cards.
(iii) Identifying the value of picture cards (i.e. jack, queen and king).
(iv) Integrating the techniques developed for (i) to (iii).

Figure 7.1. Recognising playing cards by counting bays. There are four bays for a club (♣), two for a spade (♠), one for a heart (♥) and none for a diamond (♦). Any other score indicates that there has been an error, or that the object is a not a playing-card suit symbol.

Although identifying playing cards might seem to be a frivolous application, it does possess many features in common with important industrial image processing tasks. For example, a printed company logo, an alignment arrow, or a component-identification mark might resemble a playing-card suit symbol. Recognising complex printed/engraved/embossed patterns, such as a trade-mark, has many points of similarity with the task of identifying a "picture" playing card.

Identifying the Suit

Suppose that one of the suit symbols, hereafter called a "pip" for convenience, has already been isolated. (This is based upon the operator **ndo**.) The pip is assumed to be white, against a black background. Although lighting is outside our present remit, it is perhaps of interest to note that cards from the two red suits (i.e. hearts and diamonds) yield a high contrast image, if green light is used for illumination; there should be very little difficulty. The following Prolog+ program will identify the suit (see Figure 7.1).

```
suit(S) :-
        chu,                    % Convex hull
        max,                    % OR with the original (i.e. binary) image
        blb,                    % Fill the holes
        xor,                    % Exclusive OR to show the holes
        3•skw,                  % Shrink white areas to eliminate noise
        3•exw,                  % Expand white areas back to original size
        ndo(1,N),               % Count the white regions
        suit1(N,S).             % Find suit from N
```

Figure 7.2. A non-picture card. Notice the small pip and the numeral at the top-left and bottom-right corners of the card. Hence, the value of this type of card can be obtained by counting the blobs and then subtracting 4.

```
suit1(0,diamond).
suit1(1,heart).
suit1(2,spade).
suit1(4,club).
suit1(_,error).
```

This simple procedure is size, orientation and position invariant. However, certain objects which do not appear as printed figures on playing cards might be mistaken for pips; the procedure does not purport to be a pip recogniser.

Identifying the Value of a Non-Picture Card

This time, we shall assume that the image of the playing card has been pre-processed, so that the pips and the numerals are black against a white card. The background is also black (See Figure 7.2). If the lighting is uniform, this can be achieved using simple thresholding. The following program will identify the value of the card:

```
value(N) :-
    blb,            % Fill holes to eliminate pips
    xor,            % Exclusive OR, to show pips
    blb,            % Fill holes in numeral
    eul(M),         % Count the blobs
    N is M-4,       % There are four extra ones!
    N ≥ 1,          % Is the result within acceptable range?
    N ≤ 10.         % Is the result within acceptable range?

value(error).
```

Identifying the Value of a Picture Card

Identifying the value of a picture card is a more difficult task than those discussed so far, although there are many different approaches, as we shall see. While it is theoretically possible to use correlation to recognise a given picture card, this approach is computationally expensive; it would be necessary to use three-dimensional correlation, if the position and orientation of the card were unknown. Moreover, this must be done for each of 12 reference images, one for each of the picture cards (jack, queen and king), in each of the four suits. The heavy processing load is compounded by the requirement for large amounts of computer memory. The computational effort could be reduced very considerably, by calculating the position and orientation parameters first, then shifting and rotating the card image, before applying a simple template match. A simple normalisation procedure, which relies upon the primitive command **lmi**, computes the orientation of the axis of the minimum second moment, as well as the centroid coordinates.

```
normalise_card :-
    grb,                    % Digitise an image
    wri(temp),              % Save image on disc
    sed,                    % Sobel edge detector
    thr(16,255),            % Threshold
    edg(0,16),              % Remove the edge
    3•exw,                  % Noise removal to help next operation a bit
    blb,                    % Fill holes
    3•skw,                  % Restore white areas to original size
    lmi(_,X,Y,Z),           % Centroid coordinates and orientation, Z
    rea(temp),              % Read image
    min,                    % Eliminate background
    Z1 is –Z,               % Negative of Z
    int(X,X1),              % Convert to integer
    int(Y,Y1),              % Convert to integer
    tur(Z1,X1,Y1),          % Rotate image
    X2 is 256 – X1,         % Calculate X-shift parameter
    Y2 is 256 – Y1,         % Calculate Y-shift parameter
    psh(X2,Y2).             % Shift the card to centre of image
```

Once **normalise_card** has been applied, it is a simple matter to perform a template match. The following program uses 12 stored images, each of which has previously been normalised in this way. A similarity measure is computed, by the as yet unspecified predicate **difference**, which compares the image of the test card with each of the 12 stored reference images.

```
identify_card(A) :-
    normalise_card,
    wri(temp),
    remember(dist,[1000000.0,none]),
    template_match(A,_).
```

```
template_match(_,_) :-
    rea(temp),
    member(A,[ jack_clubs, queen_clubs, king_clubs,
    jack_spades, queen_spades, king_spades,
    jack_hearts, queen_hearts, king_hearts,
    jack_diamonds, queen_diamonds, king_diamonds] ),
    rea(A),
    difference(S),
    recall(dist,[_,B]),
    S < B,
    remember(dist,[S,A]),
    fail.

template_match(A,B) :-
    remember(dist,[B,A]).
```

Here is one naive suggestion for the predicate **difference**.

```
difference(A) :-
    dif,                    % Absolute intensity diffference
    gli(_,A).               % Distance measure is maximum intensity difference
```

Here is another:

```
difference(A) :-
    dif,                    % Absolute intensity diffference
    pct(1,A).               % A is threshold which makes one percent white
```

A third possibility is to compute the average intensity difference over the surface of the playing card.

```
difference(A) :-
    dif,                    % Absolute intensity diffference
    set_window,             % Adjust processing window to cover card only
    avr(A),                 % A is the average
    restore_window.         % Restore the processing window to full image
```

There are many other possibilities. No doubt, the reader can suggest several.

A number of alternative approaches to recognising picture playing cards can be devised. For example, we might sample the intensity along a line (called a *scan line*) which joins two well-defined points, such as diagonally opposite corners of the card. The intensity, expressed as a function of position along this line, might be likened to a bar-code and could be matched to a set of 12 stored reference vectors. Here is a simple program[1] for calculating the "bar-code" across the centre of the card (see Figure 7.3).

```
bar_code(X) :-
    normalise_card,
    raf(11,11),             % Blur to eliminate minor detail
    thr(128,255),           % Threshold
    vgt(256,1,256,512,X).   % Instantiate X to intensity vector
```

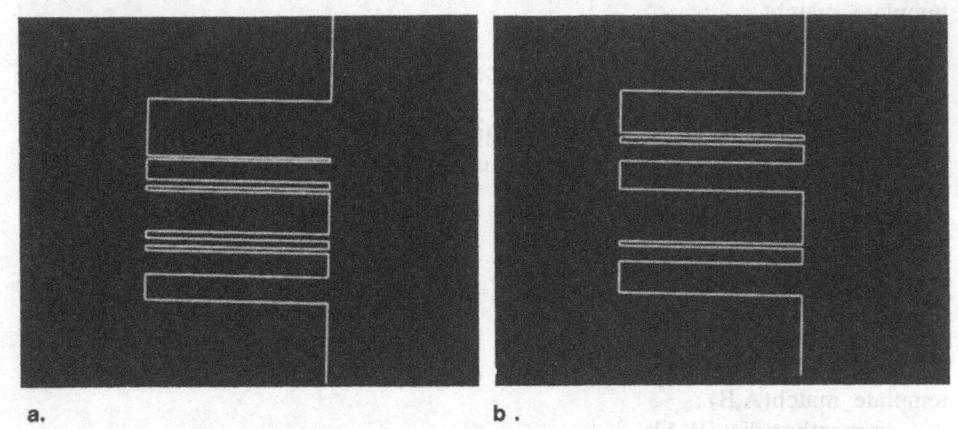

Figure 7.3. "Bar codes" for two picture cards compared. (a) Queen of clubs. (b) Queen of spades.

Matching could be accomplished by minimising the Hamming distance between the observed vector and a number of stored reference vectors. The Hamming distance can be computed in the following way:

```
hamming_dist([],[],H,H).

hamming_dist([X|Y1],[X|Y2],H1,H2) :-
    hamming_dist(Y1,Y2,H1,H2).

hamming_dist([_|Y1],[_|Y2],H1,H2) :-
    H3 is H1 + 1,
    hamming_dist(Y1,Y2,H3,H2).
```

If necessary, the Hamming distance could be computed in this way for each of several scan lines, thereby yielding a number of numeric descriptors. These data could form the input to a pattern classifier of the type described in Section 7.9 so we shall defer discussion of this until then.

Another method for recognising picture playing cards could be based upon the recognition of the printed letters "J", "Q" and "K" (representing Jack, Queen and King), which appear at the top-left and bottom-right corners. This could be achieved using the following measurements:

(a) The number of holes. (This is 1 or 2 for "Q" and 0 for "J" and "K".)

(b) The number of limb-ends. (This is 1 for "Q", 3 for "J" and 4 for "K".)

Additional methods for recognising both printed and hand-written letters will be discussed again later (section 7.8).

Yet another possible method of identifying picture cards makes use of differences in the radial distribution of ink on the cards.[2] The program is quite simple (see Figure 7.4).

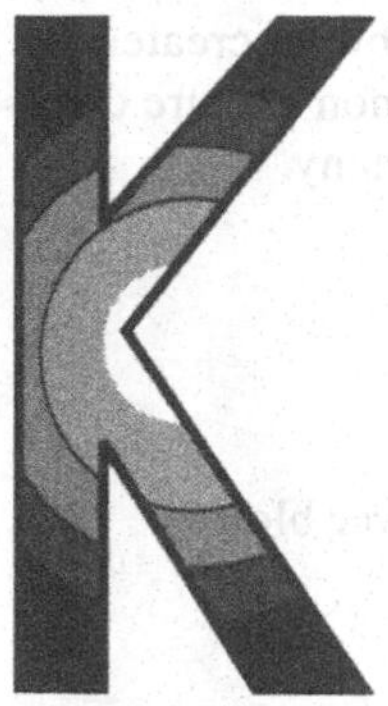

Figure 7.4. Calculating the radial distribution of white pixels about the centroid of a blob. An intensity cone is placed with its brightest point at the centroid. The blob is used to mask the cone. The intensity histogram is then computed.

```
radial_distribution(X) :-
    rea(card_image),        % Read image from the disc
    card_silhouette,        % Ignore the printing for the moment
    cgr(X,Y),               % Find centroid of the card
    rea(card_image),        % Read image from the disc
    highlight_printing,     % Ignore the silhouette of the card
    hic(X,Y),               % Intensity cone
    min.                    % Use printing as a mask over the cone
    hgi(X).                 % Get histogram vector
```

The resulting vector (X) is then compared with each of the stored reference vectors, in a similar way to that defined in **hamming_dist**.

Additional Tasks

Of course, the program segments given above form only a part of a complete card recognition program. An important component of this is the predicate **card_type** which decides whether a given playing card is a picture or non-picture type. The following predicate simply decides whether the image is complex, containing a lot of detail or relatively simple. (A general predicate, called **complex**, can be defined in a similar way. See Appendix III.)

```
card_type(non_picture) :-
    frz,          % Digitise picture of card
    filter,       % Many different types of high-pass filter could be used
    thr(128),     % Threshold. Level found experimentally
    blb,          % Fill holes
    xor,          % Exclusive OR, shows holes
    ndo(1,N),     % Count blobs
    N ≤ 15.       % Card produces small number of holes

card_type(picture).
```

Another possibility is to base the decision upon the number of blobs created by high-pass filtering, followed by thresholding. A simple image (i.e. non-picture card) will generate few blobs, whereas a more complex one will create many.

```
card_type(non_picture) :-
    high_pass_filter,
    threshold,
    ndo(3,A),
    A < 15.                % Card is non-picture type if 14 or fewer blobs

card_type(picture).
```

Playing card games is an obvious application for an AI language, such as Prolog and its derivatives. For example, Prolog is well suited for representing the rules of a game such as poker or whist. This involves straightforward Prolog programming techniques, so we shall not pursue the matter further.

Of course, recognising and playing cards is a flippant application of Prolog+. However, as so often happens, it models some interesting industrial tasks. For example, the suit recognition program, **suit**, could be applied to the recognition of flat, i.e. laminate, objects, or to optical character recognition. In the latter case, it might form a useful, if somewhat limited method for recognising icons, logos or whole printed words, such as pharmaceutical product names. The recognition of printed packaging is particularly important in pharmacy, since for many drugs, it is critical that the packaging should properly indicate the contents. The technique embodied in the predicate **picture_card** was first developed for verifying that pills had the correct label stamped on them. Recognising picture cards is a model of the more significant industrial task of identifying boxes or other packaging, without bothering to read the words printed on them.

7.3 Stacking

Figure 7.5 shows, in diagrammatic form, a simple scene consisting of several blocks, each possessing a machine-recognisable label. The task to be performed is that of stacking the blocks in some desired order, specified by the user. In Figure 7.5, for example, the desired ordering might be based upon alphabetic ranking. To obtain a more general ordering, the user of the machine vision system might specify the desired stacking in terms of a simple set of statements like that shown below:

```
upon(i,g)
upon(a,floor)
upon(g,f)
etc.
```

Obviously, this is a possible application for the techniques discussed in Section 7.2, although we shall not dwell on details, since this will distract us from our main topic (see Sections 5.3 and 5.4).

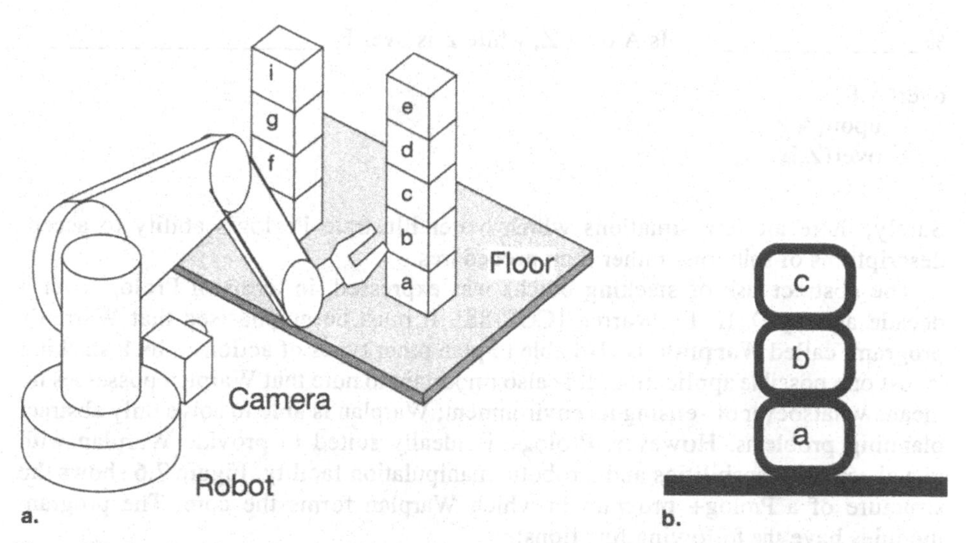

Figure 7.5. Stacking blocks in some desired order. (a) Showing the camera and robot which will be used to move the blocks around. Notice that the blocks possess machine-recognisable labels. (b) The faces of the blocks are assumed to have black borders and black labels but are otherwise white.

Checking that the specification of the desired state is non-contradictory and physically sensible is trivial in Prolog. Here are a few ways of detecting errors:

```
% _______________The floor must NOT be on top of anything else__________

bad_spec :- upon(floor,_).

% ________________If A is above B then B cannot be above A__________

bad_spec :-
    over(A,B),
    over(B,A).

% ______________A cannot be on top of itself?_______________

bad_spec :- upon(A,A).

% ___________Each block (A) must apply downward pressure on the floor_____

bad_spec :- not(above(A,floor)).

% _______________We have failed to prove that there is an error__________

bad_spec :-
    !,
    fail.                       % So we force a failure

% ______________________Subsidiary predicates______________________

over(A,B) :-upon(A,B).   % A is over B if A is resting directly on B
```

```
% ____________________Is A over Z, while Z is over B?____________________

over(A,B) :-
    upon(A,Z),
    over(Z,B).
```

Surely, there are few situations which better illustrate Prolog's ability to accept descriptions of solutions rather than procedures.

The abstract task of stacking blocks was expressed, in terms of Prolog, over a decade ago by D. H. D. Warren [COE-88]. It must be emphasised that Warren's program, called **Warplan**, is also able to plan other types of actions; block stacking is just one possible application. It is also important to note that Warplan possesses no means whatsoever of sensing its environment; Warplan is able to solve only abstract planning problems. However, Prolog+ is ideally suited to provide Warplan with visual sensing capabilities and a robotic manipulation facility. Figure 7.6 shows the structure of a Prolog+ program in which Warplan forms the core. The program modules have the following functions:

(a) *Pre-planner*: Digitise an image from a camera. Identify the blocks, determine their positions. Format the data ready for Warplan.

(b) *User dialogue and error checker*. The latter is based on **bad_spec**.

(c) *Warplan*. Plans the action of the robot, calculating a series of moves of the form:

    move(Object, From, To)

    This represents a decision to move the block labelled *Object* from the block labelled *From*, onto the block labelled *To*. Notice that Warplan does not have any knowledge about the coordinate positions of the blocks, since it "manipulates" blocks in purely symbolic form.

(d) *Post planner*. The post planner translates the symbolic movements dictated by Warplan into a series of movements of a real robot. This is performed in three stages. First, the robot movements are calculated in terms of the coordinate axes used by the vision system. Second, the vision system coordinates are transformed into those used by the robot. Third, the robot is ordered to move, pick up the blocks, one at a time, and transfer them to new positions, as appropriate.

It would be a straightforward matter to add further program code to ensure that the stacking takes place as expected. This would merely require invoking the appropriate part of the *pre-planner* again, after each move has been completed. The predicted scene description would then be compared with that observed by the camera. In this way, it would be possible for the stacking system to monitor unexpected changes in the scene, due for example, to blocks toppling over, or to the user "stealing" or moving blocks during the stacking process.

A number of additional refinements could be added. For example, it is possible to allow the user to provide *non-specific definitions* of the desired state. Suppose for example that the blocks possessed numeric labels. Then, one possible stacking specification might be to pile them in ascending order. Alternatively, blocks whose

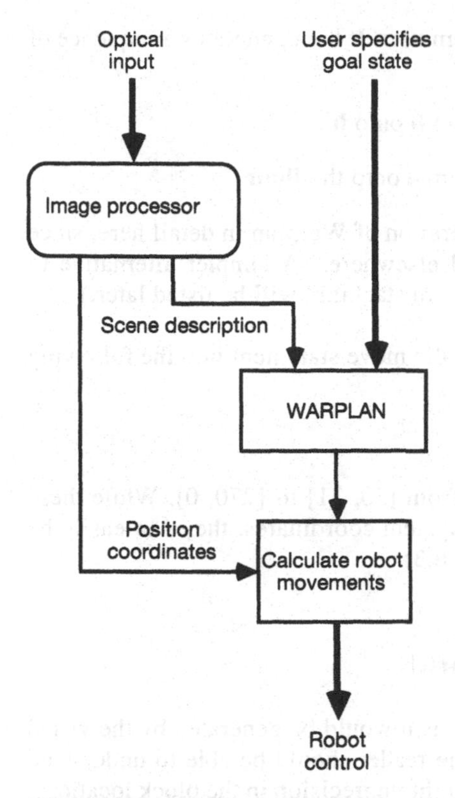

Figure 7.6. Program structure for stacking.

label numbers are greater than some user-specified limit might be placed in a reject bin, while the remainder should be stacked in order. This is a simple model of stacking goods on a supermarket shelf; goods with unexpired date-labels are stacked, while those which have passed their "sell by" dates are discarded.

The above ideas have been implemented in Prolog+, in a three-part program, which is listed later.

Part 1. This part of the program generates a set of statements of the form:

given(start, on_top_of(x,y)). % In the state called "start", x is on top of y.

and another set of of the following type

given(start, clear_top(x)). % In the state called "start", x has a clear top.

Notice that Part 1 is also required to pass block-position parameters to the robot controller (Part 3), which must be able to keep track of the positions of the blocks, as they are being moved around.

Part 2 This is the original program written by Warren and is listed in several more recent publications [COE-88]. Warplan receives a description of the initial stacking of the blocks and the goal state from the system user. It then generates a sequence of robot moves, of the form:

```
move(f,b,h).           % Move f from b onto h

move(d,e,floor).       % Move d from e onto the floor
```

We do not attempt to explain the internal operation of Warplan in detail here, since it is both complicated and well documented elsewhere. (A simpler alternative to Warplan, that is specifically designed to perform stacking, will be listed later.)

Part 3 This part translates this form of symbolic move statement into the following form:

```
moving(d, [30, 31], [270, 0]).
```

This indicates that block **d** is being moved from [30, 31] to [270, 0]. While these positions are defined in terms of the vision system coordinates, they can easily be translated into the robot coordinates (Section 6.3).

Initial Description of the Blocks World

Here is the description of the initial stacking, as it would be generated by the visual analysis section of Part 1 of the program. The reader should be able to understand the scene without any difficulty. Notice the slight imprecision in the block locations, which arises as a result of quantisation effects, slight variations in block sizes and sloppy stacking.

```
block(d,30,31).        % There is a block, labelled d, at position [30,31]
block(e,31,21).        % There is a block called e at position [31,21]
block(b,29,10).
block(a,29,1).
block(c,10,1).
block(h,50,20).
block(f,49,0).
block(g,51,10).
```

Of course, the user can easily infer that block **h** is on block **g** is on block **f** is on the **floor**, but this must be stated explicitly for Warplan. Notice too that there is no statement about which block(s) is (are) resting on the floor. Nor is there any indication as to which blocks have clear tops, i.e. have no other blocks resting on them. Hence, Part 1 of the program must translate these absolute positions into the following format.

```
given(start, clear_top(d)).        % In state "start", d has a clear top (nothing is on
                                   top of d)
given(start, on_top_of(d, e)).     % In state "start", d is on top of e
given(start, on_top_of(e, b)).     % In state "start", e is on top of b
```

```
given(start, on_top_of(b, a)).
given(start, on_top_of(a, floor)).
given(start, clear_top(c)).
given(start, on_top_of(c, floor)).
given(start, clear_top(h)).
given(start, on_top_of(h, g)).
given(start, on_top_of(g, f)).
given(start, on_top_of(f, floor)).
```

Warplan requires data in the above format.

User Specification of the Desired (i.e. Goal) State

Warplan uses the following relatively crude method of specifying the goal state.

```
desired_stacking(
        on_top_of(a, b) &
        on_top_of(b, c) &
        on_top_of(c, d) &
        on_top_of(d, e) &
        on_top_of(e, f) &
        on_top_of(f, g) &
        on_top_of(g, h) ).
```

Making the specification of the goal state more elegant is, of course, straightforward but need not concern us here.

Additional Notes

(i) The following dummy clauses must be put in the database initially, to ensure that **assert** works properly. (In LPA MacProlog, **assert** needs this information to ensure that it knows where to put the new clauses.) They play no active part whatsoever in the program execution.

given.

move.

moving.

(ii) Warplan is invoked by attempting to satisfy the goal

plans(X, start)

where X is the desired state.

(iii) The following clauses are inserted in the database by Warplan.

```
move(d, e, floor).      % Move d from (on top) of e onto the floor
move(e, b, floor).      % Move e from b onto the floor
move(b, a, floor).
```

```
move(h, g, floor).
move(g, f, floor).
move(g, floor, h).
move(f, floor, g).
move(e, floor, f).
move(d, floor, e).
move(c, floor, d).
move(b, floor, c).
move(a, floor, b).
```

(iv) The goal

continue.

invokes Part 3 of the program, which generates output in the following format:

```
moving(d, [30, 31], [270, 0]).        % Move d from [30,31] to [270,0].
moving(e, [31, 21], [290, 0]).        % Move e from [33,21] to [290,0].
moving(b, [29, 10], [310, 0]).
moving(h, [50, 20], [330, 0]).
moving(g, [51, 10], [350, 0]).
moving(g, [350, 0], [330, 10]).
moving(f, [49, 0], [330, 20]).
moving(e, [290, 0], [330, 30]).
moving(d, [270, 0], [330, 40]).
moving(c, [10, 1], [330, 50]).
moving(b, [310, 0], [330, 60]).
```

This defines the sequence of moves required by the robot. The movements are expressed in terms of the vision system coordinates but transforming them into robot coordinates is straightforward, as we showed in Section 6.3.

(v) The following clause defines the X-coordinate of a point in the robot work space, expressed in terms of vision system coordinates.

max_x(250).

This clause specifies that it is safe for the robot to move blocks anywhere to the right of the position X = 250; all existing stacks are to the left of this point.

Program Listing

```
% ______________________Defining operators__________________________

op(900, xfx, is_about_equal_to).
op(650, yfx, =>).
op(700, xfy, &).
op(650, fx, push).

%_________________"go" is the top-level predicate_______________________
```

```
% ____________________Part 1 of the program____________________

go :-
    initialise,                     % Initialise the program
    label_blocks,
    find_stack(L),
    save_data(L),
    fail.

% ____________________Parts 2 and 3 of the program____________________

go :-
    desired_stacking(X),            % Instantiate X to desired stacking
    plans(X, start),                % Part 2, Warplan
    continue.                       % Part 3
```

/* Recognise the block labels and locate them in (X,Y)-space. The blocks are assumed to have white faces, with a black edge. The labels are also black. The predicate "locate_and_identify" is programmed, as appropriate, to read the block labels. */

```
label_blocks :-
    frz,                            % Digitise image
    thr(0,128),                     % Naive method of segmenting the image
    blb,                            % Eliminate borders of the blocks...
    xor,                            % ... and leave only the labels
    remove_noise,                   % Remove minor noise spots
    ndo(1,N),                       % Shade blobs from top to bottom
    wri(temp),                      % Save image on disc
    separation_loop(N).             % Separate the blobs, identify and locate them

separation_loop(0).                 % Stop when there are no more blocks to consider

separation_loop(N) :-
    rea(temp),                      % Read shaded image
    M is N - 1,
    hil(N,N,0),                     % Remove brightest blob
    wri(temp),                      % Save picture of remaining blocks
    swi,                            % Restore image of N blocks
    thr(N,N),                       % Isolate one of the blocks
    locate_and_identify(S,X,Y),     % Block label is S. Block position is (X,Y)
    assert(block(S,X,Y)),           % Save suit and position in the database
    !,                              % Avoid back-tracking
    separation_loop(M).             % Repeat process until all blobs processed

% ____________________Initialise the database____________________

initialise :-
    retractall(block(_,_)),
    retractall(given(_,_)),
    retractall(move(_,_,_)),
    retractall(moving(_,_,_)).
```

```
/* Identify any existing stacks of blocks, based upon a block, A, which is standing
on the floor. */

find_stack(L) :-
        upon(A,floor),
        build_stack(A,[A],L).

% A stack has been found, now identify all of its members

build_stack(A,L1,L2) :-
        upon(B,A),
        append([B],L1,L3),
        build_stack(B,L3,L2).

build_stack(A,L,L) :-
        not(upon(_,A)).

% Is block Ka, which is at [Xa,Ya], resting on block Kb, which is at [Xb,Yb]?

upon([Ka,Xa,Ya],[Kb,Xb,Yb]) :-
        block(Ka,Xa,Ya),
        block(Kb,Xb,Yb),
        Xa is_about_equal_to Xb,
        Y is Ya – Yb,
        Y is_about_equal_to 10.

% Is block Ka (at [Xa,Ya]) resting on the floor?

upon([Ka,Xa,Ya],floor) :-
        block(Ka,Xa,Ya),
        Ya is_about_equal_to 0.

% Is A about equal to B, i.e. within ±3?

A is_about_equal_to B :-
        S is A – B,
        abs(S,T),
        T < 3.

% ________________Retain knowledge gained so far____________________

% This clause looks after those blocks which are at the top of a stack

savc_data([[H,_,_]|T]) :-
        assert(given(start,clear_top(H))),
        keep_data([[H,_,_]|T]).

% List of blocks is empty – so do nothing

keep_data([]).

% If the block is on the floor, add a fact to this effect in the database
```

```
keep_data([[A,_,_]]) :-
      assert(given(start,on_top_of(A,floor))).

% Blocks are neither on the floor nor at the top of a stack

keep_data([[A,_,_],[B,_,_]|T]) :-
      assert(given(start,on_top_of(A,B))),
      keep_data([[B,_,_]|T]).

/* ______________________Part 2______________________*/

/* This is the standard Warplan program, except that the predicate "output" has been
altered to permit easier interfacing with Part 3. */

output(Xs => X):-
      !,
      output1(Xs).

output1(Xs => X):-
      !,
      output1(Xs),
      assert(X).

/*____________________________________________________
Part 3 - Facts of the form

move(A,B)

are generated by "output" in Part 2. A new fact
moving(A, [Xold, Yold], [Xnew, Ynew])

is created here.
____________________________________________________*/

continue :-
      move(A,_,B),                      % Find a clause to translate
      retract(move(A,_,B)),             % Remove it from further consideration
      new_position(A,B),                % Convert to Cartesian coordinates
      continue.                         % Repeat

continue.

% A is going onto floor. We must place each block at new position on floor.

new_position(A,floor) :-
      max_x(X1),                        % Find where last block was put on the floor
      X2 is X1 + 20,                    % Increment position counter
      assert(max_x(X2)),                % Save for next block to be put on floor
      retract(max_x(X1)),               % Retract old position
      block(A,B,C),                     % Find where block A is (Cartesian coords)
      assert(block(A,X2,0)),            % Save block position
      retract(block(A,B,C)),            % Retract block position
      assert(moving(A,[B,C],[X2,0])).   % Save robot movement
```

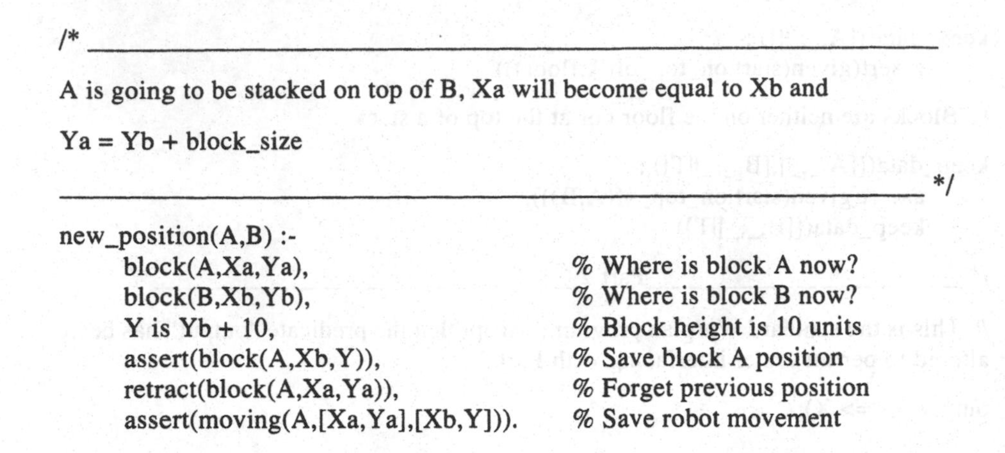

```
/* ____________________________________________________________________

A is going to be stacked on top of B, Xa will become equal to Xb and

Ya = Yb + block_size
____________________________________________________________________ */

new_position(A,B) :-
        block(A,Xa,Ya),                     % Where is block A now?
        block(B,Xb,Yb),                     % Where is block B now?
        Y is Yb + 10,                       % Block height is 10 units
        assert(block(A,Xb,Y)),              % Save block A position
        retract(block(A,Xa,Ya)),            % Forget previous position
        assert(moving(A,[Xa,Ya],[Xb,Y])).   % Save robot movement
```

Additional Remarks About Warplan

As we have stated before, Warplan is a very well-known program, written in Prolog+, and is popularly acknowledged to be one of the more significant achievements of AI research. It can perform tasks other than stacking, including assembly, planning meals, route finding etc. [COE-88] Warplan operates in an abstract world and possesses no sensing or manipulative ability. The Prolog+ program listed above demonstrates, in principle, how Warplan can be integrated with these facilities, enabling it to operate in the real world, stacking real blocks, with a real robot.

The calibration of the vision system and robot coordinates (see Section 6.2) has not been included in the above program. Calibration would be needed to a make the program useful in practice. The robot drive software would also be required.

It should be noted that Warplan is not able to guarantee that an optimal stacking plan will be found and it may take a very long time to plan complex tasks. For this reason, a simplified stacking program has been devised and is listed below.

Simplified Stacking Program; an Alternative to Warplan

stack, moves a set of blocks around in an abstract world. The objective is to build a goal, or desired, stack, which is defined by the user. Note:

(i) The user defines the initial stacking using the predicate **is_on**. An example is presented below.

(ii) At any time, except of course in the final state, there may be more than one stack of blocks on the **floor**. In particular, there may be any number of stacks of blocks defined for the initial state.

(iii) There may be more than one block resting on another block in the initial state, although a block cannot simultaneously rest on two other blocks.

(iv) The **floor** cannot rest on a block.

(v) There is just one desired stack, specified by the user in the form of a list. This is illustrated below.

Program Listing

```
% ________________Top-level predicate for stacking________________

stack([A|B]) :-
      writenl('INITIAL STACKING:'),
      stackprint,                        % Print initial stacking
      nl,
      (is_on(A,floor);                   % Is the bottom block (A) of desired
                                           stack already on floor?
      (clear(A),shift(A,floor))),        % No! So put A on floor
      stack1([A|B]),                     % Now stack the rest
      writenl('FINAL STACKING:'),
      stackprint,                        % Print final stacking
      nl.

stack1([A,B]) :-
      move(B,A).                         % Move the last remaining block³

stack1([A,B|C]) :-
      move(B,A),                         % Move B onto A
      !,
      stack1([B|C]).                     % Repeat for the rest of the desired stack

move(A,B) :-
      is_on(A,B).                        % Do nothing if A is already on B

move(A,B) :-
      clear(A),                          % Remove all blocks from above A
      clear(B),                          % Remove all blocks from above B
      shift(A,B).                        % Shift block A onto B

shift(A,B) :-
      write(A),
      write(' ——→ '),
      write(B),
      nl,
      retract(is_on(A,_)),               % Forget previous position of A
      assert(is_on(A,B)),                % Remember A's new position
      stackprint,                        % Print new stacking
      nl.

clear(A) :-
      not(is_on(_,A)).                   % Nothing is on A

clear(A) :-
      is_on(B,A),                        % B is on A
      clear(B),                          % Clear B first
      shift(B,floor).                    % Now we can shift B onto the floor
```

```
stackprint :-
        is_on(A,_),               % Choose a block A
        not(is_on(_,A)),          % Is A the top of a stack?
        sp1(A,[A],L),             % Find blocks in stack with A at top
        writenl(L),               % Print stack list
        fail.                     % Do same for all existing stacks

stackprint.                       % Force stackprint to succeed

sp1(A,L,L) :-
        is_on(A,floor),           % A is on floor
        !.                        % Stop resatisfaction of this goal

sp1(A,L1,L2) :-
        is_on(A,B),               % A is on B
        sp1(B,[B|L1],L2).         % Add B to L1 and repeat for whole stack
```

Sample Query

The operation of **stack** will be demonstrated using the following query

stack([a, b, c, d, e, f, g, h, i])

The desired stack is [a, b, c, d, e, f, g, h, i]; **a** is resting on the floor and **i** is at the top.

Database Prior to Query

```
is_on(g, f).                      % States that g is on f
is_on(h, floor).                  % h is on the floor.
is_on(i, g).                      % i is the top of this stack, since there is no block on i
is_on(f, h).
is_on(a, floor).                  % There are two initial stacks
is_on(b, a).
is_on(c, b).
is_on(d, c).
is_on(e, d).
```

Database After Query

```
is_on(a, floor).
is_on(b, a).
is_on(c, b).
is_on(d, c).
is_on(e, d).
is_on(f, e).
is_on(g, f).
is_on(h, g).
is_on(i, h).
```

Output Printed by "stack"

INITIAL STACKING:
[h, f, g, i] % **i** is on **g** is on **f** is on **h** is on the **floor**
[a, b, c, d, e] % **e** is on **d** is on **c** is on **b** is on **a** is on the **floor**

i ⟶ floor % Move **i** onto the floor
[h, f, g] % **i** has been taken from the top of this stack
[a, b, c, d, e] % This stack is unchanged
[i] % New stack consists only of **i**

g ⟶ floor
[h, f]
[a, b, c, d, e]
[i]
[g]

f ⟶ e % **f** is put onto **e**, i.e. onto stack **[a, b, c, d, e]**
[h]
[i]
[g]
[a, b, c, d, e, f]

g ⟶ f
[h]
[i]
[a, b, c, d, e, f, g]

h ⟶ g
[i]
[a, b, c, d, e, f, g, h]
i ⟶ h
[a, b, c, d, e, f, g, h, i]
FINAL STACKING:
[a, b, c, d, e, f, g, h, i] % Desired stack has been built
№1 yes % **stack** always succeeds

Notice that in the initial state there is a stack [a, b, c, d, e] which is a subset of the desired stack. This sub-stack is not altered by stack. Furthermore, if the initial stacking is the same as the desired stacking, then no blocks will be moved.

7.4 Packing

The next application of Prolog+ to be discussed is that of packing arbitrary two-dimensional objects into a two-dimensional space, also of arbitrary shape. Packing has important industrial applications:

Figure 7.7. The epitome of the human ability to pack irregular three-dimensional objects. In parts of Britain, there are many miles of dry stone walling like this (Brecon, Wales) which have already stood intact for several centuries and show very few weak points. Sadly, the skill needed to build such walls is a dying art.

(a) Calculating what coins and bank-notes to give as change, when a specified sum of money is to be paid (one-dimensional problem).
(b) "Packing" television advertisements into a given commercial break[4] (one-dimensional problem).
(c) Cutting steel rod or glass tubing to specified length (one-dimensional problem).
(d) Cutting rectangular pieces of carpet, sheet glass, or any other web product (two-dimensional problem).
(e) Cutting pieces of leather from animal hides, for shoes, handbags, clothing, etc. (two-dimensional problem).
(f) Packing agricultural, food and other unpredictable and irregularly shaped products into cases (two or three-dimensional problem).
(g) Packing postal parcels into containers, prior to shipping (three-dimensional problem).

Large numbers of people are employed as packers in the postal services, manufacturing industry and trade (supermarket check-out staff). If the packing problem could be solved mechanically, these people could be employed on less tedious work.

In certain parts of Britain there are mortarless walls separating fields. These are called dry stone walls and are built from irregularly shaped pieces of natural rock, very carefully packed together (Figure 7.7). Despite this apparently crude method of construction, dry stone walls are very robust and many have survived for several centuries. Walls of this type provide evidence of the ability of human beings to pack irregular three-dimensional shapes very efficiently, even though there is no mathematical formula for doing so. It comes as a surprise to most people to learn that the supposedly densest hexagonal packing of equal-sized circles on a plane has not yet been *proved* to be optimal.

Prolog+ is an ideal language for experimenting with various procedures for packing, for a number of reasons:

(i) It allows visual sensing of the shapes of both two- and three-dimensional objects (Section 6.8). Information about both the shapes to be packed together and the space into which they will be placed can be obtained by visual sensing.
(ii) Since there is no known algorithmic solution to the packing problem (in more than one dimension), the use of heuristic procedures is essential. Prolog+ is very well suited for defining rule-based procedures, as is its progenitor, Prolog.
(iii) Back-tracking is ideally suited for trial-and-error processes, such as those required in packing, since the programmer specifies the nature of the solution, not the method of reaching it.
(iv) Prolog+ combines visual sensing of the environment, intelligent decision making and physical manipulation in one well integrated language. These are *all* essential for packing. That the last of these is of vital importance may not at first seem obvious. It is clear, however, that packing is a recursive process; one piece/object is put in place, the scene is then viewed and then the next one is fitted. In other words, it is necessary to fit the objects together, one at a time, in order to make the next decision.

In order to progress one step at a time, it is helpful to consider one-dimensional packing first. Here then is a program for sensing the lengths of metal rod, which is to be cut up into sections to suit customer needs. The cut sections of rod are then checked visually to make sure that they are of the appropriate length.

```
cut_rod :-
    get_next_rod,                    % Move rod for measurement
    rod_length(X),                   % Use vision to measure the rod length X
    customer_orders(L1),             % Consult the database for orders for rod
    pack1(X,L1,Y,L2),                % Enter the packing routine
    nl,                              % Information for the user
    write('There is'),
    write(Y),
    write('metres of rod remaining unused'),
    retract(customer_orders(L1)),    % Update the database
    assert(customer_orders(L2)),     % Update the database (continued)
    cut_rod.                         % Continue processing the next rod

pack1(X,[],X,[]).                    % No more customer orders left. X is the
                                     remanent of rod left

/*  X is the length of rod before cutting the next piece
    P is the length of rod to be delivered to customer C
    Q is the set of remaining customer order
    R is the remanent of rod left over after cutting */

pack1(X,[[P,C]|Q],Y,R) :-
    Z is X – P,                      % Find how much rod will be left over
    Z ≥ 0,                           % Check that this is a positive amount
    cut_and_check(Z),                % Cut length of rod Z and check it visually
    nl,                              % Information for the user
    write('After satisfying the order for customer'),
```

```
    write(C),
    write('there is',
    write(Z),
    write('metres of rod left'over'),
    !,
    pack1(Z,Q,Y,R).             % Do it again until no more rod or orders left

pack1(X,L,X,L). % Remanent too short to satisfy any more orders
```

Vision is used in two predicates: **rod_length** and **cut_and_check**. The function of the former is fairly obvious. However the latter is a little more tricky. If the length of rod cut off is not of the length specified by the customer, there is a machine fault, in which case continuing the processing would be foolish. So, **cut_and_check** might have the following form:

```
cut_and_check(A) :-
    measure_length(X),          % Use vision to measure rod length cut
    Y is A – X,
    abs(Y,Z),
    tolerance(P),
    Z < P.                      % Rod length is within allowable tolerance

cut_and_check(_) :-
    warn_engineer,
    switch_off_power.           % Switch off the power to the rod cutter
```

We shall not, at this stage, consider recovery from this error state, except to say that clearly there are several different ways of restarting the cutting process.

Of course, **cut_rod** represents a very simple approach to packing, but there is a hidden trick that we have not discussed. The list L1, which forms the "input" to **pack1**, can be put into some pre-defined order before **cut_rod** is first called. This has the effect of causing **pack1** to operate in quite different ways. If the members of L1 are put into descending order, the effect is to satisfy orders for big sections of rod first. We all know that, when calculating change, we choose the high-value coins/notes first. This is a useful maxim in all packing problems. When we are packing suitcases for example, we put big items in first. We shall also use this principle when packing two-dimensional objects.

Packing Rectangles

Our approach to two-dimensional packing is most easily explained by first considering the simpler task of packing rectangles into a rectangular space. Consider Figure 7.8. The shaded areas are rectangles which have already been fitted in place. A procedure is then applied in which the white areas are run-coded (**rlc**). That is, each white pixel is given an intensity value which indicates how many white pixels lie to its left in a continuous sequence of white pixels. Black pixels remain black. Suppose that, in the run-coded image, the $(i,j)^{th}$ pixel has an intensity value of $X_{i,j}$, then it is known that a rectangle of width A pixels, where

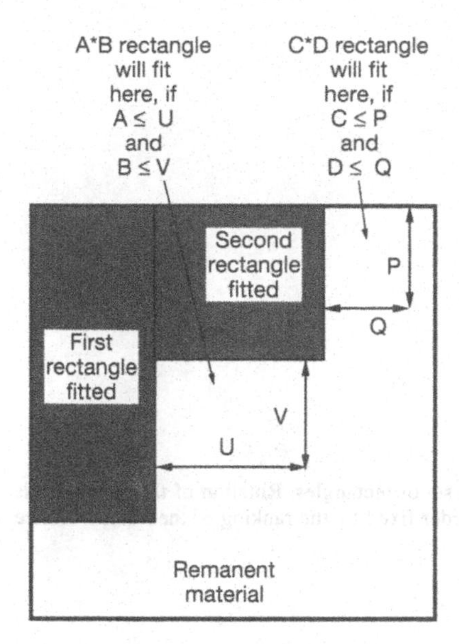

Figure 7.8. Packing rectangles.

$A \leq X_{i,j}$

can be fitted into the space to the left of (i,j). This procedure is repeated, this time calculating the run code along the vertical axis. In this way, it is possible to determine whether a rectangle of height B pixels could be fitted into the space above (i,j). In order to find whether a rectangle of size (A,B) pixels could be placed, we merely perform the following steps:

(i) Run code along the horizontal axis.

(ii) Threshold the resulting image, at level A.

(iii) Run code the image output from step (ii), along the vertical axis.

(iv) Threshold at level B.

(v) Locate the top-left-most point of the white region resulting from step (iv). This determines where the bottom-right-most corner of the (A,B) rectangle should be placed.

(vi) If there is no white area resulting from step (iv), the rectangle of size (A,B) cannot be fitted anywhere in the available space. It is then necessary either to try and fit the rotated rectangle (i.e. of size (B,A) pixels), or select a completely new (i.e. smaller) rectangle for packing. (Of course, in some applications, such as cutting shapes from a material, such as fabric, leather or wood, which has a "grain", rotating the shapes may not be permitted.)

A program based upon these rules has been written in Prolog+ and is listed later. This program could *not* be coded in standard Prolog, since it relies heavily upon the use of the image processor. The result of applying this program to a simple problem involving the packing of rectangles, without rotation, is shown in Figure 7.9.

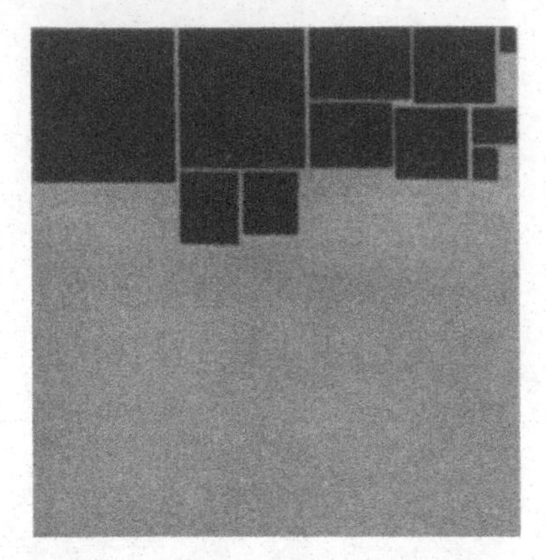

Figure 7.9. Result of applying the packing program to a set of rectangles. Rotation of the rectangles is not permitted. The rectangles were fitted in place in an order fixed by the ranking of their areas. Notice how dense the packing is.

Packing Complex Shapes

More complex shapes can be packed in a similar way (Figure 7.10). First, it is necessary to digitise and analyse one or more images, so that the shapes to be packed can be determined. It is also necessary to define the space into which these shapes are to be packed. Once again, this requires the use of vision. Each of the shapes to be packed is then described by a pair of numbers, representing the size of

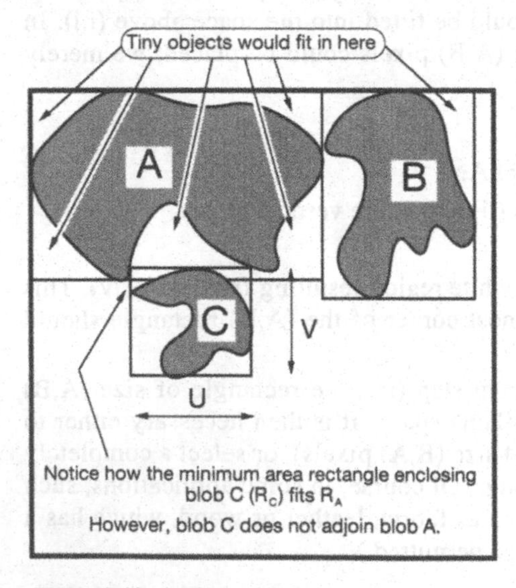

Figure 7.10. Packing arbitrary shapes. Objects A and B have already been fitted in place and we are now trying to find where to place object C. The program actually decides where to place the minimum-area rectangle which can be drawn around C (denoted by R_C). However, once C has been put into place, R_C is forgotten; any further objects can be placed completely or partially within R_C, as long as they do not overlap C itself. An example of this is seen here: both C and R_C overlap R_A.

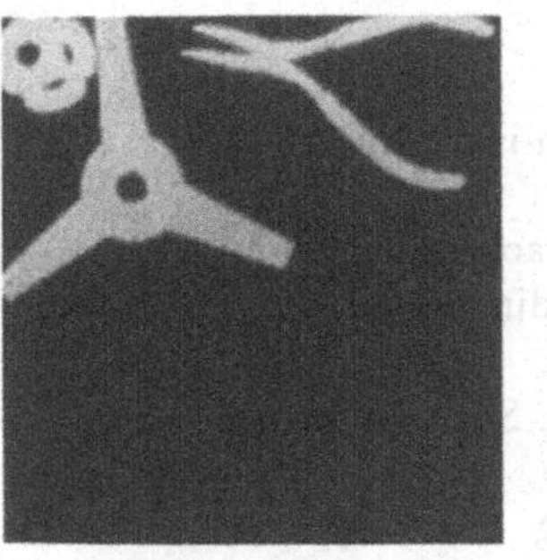
a.

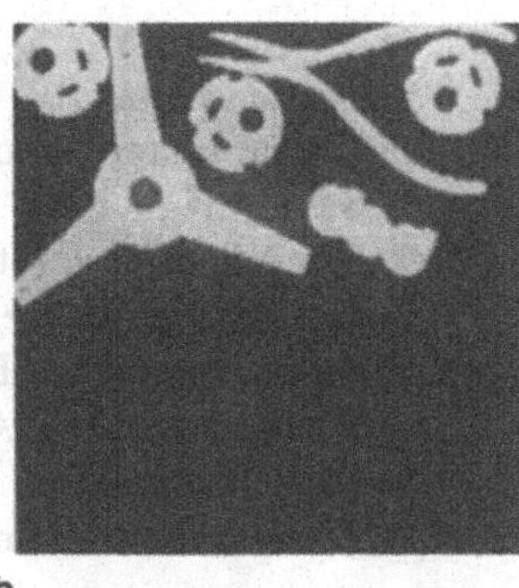
b.

c.

Figure 7.11. Packing irregular shapes whose silhouettes were derived from a digital image. (**a**) Three shapes packed into a rectangular space. The three-pointed star was fitted first, since its minimum-area rectangle is the largest. Next, the pliers were placed, followed by the circular object. (**b**) Six shapes packed. Although we can see that this is a sub-optimal result, the program has achieved a reasonably good packing, within the limitations of its naive rule-set. If a sufficiently large number of small objects were to be packed in addition to the six shown here, the interstitial spaces would be filled quite efficiently; the packing density depends upon the statistics of the object sizes. A wide range of object sizes usually yields a high packing density. (**c**) Packing the same six objects into a space devised to resemble an animal skin. This time, the pliers were too big to be fitted. (Of course, by rotating the pliers, they could be fitted into this space.) The reader should not be too critical of the loose packing here, since the intention was to demonstrate the general principles of two-dimensional packing and the expressional efficiency of the Prolog+ code, rather than to obtain a "good" solution to the general two-dimensional packing problem. In order to appreciate this point, try to code the packing procedure using an imperative language, instead of one that is based upon Prolog.

the minimum-area rectangle which can be placed around it. The program then packs these rectangles, as before. Clearly, this is not ideal, since spider-like shapes will not be fitted close together and a sloppy packing results. However, it should be noted that once a shape has been fitted in place, a new, smaller shape may be placed very close to it, within the smallest rectangle surrounding the first one (Figure 7.10). As evidence of this, notice that in Figure 7.11(b) a circular object has been fitted in the space between the handles of the pliers. In Figure 7.11(c), the same shapes have been packed into a space resembling the silhouette of an animal skin.

Packing Program, Partial Listing

The predicate **pack_rectangles** makes use of **rlc** (run-length coding). Consider Figure 7.8. The image generated by **rlc** indicates how far the nearest edge point is, when measured in the left-ward direction. Suppose, for the moment, that we wish to pack a rectangle of size A × B pixels into the non-black space. If we threshold the image generated by **rlc** at level A, we can find those horizontal chords which are long enough for this to be done. Of course, we also have to do this by comparing B with the vertical run-lengths (i.e. found by **rlc** after applying **yxt**). Here then is the Prolog+ listing of an application for packing rectangles in a rectangular window.

```
pack_rectangles :-
        zer,                            % Black image
        neg,                            % Create image which is all white
        wri(remanent),                  % Save image
        !,                              % Cut – Limit back-tracking
        rectangle(A,B),                 % Get next rectangle dimensions from database
        (fit_rectangle(A,B,X,Y) ;       % Fit rectangle A × B
        fit_rectangle(B,A,X,Y)),        % Fit rectangle B × A. See note below
        retract(rectangle(A,B)),        % Modify database
        fail.                           % Force back-tracking

pack_rectangles.                        % Force predicate to succeed

fit_rectangle(A,B,X,Y) :-
        rea(remanent),                  % Read image file storing remanent
        rlc,                            % Run-length code, to the left
        thr(A),                         % Find horizontal runs of length ≥ A pixels
        wri(temp),                      % Save the result
        rea(remanent),                  % Read remanent image file again
        yxt,                            % Interchange X and Y axes
        rlc,                            % Run-length code
        yxt.                            % Interchange X and Y axes
        thr(B),                         % Find vertical runs of length ≥ B pixels
        rea(temp),                      % Read intermediate result back again
        min,                            % AND white regions
        cwp(N),                         % Count white points
        !,                              % Limit back-tracking
        N > 0,                          % Can rectangle A × B fit into remaining space?
        top_left(X,Y),                  % Find X & Y values for top-left-most white point
        wri(remanent),                  % Read remanent image again
        box(X,Y,A,B,0),                 % Draw black rectangle rectangle A × B at (X,Y)
        wri(remanent).                  % Save modified remanent image
```

Notes on the Packing Program

(i) Two-dimensional packing can only be programmed with difficulty using standard Prolog, or an image processing language, such as VCS; Prolog+ is a natural choice for this particular application, since both image processing and symbolic processing are needed.

(ii) The order in which the various **rectangle** facts are listed in the database is crucial. In order to try and pack big rectangles first, the rectangles with the largest areas should be near the top of the database. (Although sorting is not included as a built-in predicate in Edinburgh Prolog, MacProlog does have the **sort** predicate which makes this a trivial matter.)

(iii) Notice that **pack_rectangles** contains the ORed condition

```
(fit_rectangle(A,B,X,Y) ;        % Fit rectangle A × B
fit_rectangle(B,A,X,Y)),         % Fit rectangle B × A. See note below
```

This permits the rectangles which are to be packed together to be inserted in either orientation. As we said earlier, there are many situations in which it is not desirable to allow this to happen. In this instance we can use the following modified version of **pack_rectangles**:

```
pack_rectangles :-
    zer,                          % Black image
    neg,                          % Create image which is all white
    wri(remanent),                % Save image
    !,                            % Cut – Limit back-tracking
    rectangle(A,B),               % Get next rectangle dimensions from database
    fit_rectangle(A,B,X,Y),       % Only one call now!
    retract(rectangle(A,B)),      % Modify database
    fail.
```

(iv) A naive approach to packing arbitrary shapes would be to enclose each one within its minimum-area rectangle and then pack these into the given space. The following predicate may be used to derive the rectangle parameters from an image containing a set of blobs. The facts are inserted in the database in order of the areas of the blobs.

```
build_database:-
    ndo,                          % Shade blobs according to their areas
    !,                            % Limit back-tracking
    gli(_,Z),                     % Find maximum intensity
    Z > 0,                        % Is it > 0?
    thr(Z,Z),                     % Select brightest (i.e. biggest) blob
    dim(X1,X2,Y1,Y2),             % Get blob dimensions
    X is X2 - X1,                 % Find X-dim. of min. enclosing rectangle
    Y is Y2 - Y1,                 % Find Y-dim. of min. enclosing rectangle
    assertz(rectangle(X,Y)),      % Add fact to the end of the database
    hil(Z,Z,0),                   % Eliminate the blob just coded
    fail.                         % Go back and do the rest
build_database.                   % Force success of this goal
```

Packing Arbitrary Shapes

It requires only a minor change to the above program, in order to pack arbitrary shapes. As we have already stated, a sub-optimal policy is to use the minimum-area rectangles surrounding the shapes when deciding which shape to move and where to put it. However, the shape itself can be moved into place, rather than its surrounding rectangle. This does not always achieve a very tight packing for two obvious reasons:

(i) The minimum-area rectangle surrounding a spider-like shape, is almost empty.

(ii) There is no provision in the program for rotating and shifting a shape to obtain a better fit.

These do not matter if there is a wide disparity of sizes of the shapes to be packed with far more small ones than there are large shapes. The reason is, of course, that the small shapes will fit into the "interstitial" spaces between the larger ones, which were fitted earlier. Thus, it is impossible to be sure whether or not this or any other packing policy will be effective; only the results of trying it will prove the point.

Perhaps of greater importance is the ability to refine an initial packing by shuffling the parts around after an initial attempt at packing has been made. Unfortunately, the computational load to do this can be very large indeed, since many shifts and rotations may be needed to move a given shape into a given slot, close to its "optimal" position. (Here, "optimal" is judged by a human being.) The most promising approach to obtaining a tighter packing in a multi-stage packing program seems to rely upon the use of heuristic rules for refinement.

One such heuristic rule is described below. It is clearly typical of that type of rule that is difficult to justify in any way, other than by the results it produces and which is awkward to express in a conventional computing language.

Step 1. Apply a (crude) packing procedure, but stop it prematurely.

Step 2. Shade the empty spaces white, while the shapes already fitted in place are black. Call this image P.

Step 3. Select one of the larger shapes, S, which has still not been put into place and construct the structuring element for a morphological operator, using S. (The morphological operator structuring element might be defined using only edge pixels in S, to reduce the computational load.)

Step 4. Apply the morphological operator to the image P.

Step 5. Look for regions in the output image which are still white.

Step 6. Use the following heuristic procedure to identify the small spaces:

```
small_space(X,Y) :-
        ndo(3),             % Shade blobs according to size
        hil(0,0,255),       % Eliminate background from consideration
        gli(Z,_),           % Find minimum intensity
        swi,                % Switch images
        thr(Z,Z),           % Select darkest blob (i.e. smallest)
        top_left(X,Y).
```

Step 7. Fit the shape S so that the top-left corner of its minimum-area rectangle is placed at [X,Y] as found in step 6.

Step 8. Repeat steps 2 to 8.

In this instance, we shall use Prolog+ to specify only the top levels of the process, since the functions of predicates in the lower levels are quite obvious.

```
refined_packer :-
    initial_packer,          % Described above
    wri(result),             % Save the resulting packing image
    refine_packing.          % Refine the packing
```

```
refine_packing :-
    shape(S),                % Consult the database for shape S
    build_morph(S,X),        /* Use the white region in image S to build
                                structuring element (X) for a morphological
                                operator */
    rea(result),             % Read intermediate result from disc
    morph_op(X),             % Apply the morphological operator
    small_space(X,Y),        % Defined above
    shift(S,X,Y),            % Shift S to [X,Y]
    rea(result),             % Read packing result
    max,                     % Add new shape in place
    wri(result),             % Save result on disc
    refine_packing.          % Repeat recursively

refine_packing.
```

From the above skeleton, it is quite easy to write the complete program.

Here is a good example of Prolog+ being used, not only as a computational language, but also as a medium for expressing ideas about heuristics. The author can now reveal a secret: it was the ability of Prolog+ to express heuristics that allowed him to develop the ideas described in this section. These ideas came out of the author's attempts to write a Prolog+ program for packing. Conventional programming practice presupposes that the program *follows* the formulation of ideas. In this case, and in several other sections of this book, the sequence of events was the other way round. Much is claimed for the expressional power of Prolog, but here is one largely unexpected and significant practical advantage.[5] Unfortunately, we do not have space to pursue the fascinating topic of packing any further. Let it suffice to say that the author can envisage a solution, expressed in terms of Prolog+, for the task of packing (three-dimensional) cuboidal objects such as those which travel via the postal service. The procedure is a natural extension of the ideas implicit in **pack_rectangles**.

Structured lighting (Section 6.8) would be invaluable for describing the shapes of more complex (i.e. non-convex) three-dimensional objects, although the structure of a program for packing objects of this type has not yet been developed.

7.5 Shape Analysis

In this section, three quite different approaches will be discussed:

(a) Shape recognition by parameter matching
(b) Recognising shapes by matching their skeletons
(c) Analysing images which are known to be formed from overlapping convex shapes

Shape Recognition by Parameter Matching

The program on pages 60–61 identifies items of table cutlery. In the rather limited form given there, the program merely recognises forks and knives; additional clauses for **object_is** are needed if it is to identify other objects. Some further possibilities are discussed below. In each case, the object to be recognised is a represented by a white figure against a black background. The essence of this approach is that the programmer picks some distinctive feature(s) of the object that enables a Prolog+ program to recognise it and distinguish it from other objects that might be seen.

For example, the following definitions of a gear and a comb, and that of a fork on page 60 are adequate to ensure correct identification, despite the fact that they represent very sloppy recognition rules.

```
% ____A gear has ≥20 teeth and its convex hull forms an almost perfect circle___

object_is(gear) :-
    count(bays,N),          % Count number of bays
    N ≥ 20,                 % Check number of bays ≥ 20
    chu,                    % Convex hull
    blb,                    % Fill centre
    shape_factor(S),        % Calculate area/perimeter^2 (see page 50)
    S ≤ 0.785398,           % Check S is not too large
    S ≥ 0.725.              % Check S is not too small

% ____A comb has ≥20 teeth and the space between adjacent teeth is ≤10 pixels ___

object_is(comb) :-
    wri(temp1),             % Save image on disc
    5•exw,                  % Fill in gaps between teeth
    5•skw,                  % Restore silhouette to original size
    rea(temp1),             % Read image from disc
    xor,                    % Exclusive OR, leaves gap between teeth
    skw,                    % Break joins for gaps next to short teeth
    eul(N),                 % Count gaps between teeth
    N ≥ 19.                 % Number of gaps = Number of teeth–1
```

A comb fails the test for a fork because it has too many teeth, while a comb fails the test for a gear because its convex hull is not nearly circular.

Similarly, pliers and scissors can be distinguished very easily from each other, simply by counting indentations (bays) and holes (lakes).

```
% ____Pliers have no holes and three bays (pliers open) or four (pliers closed) ___

object_is(pliers) :-
    count(holes,0),
    count(bays,N),
    member(N,[3,4]).
```

% _____Scissors have two holes (for fingers) and three bays (scissors open) or four (closed)_____

```
object_is(scissors) :-
    count(holes,2),
    count(bays,N),
    member(N,[3,4]).
```

However, an open spanner (wrench) and a con-rod (Figure 7.12) are both misrecognised as pliers by these over-simple criteria. A spanner might be distinguished from pliers and a con-rod by the relative sizes (i.e. areas) of the largest and second largest bays.

The essence of this mode of programming is to select the smallest number of easily computed criteria. Always avoid being over-specific; general rules work best.[6]

Shape Recognition by Matching Skeletons

Consider the image shown in Figure 7.12(a). This is the silhouette of an automobile con-rod, while Figure 7.12(b) shows its skeleton. Figure 7.12(c) shows a piece-wise linear approximation to the skeleton. This is obtained by joining together the skeleton limb-ends and joints in an appropriate manner. A Prolog+ program has been written which can represent the skeleton of a shape such as this in the form of a list having the following form:

$[[a,b,D_{a,b}],[b,c,D_{b,c}],[a,c,D_{a,c}],[d,e,D_{d,e}],\ldots]$

This states that there is a node **a** connected to a node **b** and at a distance $\mathbf{D_{a,b}}$ from it. Node **b** is at a distance $\mathbf{D_{b,c}}$ from **c** and so on.

We shall use lists like this to represent the shapes of both the object to be identified and the reference objects(s) with which it will be compared. In order to explain the list matching algorithm, we must first define some additional notation. Let T_1 and T_2 denote the lists representing the skeletons generated from the test and (one of the) reference objects, respectively. Furthermore, let

$T_1 = \{[A_i,B_i,C_i]\}$,

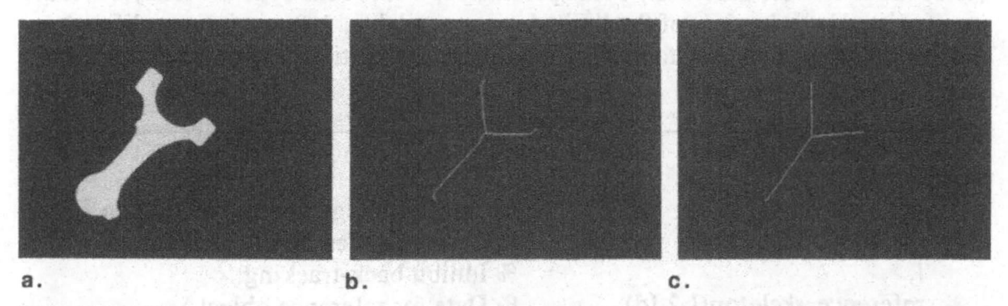

Figure 7.12. Representing a shape by a piece-wise linear approximation to its skeleton. **(a)** Silhouette of an automobile connecting rod (con-rod). **(b)** **mdl** applied to **(a)** generates the skeleton. **(c)** A piece-wise linear approximation to the skeleton shown in **(b)**.

where A_i and B_i are two nodes, denoting joints or limb ends, in the graph representing the skeleton and C_i is the distance between them. Similarly for $T_2 = \{[P_i,Q_i,R_i]\}$. The list matching procedure is defined thus:

(i) Find one member of T_1. Denote this by $[A_i,B_i,C_i]$. Delete this element from T_1.

(ii) Find one member of T_2. Denote this by $[P_i,Q_i,R_i]$. Delete this element from T_2.

(iii) If C_i and R_i are approximately equal, proceed to step (iv), otherwise back-track, replacing the elements previously deleted from T_1 and T_2.

(iv) Create a list, L, associating the following pairs of nodes $[A_i,P_i]$ and $[B_i,Q_i]$, (Alternatively, $[A_i,Q_i]$ and $[B_i,P_i]$ could be associated together instead.)

(v) Find one member of the reduced set T_1. Denote this by [A,B,C].

(vi) Find one member of the reduced set T_2. Denote this by [P,Q,R].

(vii) If C and R are approximately equal, proceed to step (viii), otherwise back-track.

(viii) Four possible conditions apply:

If [B,Q] is already a member of L, add [A,P] to L. Proceed to step (ix).

If [A,Q] is already a member of L, add [B,P] to L. Proceed to step (ix).

If [B,P] is already a member of L, add [A,Q] to L. Proceed to step (ix).

If [A,P] is already a member of L, add [B,Q] to L. Proceed to step (ix).

If none of these conditions is true, back-track. Back-tracking may proceed as far as step (i).

(ix) Delete element [A,B,C] from T_1 and [P,Q,R] from T_2.

(x) Repeat steps (v)–(ix), unless T_1 and T_2 are both empty.

Program, Partial Listing

In the listing given below, only the predicates **get_image** and **create_list** perform any image processing. The goal **get_image** simply digitises an image, processes it and generates the skeleton of the object(s) being viewed, while **create_list(L1)** analyses the image created by **get_image** and forms a list (L1) which describes the connections between pairs of features of skeletons. (These features are limb-ends and joints, where the skeleton bifurcates.) Some predicates, whose functions are fairly obvious, are not listed here, in view of the limited space available. A few minor modifications allow the Prolog+ program to match the skeletons of partially occluded components.

```
% ______________________Top-level goal ______________________

match_shape :-
        get_image,                      % Digitise image and form skeleton
        create list(L1),                % Create list describing skeleton
        !,                              % Inhibit back-tracking
        reference_skeleton(L2,Id),      % Data for reference object
        member([A,B,C],L1),             % Find one element of L1
        member([P,Q,R],L2),             % Find one element of L2
```

```
        approximately_equal(C,R),          % Are C and R approx. equal?
        delete([A,B,C],L1,L4),             % Delete [A,B,C] from L1
        delete([P,Q,R],L2,L5),             % Delete [P,Q,R] from L2
        L6 = [[A,P],[B,Q]],                % Seed list of matching node pairs
        tree_match(L4,_,L5,_,L6,L3),       % Match lists L4 and L5
        !,                                 % Inhibit back-tracking
        cull(L3,L7),                       % Eliminate any repetitions from L3
        tell_user(L7).                     % Success, so tell user about it

tree_match([],_,[],_,L,L).                 % Lists match if both empty

tree_match(L1,L2,L3,L4,L5,L6):-
        member([A,B,C],L1),                % Find one member of L1
        member([P,Q,R],L3),                % Find one member of L2
        approximately_equal(C,R),          % Are C and R approx. equal?
        node_match(A,B,P,Q,L5,L9),         % Match two more nodes
        delete([A,B,C],L1,L7),             % Delete [A,B,C] from L1
        delete([P,Q,R],L3,L8),             % Delete [P,Q,R] from L2
        tree_match(L7,L2,L8,L4,L9,L6).     % Recurse, if L7 and L8 not empty

node_match(A,B,P,Q,L1,L2) :-
        member([B,Q],L1),                  % [B,Q] have been found to match...
        L3 = [[A,P]],                      % ...so associate A and P as matching pair
        append(L1,L3,L2).                  % Append the matching node pair to L1

node_match(A,B,P,Q,L1,L2) :-
        member([A,Q],L1),                  % [A,Q] have already been found to match
        L3 = [[B,P]],                      % ...so associate B and P as matching pair
        append(L1,L3,L2).                  % Append the matching node pair to L1

node_match(A,B,P,Q,L1,L2) :-
        member([B,P],L1),                  % [B,P] have already been found to match...
        L3 = [[A,Q]],                      % ...so associate A and Q as matching pair
        append(L1,L3,L2).                  % Append the matching node pair to L1

node_match(A,B,P,Q,L1,L2) :-
        member([A,P],L1),                  % [A,P] have already been found to match...
        L3 = [[B,Q]],                      % ...so associate B and Q as matching pair
        append(L1,L3,L2).                  % Append the matching node pair to L1

approximately_equal(U,V):-
        S is U – V,                        % Difference between U and V
        abs(S,M),                          % Absolute value – modulus function
        M < 10.                            % Difference compared to arbitrary limit

% Database format of the stored reference patterns – for illustrative purposes

reference_skeleton([ [a,b,74], [c,a,68], [e,a,203], [e,d,36] ], ref_object1).
```

reference_skeleton indicates that the reference component called **ref_object1** has two skeletal features, **a** and **b**, which are connected and are 74 units apart. Features **c** and

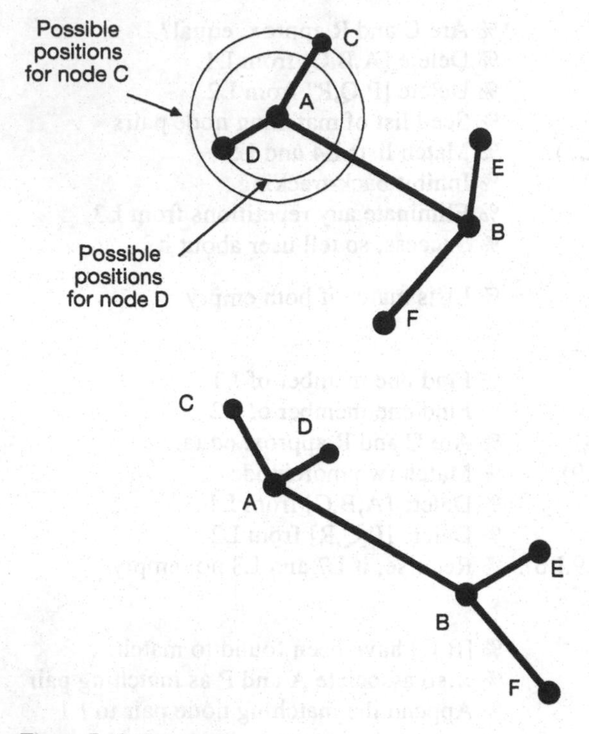

Figure 7.13. The shape matching program, **match_shape**, is unable to distinguish between objects which generate skeletons having the same connectivity and internodal distances.

a are 68 units apart, **e** and **a** are 203 units apart, while **e** and **d** are 36 units apart. In practice, there are likely to be many facts like this one in the database. There will certainly be at least one such entry in the database for each class of objects that is to be recognised. If a class of objects exhibits a large degree of variation, it may be necessary to store more than one reference list in the database. We shall see how this might arise later.

Algorithm Speed

This naive algorithm conducts an exhaustive search for possible solutions and has no rules for accelerating the speed of operation. In the worst case, the execution time is $O(N^3)$, where N is the number of arcs in the graphs to be matched. It is a simple matter to avoid wasting time, if the graphs contain different numbers of nodes. Further speed improvement might be achieved by adding rules to stop the search, if nodes are found to be unsuitable for matching because they have different connectivity scores. Other, more efficient algorithms for establishing graph isomorphism can be coded in Prolog+. Once again, it must be emphasised that our primary objective is to demonstrate the power of Prolog+, rather than to achieve a complete solution to this particular application, as desirable as that may be.

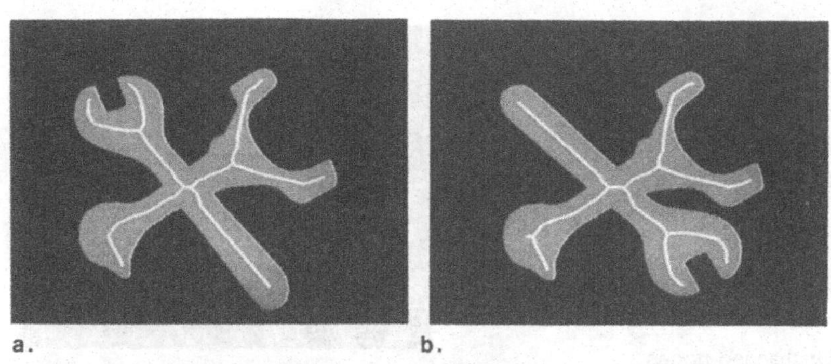

Figure 7.14. Skeletons derived from a pair of articulated lever-like objects in two different poses. (The tiny spur at the lower-left corner of the skeleton in (**b**) is an artifact caused by quantisation noise. Certain types of skeletonisation operator are more robust and do no suffer from this effect.)

Checking Geometry

The geometric relationship between nodes in the graphs being matched is ignored in the Prolog+ application listed above. Two skeletons match if they have same connectivity pattern and similar inter-nodal distances (Figure 7.13). The angles between skeleton limbs are not tested. It is a simple matter to add an extra test, to verify that the list of matching node pairs that this program discovers is reasonable. The reader may wish to exercise his/her mind in this respect.

Semi-Flexible Objects

The seemingly more complex task of recognising semi-flexible objects, such as pliers or scissors, can in fact be accommodated using the same Prolog+ program, although it will probably be necessary to store several reference lists for each class of reference objects. Notice that the skeleton generated from a simple assembly of thin levers, often changes very little, when the levers are rotated (Figure 7.14). The implication of this is that it may be necessary to store only a few reference data objects to exemplify a single class of semi-flexible objects.

Overlapping Spider-Like Objects

An even more complicated task is that of recognising objects which are touching or overlapping. The skeletons generated from non-isolated objects are likely to be fused together, with only a few nodes altered. (The remainder of the skeleton is unchanged; see Figure 7.15.) This fact can be used by a slightly modified version of the same Prolog+ program. This allows parts of skeletons to be matched, but does not demand that the complete skeleton fits one of those represented in the database. It is possible to calculate a simple score which measures how far the graphs can be matched. A Prolog+ program for matching touching/overlapping objects has been developed by modifying the program listed above slightly.

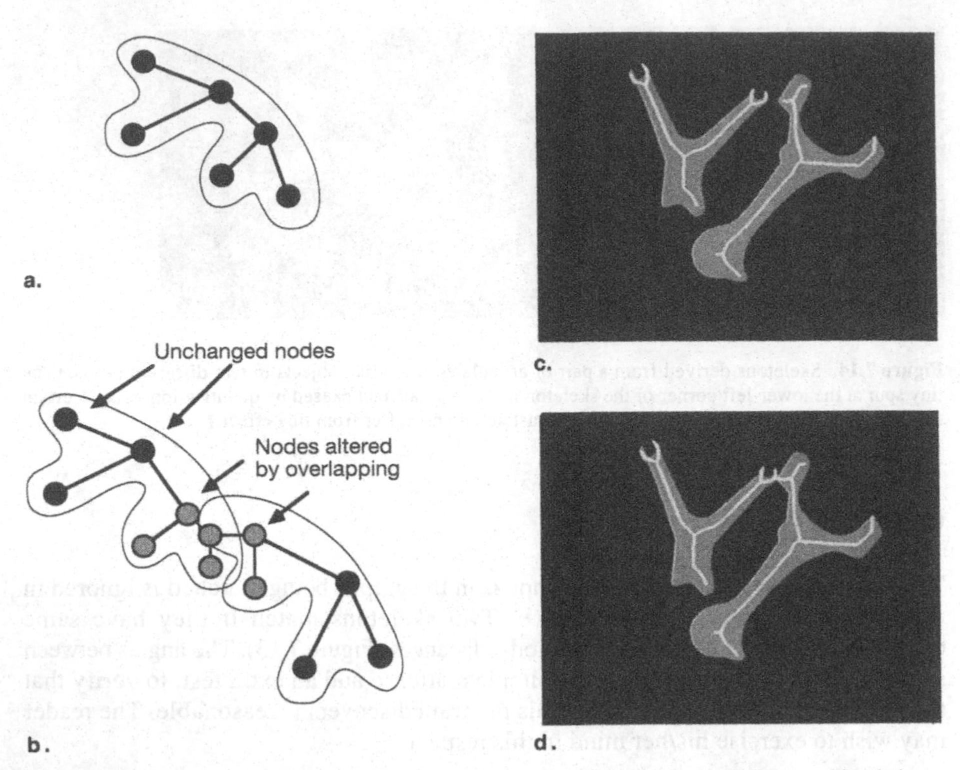

Figure 7.15. Skeletons of overlapping objects remain intact locally. (**a**) Isolated object and its skeleton. (**b**) Skeleton of two overlapping objects (diagrammatic sketch). (**c**) Two separate objects and their skeletons. (**d**) The same two objects as in (**c**) but touching. Notice that the skeleton of the combined blob contains both of the skeletons in (**c**), with only very minor changes occurring close to the point of contact.

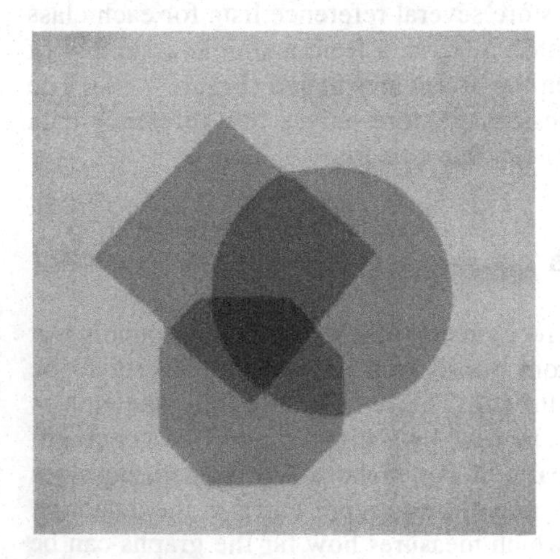

Figure 7.16. Overlapping, semi-transparent plate-like objects, viewed using back-illumination. These objects are actually pieces of coloured plastic sheeting. Compare this with Figure 7.17(a).

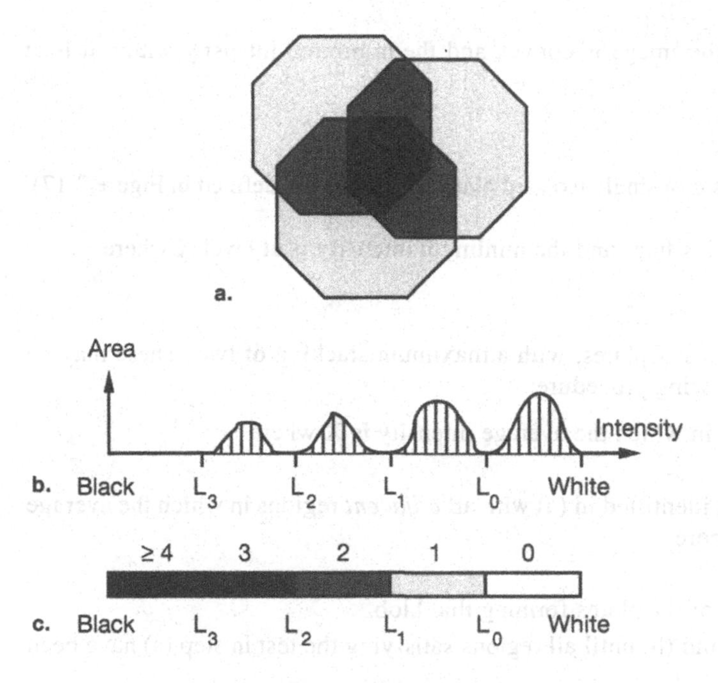

Figure 7.17. Analysing images of overlapping convex plates. (**a**) Three overlapping objects. (**b**) Intensity histogram. (**c**) Relationship between intensity and depth of stacking. Notice that the threshold parameters, L_0, L_1, L_2, L_3 indicate "troughs" in the intensity histogram.

Overlapping Convex Shapes

This task is important for the measurement of certain industrial objects. For example, the X-ray absorption of several overlapping plates may be analysed using the formula:

$$\log(\overline{A}) = \sum_{i=1}^{N} \log(A_i)$$

where A_i is the absorption of the ith plate and A is the total absorption of plates 1, 2, ..., N, stacked on top of another. An identical analysis task arises when we are trying to interpret visible-light images, obtained from semi-transparent plates stacked together. Imagine a set of thin, semi-transparent convex plates, viewed with back illumination. Where the plates do not overlap, the light attenuation is less than that where they do (Figure 7.16). In practice, it is only possible to identify overlapping reliably from the intensity in the image when the plates are stacked no more than (about) three deep.[7] In more complex situations than this, it may be permissible to signal that the stacking cannot be resolved by the program and to warn/consult a human operator and this is the approach which we shall adopt.

The following rules were devised to analyse this type of situation and can easily be coded in Prolog+. The notation is explained in Figure 7.17.

Rule 1. If a blob in the image is convex and the minimum intensity within it is at level X, where

$L_1 < X < L_0$

then that blob consists of a single isolated plate (L_1 and L_0 are defined in Figure 7.17).

Rule 2. If a blob has 2N bays and the minimum intensity is at level X where

$L_2 < X < L_1$

then the blob comprises N plates, with a maximum stacking of two. These may be identified by the following procedure:

(a) Identify a region in which the average intensity is X where
$L_1 < X < L_0$
(b) Merge the region identified in (a) with all *adjacent* regions in which the average intensity is Y, where
$L_2 < Y < L_1$
The result is one of the plates forming that blob.
(c) Repeat steps (a) and (b) until all regions satisfying the test in step (a) have been considered.

Rule 3. If the minimum intensity in a blob is L where

$L_n < L < L_{n-1}$

then the maximum plate stacking is n deep. In this situation, a blob may be decomposed into its component plates by the following recursive procedure:

(a) Set p = 1.
(b) Identify a region in which the average intensity is X, where
$L_p < X < L_{p-1}$
(c) Merge the identified region with all adjacent regions in which the average intensity is Y, where
$L_{p+1} < Y < L_p < X < L_{p-1}$
(d) Increment p and repeat steps (b)–(c) recursively, until no more regions can be added. The result is one of the plates forming that blob.
(e) Repeat steps (a)–(d) until all regions having average intensity X where
$L_1 < X < L_0$
have been considered.

Rule 4. If a blob in the image is convex and the minimum intensity within it is at level X, where

$L_2 < X < L_1$

then that blob consists of a large plate, P, which totally encloses one or more distinct

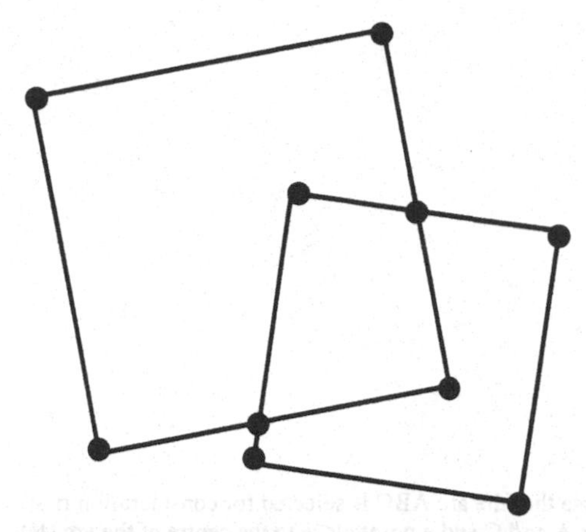

Figure 7.18. Using edges to resolve overlapping polygonal shapes (rectangles in this simple example). The edges of the transparent overlapping plates have been highlighted, using an edge-detection operator, such as **sed**. Thresholding and thinning/skeletonisation have then been applied. The resulting image then consists of a series of one-pixel wide contours. Following this, the edge contours have been broken up into linear segments, so that the techniques described in Chapter 5 can be used. (The predicate **detect_crossovers**, defined there, provides the basis for deciding what to do at points where edges cross.)

smaller plates, Q_1, Q_2,.... This situation may be analysed in the following way:

(a) To identify P, simply threshold the image at level L_0.

(b) To identify Q_1, Q_2,..., simply threshold at level L_1.

Rule 5. All blobs not satisfying Rules 1 – 4, may be referred to a human operator for analysis. This type of "catch-all" rule is included to ensure that very complex and unforeseen patterns are considered by a human operator, when the program experiences some difficulty. Of course, additional rules could be added to analyse some of those situations not covered at the moment by Rules 1 – 4.

A Prolog+ program implementing these rules has been written and tested on a real-life inspection application, which is of considerable commercial significance. The program provides a favourable demonstration of the expressional power of the language, since it is a very straightforward matter to translate the rules listed above into Prolog+.

Of course, there are many other possible procedures that we could adopt for analysing images of this type. We might, for example, employ the continuity of the edges when smoothly curved laminate objects overlap as a basis for an edge-following algorithm. The predicate **detect_crossovers** (Section 5.3) could be applied to such an image, in order to resolve the plates (see Figure 7.18). In many respects, this task is similar to that discussed in Section 5.5, where we presented a procedure for recognising tree-like objects amongst a set of similar objects, called a forest.

A completely different approach can be used if the overlapping plate-like objects are known to be of some fixed type. For example, a set of objects which are all circular is relatively easy to resolve. First, we generate an image in which the intensity

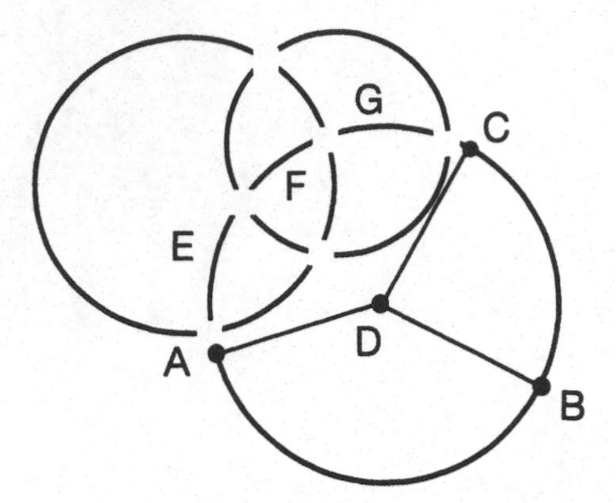

Figure 7.19. Resolving overlapping circles. Suppose that the arc ABC is selected for consideration first. (Use a procedure based upon **ndo(1)**.) Its end points A and C and a point close to the centre of the arc (B) are found. The parameters of the circle which intersects A, B and C are then computed. (In this example, the centre of this circle is at D.) A thin annulus (not shown) is then centred at D. Its inner radius is just less than |AD|, while its outer radius is just greater than this amount. All pixels which lie within this annulus are then set to black. This has the effect of removing arcs ABC, E, F and G from further consideration. This whole process is then repeated, until all arcs have been removed.

contours are broken up into circular arcs. This is easy to achieve, using a simple idea. The input image is similar to that shown in Figure 7.16.

```
break_into_arcs :-
        sed,            % Sobel edge detector
        thr(20),        % Threshold
        mdl,            % Skeletonise
        cnw,            % Count white neighbours in 3 × 3 neighbourhood
        min,            % Ignore all points not on the skeleton
        thr(1,2).       % Keep pixels on smooth arcs – remove joints
```

The remainder of the analysis can be performed using ideas discussed elsewhere in this book. The image which now has the form of that shown in Figure 7.19 is analysed one arc at a time (use **next_blob**). The coordinates of the end points and one point near the centre of an arc are then found. These values are then used by the Library predicate, **circle** (also see Figure 8.15), in order to calculate the coordinates of the centre and the radius of that circle which intersects all of these three points. All other arcs which lie either on, or very close to, this circle can then be eliminated from further consideration. The process is repeated for all arcs in the image. The set of circles so obtained traces the edges of all of the transparent circular plates.

Suppose instead that the plates were all regular polygons, for example, regular pentagons. The set of edges obtained from such an image could be analysed using the techniques outlined in Chapter 5. Alternatively, we could locate two adjacent vertices of a pentagon. Using the coordinate positions of just two adjacent vertices, we could, in theory, draw the whole pentagon. This fact could form the basis of yet another procedure for resolving an image derived from overlapping back-illuminated plates. The computational procedure would be as follows:

(i) Locate an edge.
(ii) Find two adjacent vertices lying along that edge. (In practice, this step would not be very accurate; some false vertices might be proposed.)
(iii) Predict where the other three vertices of the pentagon are.
(iv) "Draw" the pentagon.
(v) Remove all edge segments which lie along the circumference of that pentagon. This might be combined with a test to verify that step (ii) had, in fact, correctly located two appropriate vertices of a real pentagon. (If not, the procedure will back-track to step(ii).)
(vi) Repeat steps (i)–(v), for all remaining edge segments.

By now, the reader should be able to recognise the fact that a hypothesise-and-test procedure like this can be programmed very easily in Prolog+. Although step (ii) is unreliable, it can be biased to be "generous" in suggesting more possible vertices than are actually present in the image. By adding a test to verify that a real pentagon has been discovered, we can overcome the shortcomings of a weak link in the procedure.

7.6 Maze Searching

A simple maze is one whose walls are continuous (see Figure 7.20). There is a simple algorithm for guiding a person negotiating such a maze: place one hand on the wall and walk through the maze, always keeping that hand against the side wall. At a dead end, the same hand is slipped along the end wall, as the direction of travel is reversed.

This algorithm is easily coded in Prolog+ and a program is listed below. It is assumed that the maze is a simple one and is entered by moving north initially. Moreover, it is assumed that the unspecified predicate **move_robot(A,B)** moves a robot in direction A until a maze feature, B, is encountered. The robot might use vision, touch or some other sensory medium, to identify the maze features. Alternatively, the maze, or a map of it, might be viewed from above, in which case vision is clearly needed. This latter situation is akin to one which occurs during the design of printed circuit boards. More will be said about this later. First, let us discuss the maze search program.

Searching a Simple Maze

```
% ____________________Top-level predicate__________________________

search_maze :-
     go_to_maze_entrance,          % Obvious function. Not specified
     move(north).                  % Enter the maze
```

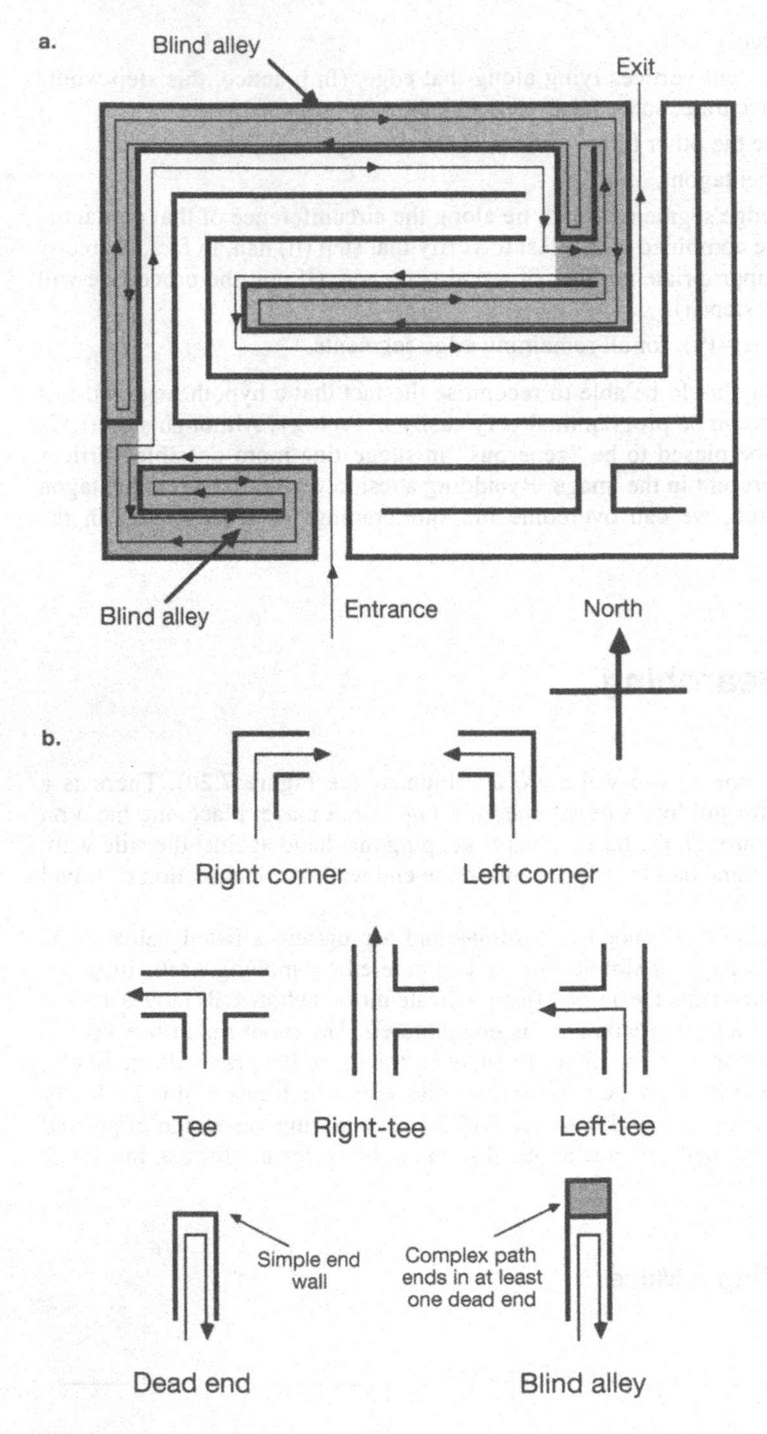

Figure 7.20. A simple maze is one in which the walls are continuous. (**a**) Map of a maze showing one path, discovered by the program **search_maze**. The associated blind alleys are also shown. (**b**) Maze features. Notice that the name of a feature depends upon the current direction of travel.

```
move(A):-
    move_robot(A,B),          % Move in direction A until feature B is found
    next_direction(B,A,C),    % Refer to database for next move, C
    assertz(maze(B,A,C)),     % Remember the move (optional)
    move(C).                  % Recurse

move(_).                      % Halt when the maze has been traversed

% Rules for finding the next move when different types of feature are encountered

% At a T-junction, turn left

next_direction(tee,north,west).
next_direction(tee,west,south).
next_direction(tee,south,east).
next_direction(tee,east,north).

% When there is a "side-road" on the left, turn left.

next_direction(left_tee,X,Y):- next_direction(tee,X,Y).

% When there is a "side-road" on the right, carry straight on

next_direction(right_tee,X,X).

% At at dead end, reverse the direction of travel
next_direction(dead_end,north,south).
next_direction(dead_end,east,west).
next_direction(dead_end,south,north).
next_direction(dead_end,west,east).

% At a corner, there is no choice, so follow the wall

next_direction(right_corner,west,north).
next_direction(right_corner,north,east).
next_direction(right_corner,east,south).
next_direction(right_corner,south,west).
next_direction(left_corner,X,Y):- next_direction(tee,X,Y).

% Force failure at end of the maze

next_direction(_,_,_) :-
    !,
    fail.
```

After running **search_maze**, the database contains a series of facts of the form:

```
maze(tee,north,west).
maze(dead_end,west,east).
maze(right_tee,west,west).
etc.
```

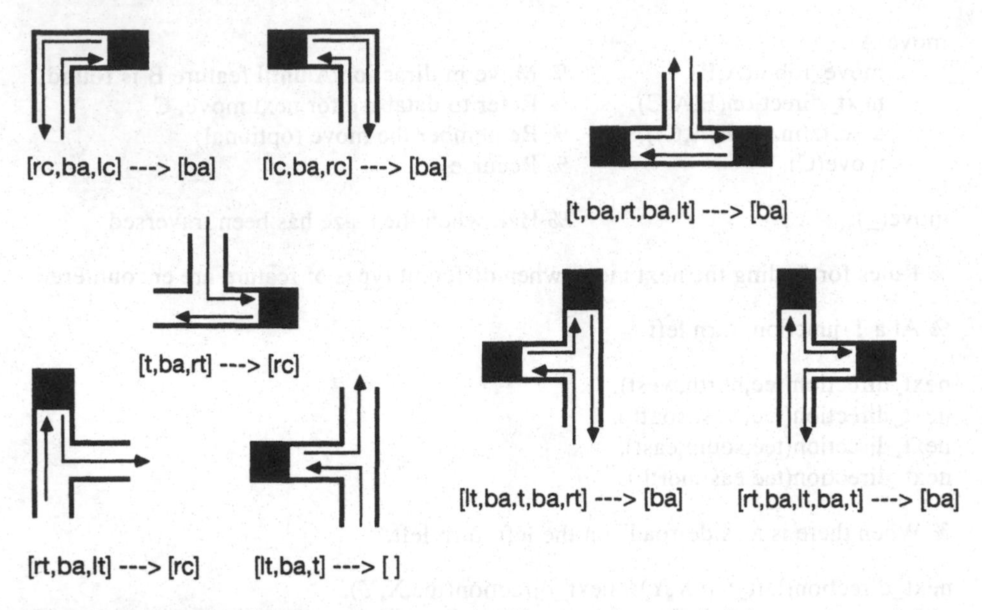
[rc,ba,lc] ---> [ba]
[lc,ba,rc] ---> [ba]
[t,ba,rt,ba,lt] ---> [ba]
[t,ba,rt] ---> [rc]
[lt,ba,t,ba,rt] ---> [ba]
[rt,ba,lt,ba,t] ---> [ba]
[rt,ba,lt] ---> [rc]
[lt,ba,t] ---> [ ]

Figure 7.21. Reducing blind alleys requires recursion. Key: **rc**, right corner; **lc**, left corner; **t**, tee; **rt**, right tee; **lt**, left tee; **ba**, blind alley. The simplest type of blind alley is a dead end, so all dead ends should be replaced by **ba** (not shown).

Eliminating Blind Alleys

Figure 7.21 illustrates the principle of operation of a program which eliminates blind alleys. Notice that recursion is essential for understanding and helpful for recognising a blind alley. A program may be written in terms of *three* different types of facts:

```
% A dead end is a simple type of blind_alley. This is the only fact of this type

prune_blind_alley([dead_end,D1,D2], [blind_alley,D1,D2]).

% A right corner replaces a blind alley after a tee junction. There are five facts like
this

prune_blind_alley([tee,D1,_], [blind_alley,_,_], [right_tee,_,D2],
            [right_corner,D1,D2]).

% A blind alley defined in terms of two simpler blind alleys. There are three facts
like this

prune_blind_alley([left_tee,D1,_], [blind_alley,_,_], [tee, _,_], [blind_alley, _,_],
            [right_tee,_,D2], [blind_alley,D1,D2]).
```

These facts represent a set of nine simplification rules which operate upon a list[8] of moves through a maze. Here is the top-level predicate for eliminating blind alleys:

```
% Terminate recursion

eliminate_blind_alleys([],[]).

% Rename dead ends as blind alleys

eliminate_blind_alleys([A | T],[B | T]) :-
        prune_blind_alley(A,B).

% Reduce single blind alleys at tee junctions and at corners

eliminate_blind_alleys([A,B,C | T1],T2) :-
        prune_blind_alley(A,B,C,D),
        eliminate_blind_alleys([D | T1],T2).

% Reduce double blind alleys at tee junctions

eliminate_blind_alleys([A,B,C,D,E | T1],T2) :-
        prune_blind_alley(A,B,C,D,E,F),
        eliminate_blind_alleys([F | T1],T2).

% No blind alley at the head of the list so strip it off and repeat for the tail

eliminate_blind_alleys([A | T1],[A | T2]) :-
        eliminate_blind_alleys(T1,T2).
```

Non-Simple Mazes

Now, let us consider more complex mazes, where these ideas do not apply. A complex (i.e. non-simple) maze is one in which the walls are not continuous. The following test may be applied to decide whether or not a maze is simple; the predicate **simple_maze** succeeds if the maze is in fact of the simple type. It is assumed that the plan of the maze is represented by a binary image.

```
simple_maze :-
        ndo(1,X),              % Count the walls
        X = 2).                % Simple maze has two separate walls
```

There is an important practical problem relating to non-simple mazes: suppose that we want to draw a conductor between two given points on a printed circuit board. Figure 7.22 shows a simple diagrammatic representation of such a board, with components shown by cross hatching and previously fitted conductors are drawn in black. Sterling & Shapiro [STE-86] describe Lee's algorithm and its implementation in standard Prolog. They represented a *very* coarsely sampled image within Prolog but it is unrealistic to expect Prolog to be able to manipulate high-resolution images in this way. However, when programming in Prolog+, the image processor is available to perform certain operations which Prolog finds difficult. Sterling & Shapiro did not explain how the component layout of a PCB can be entered into the database. Of course, using Prolog+, information about the component layout is very easily acquired, by simply digitising either a map of a PCB, or a view of a real PCB.

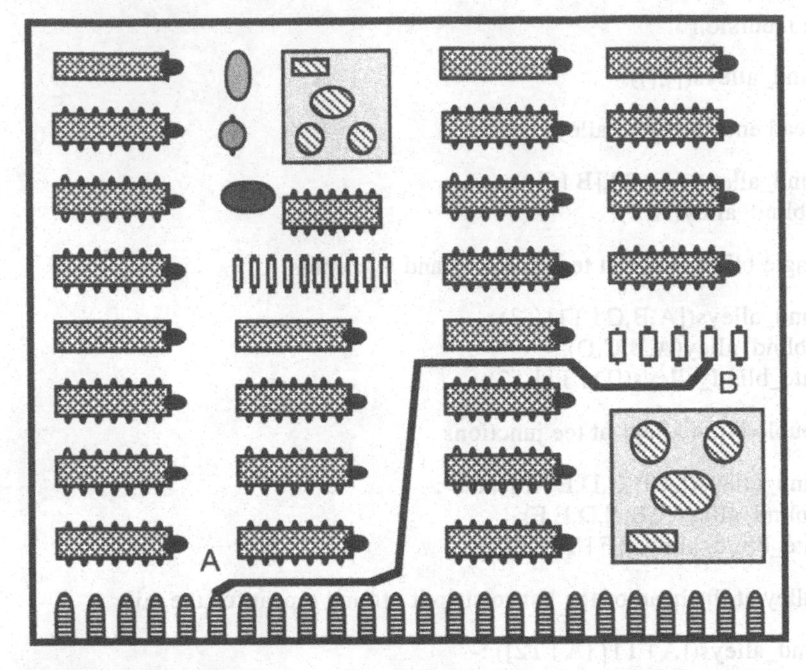

Figure 7.22. Diagrammatic layout of a simple printed circuit board. Planning the optimal route of even a single conductor, say from A to B, is far from straightforward, since there are many obstacles and numerous alternative paths to be considered. Planning the routes of a large number of conductors is so complicated that algorithmic methods are unlikely to succeed.

Consider Figure 7.23. The shortest path between P and Q is an arc which lies along the edge of the convex hull (operator **chu**) enclosing the points P and Q, as well as those obstacles which lie along the line joining P and Q. Here is the procedure coded in Prolog+:

```
shortest_route1([X1,Y1],[X2,Y2]) :-
        rea(obstacle_map),          % Assumed to have been digitised earlier
        exw,                        % Expand obstacles a bit
        swi,                        % Switch images
        zer,                        % Black image
        vpl(X1,Y1,X2,Y2,255),       % Draw line between [X1,Y1] and [X2,Y2]
        touching,                   % Library predicate. See Appendix II
        chu,                        % Convex hull
        pfx(X1,Y1,0),               % Cut the convex hull at [X1,Y1]
        pfx(X2,Y2,0),               % Cut the convex hull at [X2,Y2]
        ndo,                        % Shade both arcs according to their lengths
        thr(254,254),               % Get the shorter one
        wri(shortest_route).        % Save the result
```

This works best if the density of obstacles is low. However, Figure 7.24 demonstrates that this procedure does not always find the optimal (i.e. shortest) route. Hence it is heuristic, not algorithmic and its name is slightly misleading.

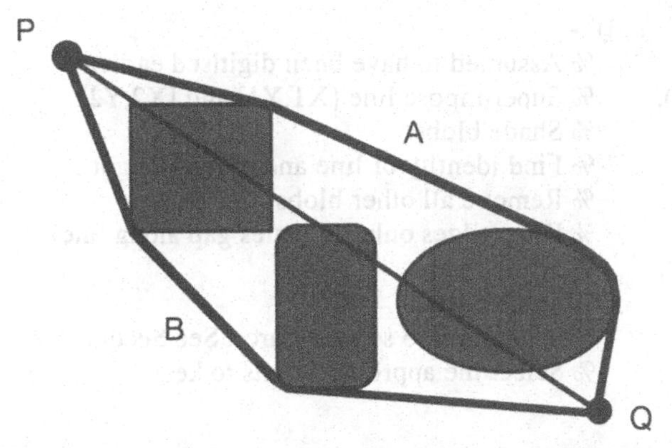

Figure 7.23. One possible method for planning routes when there are very few obstacles. The route between P and Q is to be computed. Those obstacles which lie on the line PQ are represented as blobs in a binary image. The line PQ is also drawn. The convex hull is then drawn around the blob so constructed. This is then divided into two parts, A and B. The shorter one of these then defines the desired path.

There is a simple a posteriori test which can decide whether **shortest_route1** has, in fact, found the shortest possible path:

```
check_route :-
    rea(obstacle_map),
    skw,                        % Shrink white regions
    rea(shortest_route),        % Read the image of the supposed shortest route
    min,                        % AND the images together
    cwp(0).                     % Path is OK if there are no white pixels
```

If **check_route** fails, then a slower, more refined procedure could be invoked to generate alternative paths, around obstacles like those shown cross-hatched in Figure 7.24. Another sub-optimal route-finder may easily be coded in Prolog+:

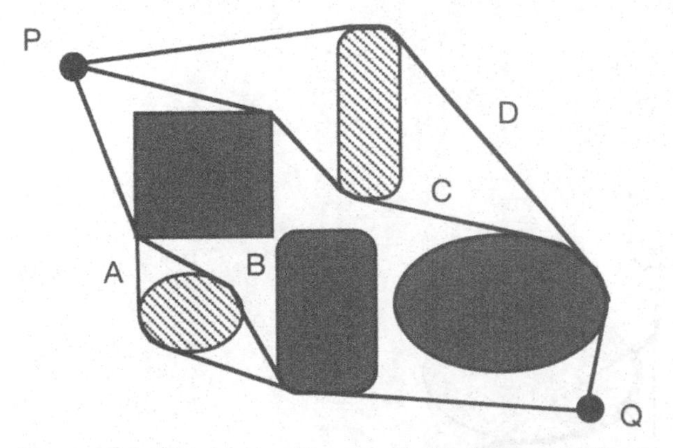

Figure 7.24. Why the procedure described in Figure 7.23 does not always achieve the shortest route if there are other obstacles nearby. The cross-hatched obstacles do not lie on the line PQ but do intersect the convex hull computed by the procedure of Figure 7.23. Hence there are some additional paths to be considered. In this case there are four: A, B, C and D.

```
shortest_route2([X1,Y1],[X2,Y2]) :-
        rea(obstacle_map),             % Assumed to have been digitised earlier
        vpl(X1,Y1,X2,Y2,255),          % Superimpose line [X1,Y1] and [X2,Y2]
        ndo,                           % Shade blobs
        pgt(X1,Y1,Z),                  % Find identity of line and blobs along it
        thr(Z,Z),                      % Remove all other blobs
        bed,                           % Keep edges only. (Creates gap along line)
        vpl(X1,Y1,X2,Y2,255),          % Fill the gap
        mdl,                           % Skeletonise
        acquire_data(L1),              % Segment into separate arcs. See Section 5.2
        route_finder(L1,L2).           % Select the appropriate bits to keep

route_finder(L1,L2) :-
        member([A,B,C],L1),
        member([D,E,F],L1),
        near(A,D),
        near(B,E),                     % Arcs have both ends close together
        not(A = D),                    % Make sure arcs are not the same
        C < F,                         % C is the length of the shorter arc
        delete([D,E,F],L1,L3),         % Delete longer of the two arcs
        route_finder(L3,L2).

route_finder(L1,L2) :-
        member([A,B,C],L1),
        member([D,E,F],L1),
        near(A,E),
        near(B,D),
        not(A = E),
        C < F,                         % C is the length of the shorter arc
        delete([D,E,F],L1,L3),
        route_finder(L3,L2).
```

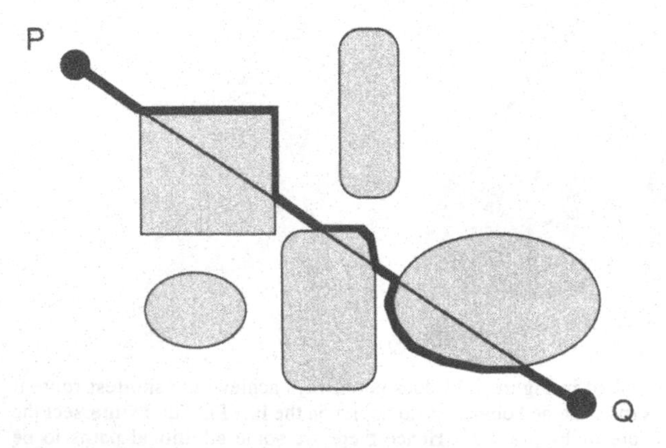

Figure 7.25. Path found by **shortest_route2.**

```
route_finder(L1,L2) :-
        member([A,B,C],L1),
        member([D,E,F],L1),
        near(A,D),
        near(B,E),
        not(A = D),
        C > F,                          % F is the length of the shorter arc
        delete([A,B,C],L1,L3),
        route_finder(L3,L2).

route_finder(L1,L2) :-
        member([A,B,C],L1),
        member([D,E,F],L1),
        near(A,E),
        near(B,D),
        not(A = E),
        C > F,                          % F is the length of the shorter arc
        delete([A,B,C],L1,L3),
        route_finder(L3,L2).

route_finder(L,L).
```

Figure 7.25 shows a path found by **shortest_route2**.

Notice that **shortest_route1** and **shortest_route2** find different types of solutions. The former attempts to minimise the Euclidean distance, while the latter finds an arc, which as far as possible, follows the line joining the two given points. Lee's algorithm, which is coded in standard Prolog by Sterling & Shapiro [STE-86], minimises the City Block distance and produces a solution consisting of lines at 0° and 90°.

Having briefly mentioned Lee's algorithm several times, we shall now consider it in a little more detail. Sterling & Shapiro present a Prolog program which processes a very coarsely quantised space (i.e. the PCB surface). The process depends upon

(a) The generation of a surface which defines the distances from each point in this image to one of the end points of the path to be found. This surface has "wells" in it, corresponding to the obstacles on the PCB. The surface generation process described by Sterling & Shapiro depends upon the propagation of arithmetic values from pixel to pixel. Many different propagation processes can be performed easily in Prolog+. For example, the grass-fire transform (**gft**) is an archetypal propagation procedure and is a built-in predicate in Prolog+.
(b) The other essential element of Lee's algorithm is **hill_climbing**. Once again, Prolog+ contains facilities which make it far more powerful and versatile than standard Prolog. For example, the image processing primitive **dbn** generates an image in which the *direction* of the brightest point in each 3 × 3 neighbourhood is coded.

Hill Climbing

Hill climbing in a two-dimensional space (i.e. an image) is of more general significance than track routing on PCBs. For this reason, we shall define a general two-dimensional hill-climbing program.

```
route(X1,Y1,X2,Y2) :-
        make_hills(X1,Y1,X2,Y2),        % Create hill surface to be climbed.
        rea(obstacle_map),              % Obstacles are black, background white
        min,                            % Mask hills image with obstacles image
        dbn,                            % Direction of brightest neighbour
        hill_climber(X1,Y1,X2,Y2).      % Enter hill-climbing process

hill_climber(X1,Y1,X2,Y2) :-
        near([X1,Y1], [X2,Y2]).         % Near hill top, stop recursion

hill_climber(X1,Y1,X2,Y2) :-
        pgt(X1,Y1,Z),                   % Get intensity at [X1,Y1]
        direction_code(Z,Dx,Dy),        % Consult db for direction to move next
        X3 is X1 + Dx,                  % Increment X
        Y3 is Y1 + Dy,                  % Increment Y
        swi,                            % Switch images
        sof,                            % Avoid overwriting output of dbn
        pfx(X1,Y1,255),                 % Plot locus as hill is climbed
        son,                            % Return to normal operating mode
        swi,                            % Switch images
        !,                              % Cut makes recursion more efficient
        hill_climber(X3,Y3,X2,Y2).      % Repeat process

/* What directions to move when a given direction code is found. Format:

direction_code(Direction,Dx,Dy)                 */

direction_code(0,1,1).
direction_code(1,0,1).
direction_code(2,–1,1).
direction_code(3,–1,0).
direction_code(4,–1,–1).
direction_code(5,0,–1).
direction_code(6,1,–1).
direction_code(7,1,0).
```

The predicate **make_hills** defines a surface ($F_{i,j}$) in a quantised space ($i \in [1,512]$, and $j \in [1,512]$). This surface is represented as an image. Here is a simple definition for **make_hills**:

```
make_hill(X1,Y1,X2,Y2) :-
        lic(X1,Y1),                     % Cone centred at [X1,Y1]
        lic(X2,Y2),                     % Cone centred at [X2,Y2]
        sub.                            % Subtract
```

It is of interest to note the isophotes of the resulting image are hyperbolae, whose foci are at [X1,Y1] and [X2,Y2].

The predicate **hill_climber** attempts to find a path over this surface. It traces a steepest ascent route from [X1,Y1] to [X2,Y2]; at each step in the process, a decision is made to move in the direction which maximises $F_{i,j}$. Clearly, the exact path found by **route** depends upon the surface generated using **make_hills**.

Similar techniques may be used for edge- or line-following. Serial processes like this can be used to good effect in some situations to minimise the data processing requirements.

7.7 Automatic Dissection of Plantlets

In this section, we shall discuss the automation of a task which requires intelligent visual sensing. This task is akin to pruning garden plants, such as roses, although the plants being manipulated are only 10–30 mm tall. *Micropropagation* is a technique used in horticulture for generating a monoclonal colony of plants with some desirable characteristics, such as scent, disease resistance, good fruit yield, colour, etc., quickly and cheaply. A tiny plantlet is grown in a nutrient material, such as agar jelly. When it reaches some suitable size, the plantlet is uprooted and cut into sections; each one is then replanted in agar. This whole process is repeated about 6 weeks later. Large numbers of plants, all with the same genotype, can be produced in a very short time, since the process is not limited by the annual growth, flowering and seed-formation cycle. At the moment, micropropagation is labour intensive and consequently expensive. Hence, it is only cost-effective for use on high-value plants. However, it is undesirable to use human beings in this role for two other reasons:

(i) The plants are very tiny and fragile. Hence human cutters can easily damage them.

(ii) People can infect the plants, which are growing in a material which is well able to support microbial, as well as plant, growth.

Clearly, a machine which could automatically dissect the plants would be attractive.

Of course, images of growing plants can be very complex and a high degree of intelligence is needed to interpret them. However, to make the problem tractable, we must constrain it . This is essential! An obvious simplification can be achieved by separating the plants, so that their images do not overlap. Here, as in many other instances, it is very much easier to avoid the problem than to solve it.

Consider Figure 7.26(a), which shows the silhouette of a tiny chrysanthemum plant. This image was obtained using back illumination. Micropropagation works best if a small Y-shaped piece of the plantlet is cut and replanted. An axial bud, growing in the "V" formed between the stem and a leaf stalk, will grow actively to form a new plant, when it is replanted in agar. The cutting lines shown in Figure 7.26(k) were calculated using the Prolog+ program listed below. This is able to accommodate plantlets which have an "open" structure, and which have no occlusion of the stem by the leaves.

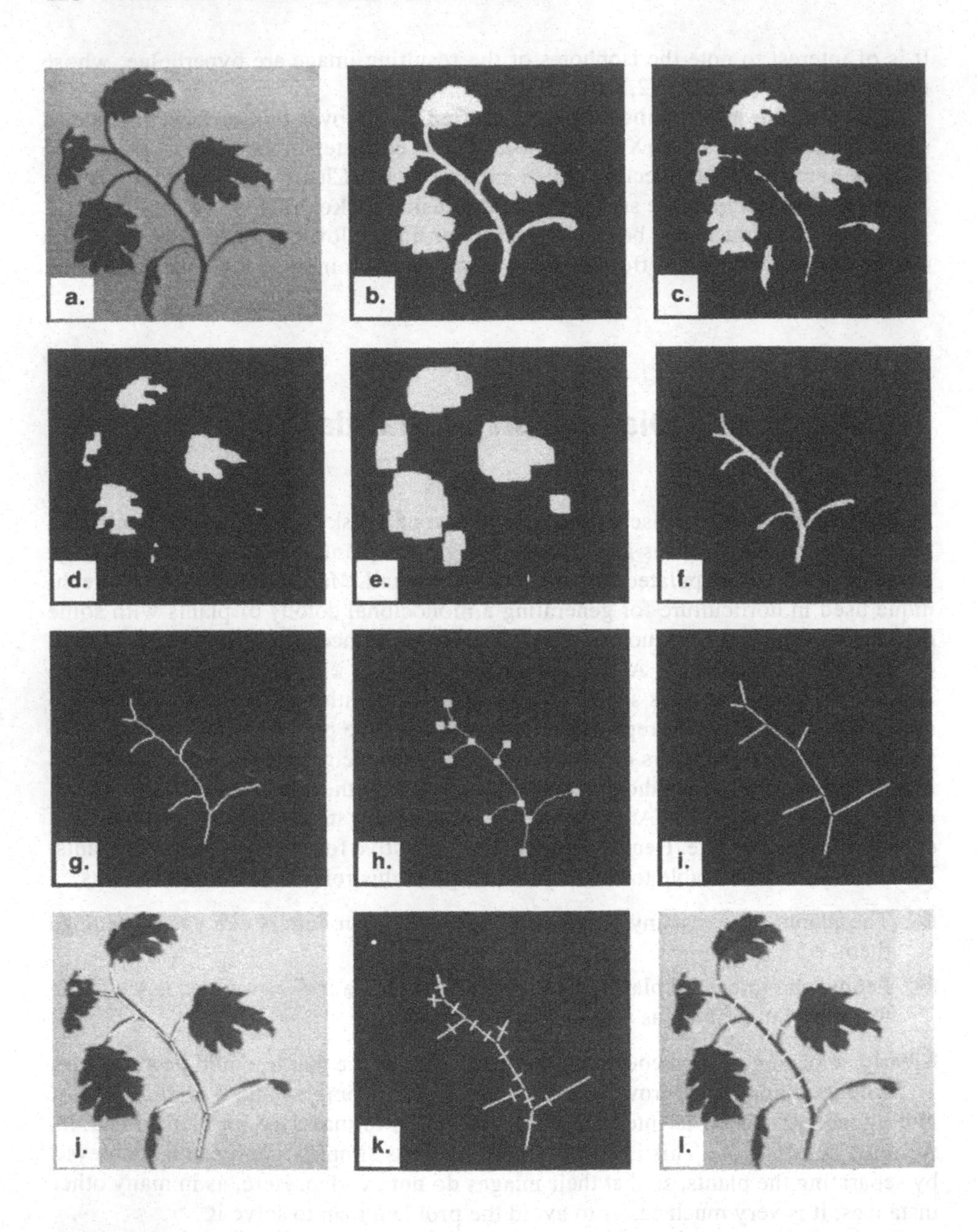

Figure 7.26. Dissecting plants with an open structure. (**a**) Original, unprocessed image of a tiny chrysanthemum plant, obtained using back illumination. (**b**) Binary image, obtained by simple thresholding. (**c**)–(**e**) Various stages in the elimination of the stem by non-linear filtering of image (**b**). (**f**) Difference between images (**b**) and (**e**), after noise clean-up. (**g**) Skeleton, derived from image (**f**). (**h**) Joints and limb-ends, obtained from image (**g**). (**i**) Piece-wise linear approximation to the skeleton, obtained by joining limb ends and joints. (**j**) Piece-wise linear approximation to the skeleton superimposed on the original image, for the purpose of comparison. (**k**) Cutting points. These were calculated from image (**j**). (**l**) Cutting points superimposed on the original image. Notice that five Y-shaped pieces can be cut from this one plant. (Courtesy of Ian Harris.)

Program, Partial Listing

The program receives as input a high-contrast, grey-level image of an "open" structure plant (Figure 7.26(a)), and ends after it has superimposed bars indicating the necessary cutting points onto that image. In practice of course, the program output would consist of a series of movement commands to a robotic device. The program has been split into several sections. The first is concerned with the processing of the plant image, in order to remove the leaves and skeletonise the remaining stem. The second section divides the skeletonised stem into its component parts and calculates the coordinates of each section of the stem. This is carried out in order to create an input for the third module, which interprets the plant description, using a number of rules, to calculate the points at which the plant needs to be cut. Finally, these points are superimposed onto the original image in the form of bars, which demonstrate to the reader where the robot would cut the plant.

```
%_____________Program for plant dissection written by Ian Harris_____________

plant :-
        get_skeleton,        % Threshold, remove leaves and skeletonise
        skeleton_breaker,    % Split skeleton into sections
        end_points,          % Calculate top and bottom points and put them in db
        track_stem,          % Track stem from bottom and calculate cut points
        plot_cuts,           % Plot the cut points on the plant
        clean_up.            % Delete all information from the database

/* "get_skeleton" thresholds the original image, performs some image processing to
remove the leaves, and finally skeletonises the remaining image. The result of this
therefore is a one-pixel wide matchstick representation of the stem and branches of
the plant. */

get_skeleton :-
        threshold,           % Threshold grey level plant image
        edg(0,25),           % Mask the edge of the digitised image
        small_box,           % Process within the smallest possible window
        wrm,                 % Remove white points (background noise)
        sof,                 % Turn off automatic switching of images
        cpy,                 % Copy current image to alternative image
        3•skw,               % Shrink white areas to remove stem
        4•exw,               % Expand white areas to return leaves to original size
        max,                 % Maximum of the current and alternative images
        xor,                 % Exclusive OR of the current and alternative images
        son,                 % Turn on automatic image switching
        edg(0,8),            % Mask edge of current image
        biggest_blob,        % Find the biggest binary blob
        red,                 % Roberts edge detector
        blb,                 % Fill holes in blob
        skw,                 % Shrink white areas
        small_box,           % Process within the smallest possible window
        mdl.                 % Skeletonise the binary image
```

/* "skeleton_breaker" break the stem skeleton into its component parts, find the coordinates of each section and assert them into a database. */

```
skeleton_breaker :-
    cnw,                          % Set intensity to number of white pixels in 3 x 3
                                    neighbourhood
    min,                          % Find the minimum of current and alternative
                                    images
    hil(4,255,255),               % Highlight the branch/stem joints
    hil(2,2,128),                 % Highlight the line ends
    hil(3,3,64),                  % Highlight the individual limbs
    hil(255,255,0),               % Set all node points to intensity 0 (black)
    ndo(2,N),                     % Shade blobs and count them
    wrw(temp),                    % Save the image in a temporary file
    find_vectors(N,1,A1).         % Find end points of each section of skeleton
```

/ * "end_points" is used to find the coordinates of the highest and lowest sections in the skeleton image, in order to obtain a point of reference for the stem tracking algorithm. */

```
end_points :-
    data(coords,A),               % Retrieve stem coordinates from the data
    y_list(A,B),                  % Construct a list of "y" coordinates
    get_low(B,L),                 % Find the lowest "y" coordinate
    find_rest(L,L1),              % Find corresponding coordinates for relevant
                                    section
    assert(data(stem_start,L1)),  % Assert values in database
    get_high(B,H),                % Find highest "y" coordinate
    find_rest(H,H1),              % Find corresponding values
    assert(data(stem_end,H1)),    % Assert values in database
    !.                            % Stop back-tracking past this point
```

/* "track_stem" is used to track the path of the stem from the bottom and calculate the positions of the cut points on each of the skeletal sections. */

```
track_stem:-
    data(stem_start,S),           % Retreive coordinates for the start of stem from db
    find_top(S,T),                % Find the highest point of the first stem section
    tracking_loop(T),             % Call the tracking_loop predicate. See below
    db_list(cut_coord,cuts),      % Construct list of cut coordinates from db
    !.                            % Stop back-tracking

tracking_loop(H,I|G]):-
    corresponding([H,I],C),       % Find end coordinates for one stem section
    C=[J,K|_],                    % Instantiate variables to coordinate values
    end_connections(H,I,E),       % Find any connected sections
    secondaries(E,H,I,J,K),       % Calculate cut point for given section
    !,                            % Stop back-tracking
    empty_list(G).                % Check for any more connected sections

tracking_loop(A).
```

Remarks

The program listed above locates the axial buds and then calculates three cutting lines for each bud. If two buds are close together, the program is able to place a single cut on the stem, mid-way between them. A rather simpler program might calculate the coordinates of each axial bud, so that a circular "pastry cutter" could be placed over each one. The advantage of this is that it is slightly faster than the technique embodied in the program above. However, it does present problems if two buds are close together, since one cutting circle might damage another bud. Obviously, a very simple test could be devised in order to avoid this problem.

Alternative approaches to detecting axial buds might be based upon

(a) a corner detector, which highlights points of high curvature.
(b) a morphological, or N-tuple, operator, which detects Y-shaped image features.
(c) a serial search along the stem. (The stem would be easy to identify initially as the thin streak-like feature which enters the growing medium.)

It might appear that we are oversimplifying the task, since plants are three-dimensional structures and our program operates in only two dimensions. However, it should be understood that there exist certain types of cutter, such as water-jet and laser, which could be used successfully with the information which our program provides. For example a laser, whose beam is coaxial with the camera's optical axis, would be able to perform the cutting using only two coordinate measurements (Figure 7.27). A water-jet cutter could also be used in the same way. Two cameras, mounted at the same height and at 90° around the plant, could provide three-dimensional data, if this really is required. This may also be useful for compensating for occlusion

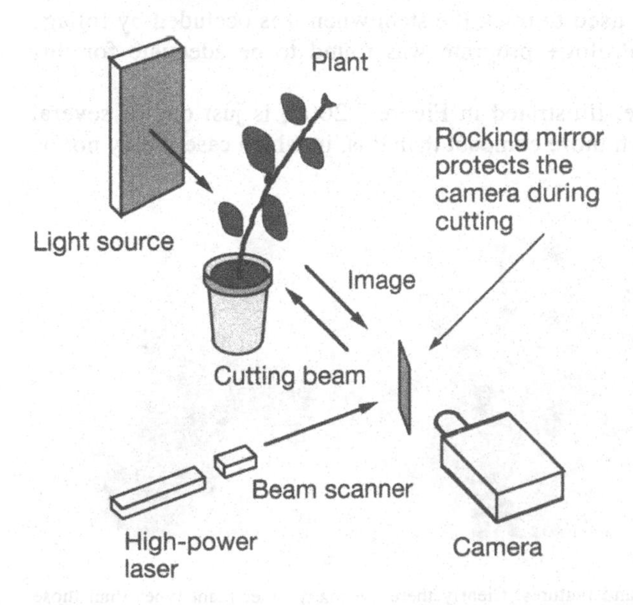

Figure 7.27. Using a laser to dissect a plant. If the laser beam is coaxial with the camera, the task is reduced to one in only two dimensions.

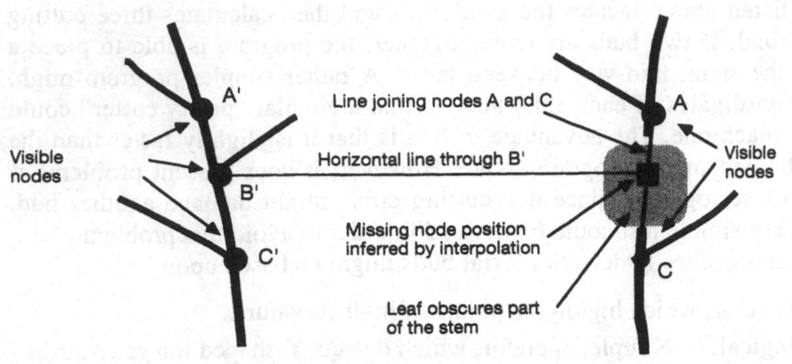

Figure 7.28. Two cameras and simple linear interpolation may be used to overcome some of the problems of occlusion of the stem by leaves.

of the stem by the leaves. The development of suitable robotic manipulators and cutting tools is the responsibility of the author's colleagues, whose work will be reported elsewhere.[9]

To date, only a very small amount of work has been done on the difficulties caused by occlusion. A preliminary investigation suggests that the problem is tractable, if the occlusion is of minor extent. Several views of the plant could be obtained, either by rotating the plant around the vertical axis or by using two, or more, cameras. The optical axis of each camera is horizontal. The presence of an axial node in any one image in a series provides evidence to decide that this is a suitable site for cutting. Interpolation may be used to track the stem when it is occluded by foliage (see Figure 7.28). A small Prolog+ program was found to be adequate for this interpolation.

The "open" plant structure, illustrated in Figure 7.26(a), is just one of several types. Some plantlets are much more compact than this, in which case it may not be

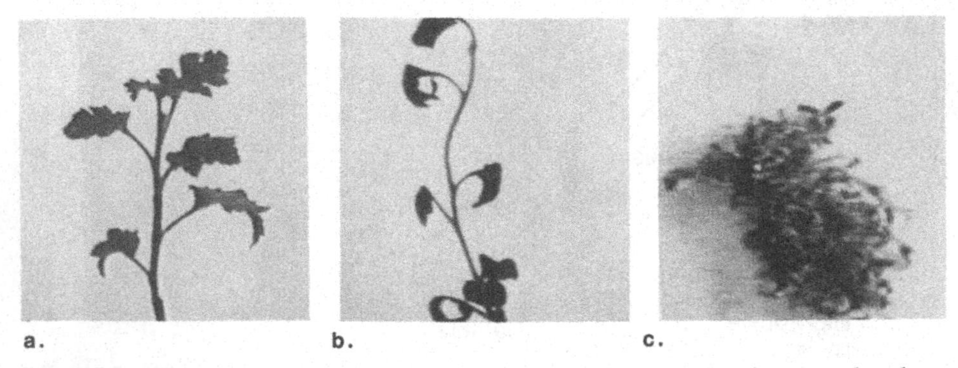

Figure 7.29. Different plant types and features. Clearly there are many other plant types than those illustrated here. (**a**) Open structure plant. There is no occlusion of the stem. (**b**) Semi-open plant structure; there is some occlusion of the stem by leaves. (**c**) View of a plant with a dense foliage structure.

worthwhile trying to distinguish different leaves and stems. It is not unusual, for example, to find that the foliage of certain species of plants forms a tiny, hemispherical bush. The approach to the image processing is then quite different; it may be better to view such a plant from above and cut it in a vertical plane. There are of course plantlets whose structure is intermediate between these extremes (Figure 7.29(b)). The development of more advanced programs with better judgement is continuing.

It is also necessary to find suitable grasping points, so that a robot can replant the cuttings. A Prolog+ program to do this has also been developed.

Several of the steps in the program cannot conveniently be performed in a traditional image processing language, although Prolog+ is well able to perform these functions with ease. Since Prolog+ is also able to control electro-mechanical devices, such as an (X,Y,θ)-table and (with minor software changes) the robotic cutter, it is an ideal language for this application. The plant-dissection task is rather unusual, since the program intelligence, rather than processing speed, is of prime importance.

7.8 Optical Character Recognition

Optical character recognition (OCR) was one of the earliest successful applications for image processing. It was developed commercially as early as the mid-1960s, and it is possible now to buy very effective OCR software for a few hundred pounds. This can be used in conjunction with a desk-top computer and a suitable document scanner or video camera. The principal objective of OCR is to transfer typed or printed text, originally appearing on paper, into machine readable (i.e. electronic) format, such as a disc file containing a sequence of ASCII characters. Since OCR does not require human intervention, tasks such as retyping and type-setting can achieve large savings of both time and money.

There are several other important and closely related tasks, including:

(a) *Verification*. A manufacturer of pharmaceutical products may need to verify that cartons containing medicines have appropriate printing to indicate their contents. A given manufacturer may well produce several drugs with very similar brand/chemical names. It is not uncommon, for example, to find that several different drug names begin with the same series of letters, indicating that the names are derived from a common Latin or Greek root. Moreover, the packets may look alike, being of the same size, shape and colour. Printed cartons, which may contain quite different products, often look very similar, since they are all made to conform to the manufacturer's house style. It is, of course, also important to be able to verify that instructions, such as the dose frequency/amount are appropriately printed.

(b) *Checking readability*. It is often important to verify that printed text is clearly readable, although there is no need for the machine to read it, since we already know what it should say. Checking readability is of obvious importance in the printing industry, commerce and manufacturing, etc.

(c) *Locating specific words*. This is important for library abstracting services, information services, military intelligence etc. How often have you scanned a page to look for one of a small number of words, such as your own name, or a company name, in order to decide whether you need to read the whole document? You can quickly judge the relevance of a document without reading it in full.

(d) *Reformatting charts and other line drawings.* An example of this is to be found in the conversion of engineering drawings, maps, navigation charts from Imperial to metric units. The task requires reading dimensions, tolerance parameters, contour heights, etc. In one sense, this is a much more closely constrained task than is normally implied by full OCR, since there is a limited set of values that a given character can assume in a certain position/context. On the other hand, the printed characters may be at any orientation and be of any size.

(e) *Checking the presence of a logo, trade-mark or other icon.* Most companies jealously protect their corporate image and a product which bears a defective company trade-mark is perceived as bringing discredit to the company. This is particularly important if the company purports to make high-quality goods.

(f) *Signature verification.* Signature verification is used in security systems. The task is to verify that a person is actually who he/she claims to be. Signature verification is just one of many possible ways of checking a person's identity.

We shall approach the task of recognising printed lettering and numerals as if OCR software did not already exist. Recall that our purpose is to demonstrate the power of the Prolog+ language, rather than trying to give complete solutions to important commercial applications.

Recognising Playing-Card Value Symbols

In section 7.2, we discussed some of the possible methods of recognising the value of a playing card. In particular, we discussed methods based upon the spatial distribution of ink over the complete surface of the picture cards and counting blobs on the non-picture cards. An alternative approach is to recognise the value symbol. This is a character printed at the top-left and bottom-right corners of each card and is a member of the set

{2, 3, 4, 5, 6, 7, 8, 9, 10, J, Q, K, A}

Notice that a ten ("10") actually consists of two separate symbols: "1" (one) and "0" (zero). The program presented below makes use of only the "0", and ignores the "1". (This is achieved using **ndo** and makes use of the fact that "1" has a smaller area than "0".)

It will be assumed that the camera views the top-left corner of the card, which is placed against a white background and is in (almost) fixed position and orientation (Figure 7.2). There may well be other printed features on the card within the camera's field of view, but the largest fully enclosed dark feature is assumed to be the value symbol. Here is a program for recognising the value symbol.

```
value(V) :-
    features(L),        % Measure eight features, explained below
    values(L,V).        % Consult database. See table below
```

```
features([A,B,C,D,E,F,G,H]) :-
        frz,                        % Digitise image
        won(2),                     % Reduce resolution to 128 × 128, for faster processing
        thr(128),                   % Fixed level thresholding
        blb,                        % Fill holes in white areas
        xor,                        % Isolate holes
        ndo,                        % Shade blobs in order of size
        thr(255),                   % Select brightest (i.e biggest)
        exw,skw,skw,exw,            % Filtering to reduce noise
        edg(0,1),                   % Remove edge of region of interest
        measure(A,B,C,D),           % Calculate four measurements
        thr(1,255),                 % Threshold to start again
        won(1),                     % Revert to 256 × 256 resolution
        yxt,                        % Rotate by 90°
        edg(0,1),                   % Remove edge
        measure(E,F,G,H).           % Calculate four more measurements

measure(A,B,C,D) :-
        dim(X1,X2,Y1,Y2),           % Dimensions of white area
        swi,                        % Switch images
        DX is X2 – X1 + 1,          % Calculate window size
        DY is Y2 – Y1 +1,
        swc(8,X1,Y1,DX,DY),         % Set window coordinates
        wgx,                        % Generate wedge
        min,                        % Mask wedge using value symbol
        thr(192),                   % Threshold
        eul(A),                     % Count blobs on right-most quarter of window
        swi,                        % Switch images
        thr(128,191),               % Threshold
        eul(B),                     % Count blobs in right-centre quarter of window
        swi,                        % Switch images
        thr(64,127),                % Threshold
        eul(C),                     % Count blobs in left-centre quarter of window
        swi,                        % Switch images
        thr(1,63),                  % Threshold
        eul(D),                     % Count blobs in left-most quarter of window
        swi.                        % Switch images

values([1,1,1,1,2,1,1,1],ace).
values([2,1,1,1,2,1,1,2],king).
values([1,1,1,1,1,2,2,1],queen).
values([1,1,1,1,1,2,1,1],jack).
......
values([2,3,3,2,1,2,2,1],2).
```

The measurements derived by the predicate **features** are explained in Figure 7.30. There are two sub-processes. In the first, the box of minimal size required to enclose the value symbol is divided horizontally into four bands of equal width. The number

a. 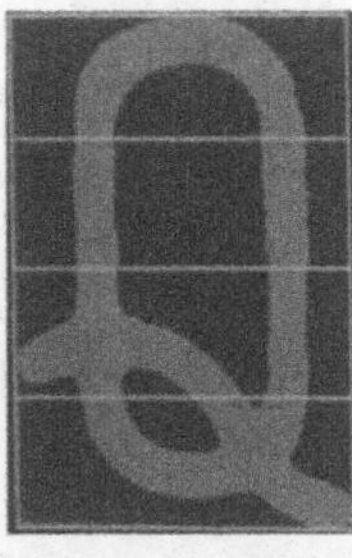

b. 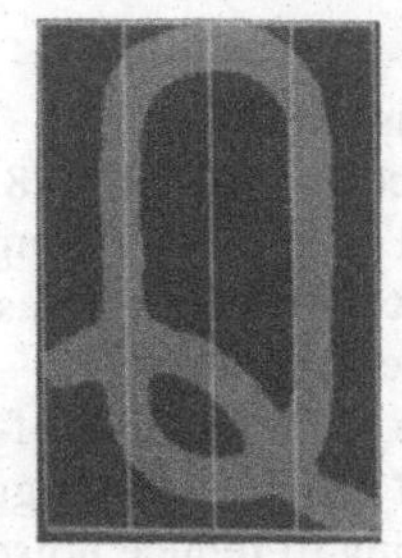

Figure 7.30. Simple measurements for recognising alphanumeric characters. (**a**) First, the box of minimal area surrounding the character is computed. Then, a set of horizontal strips are derived. In each one, the total number of white pixels is found. (This is the basis of the predicate **measure**.) Alternatively, the number of distinct white blobs in each stripe can be found. (This is the basis of **measure1**.) Both sets of measurements could be combined, providing twice as much data, for little additional computational effort. (**b**) An alternative method of analysis is to use vertical, rather than horizontal strips. Once again, this could be combined with the method of analysis explained in (**a**). Notice the situation which is apparent in the third strip from the left. There are two distinct blobs. If however, the right-most edge of this strip were placed just a little to the right, i.e. within the "ink" area of the "Q", these two blobs would then be joined, giving an Euler number of 1, not 2. For this reason, **measure1** is rather susceptible to the effects of noise.

of blobs within each band is counted using the Euler number (**eul**). This sub-process yields four measurements: [A,B,C,D]. The other four measurements, [E,F,G,H], are obtained, in the second sub-process, by dividing the same box vertically into four bands. It is necessary to use all eight measurements otherwise some confusions would arise. For example, "2" and "5" would be confused if only the first four measurements were used.

This simple recognition process is totally insensitive to size variations and is reasonably tolerant of font changes. However, it is more sensitive to orientation changes, since the minimal-area box, which is placed around the value symbol becomes larger as the characters are rotated away from the normal upright posture. Ink-to-paper contrast is likely to be high on good-quality playing cards, but in many other OCR applications the situation may be much worse. This will inevitably lead to some mistakes in the recognition process. To accommodate print-quality variations and multiple fonts, it may well be necessary to store more than one **values** fact for each value symbol. The recognition program listed above was designed simply by the author inspecting the output of the **measure** predicate. In practice, where there are many measurements and numerous fonts and/or possible output cases, it is essential to use self-adaptive learning, which is the subject of the next section.

Of course, the recognition process can make use of alternative or additional measurements, with equal ease; we merely need to modify **measure**. We conclude this section with a brief discussion of other possible measurements, which might be used for recognition instead, or in addition to the measurements described immediately above.

An Extension of the "measure" Predicate

The following is a simple extension of **measure**. Notice that two sets of parameters are calculated now, with little additional computational effort.

```
measure1(A1,B1,C1,D1,A2,B2,C2,D2) :-
    dim(X1,X2,Y1,Y2),
    swi,
    DX is X2 – X1 + 1,
    DY is Y2 – Y1 + 1,
    swc(8,X1,Y1,DX,DY),
    wgx,
    min,
    thr(192),
    eul(A1),          % As in "measure"
    cwp(A2),          % Count white pixels in right-most quarter of window
    swi,
    thr(128,191),
    eul(B1),          % As in "measure"
    cwp(B2),          % Count white pixels in right-centre quarter of window
    swi,
    thr(64,127),
    eul(C1),          % As in "measure"
    cwp(C2),          % Count white pixels in left-centre quarter of window
    swi,
    thr(1,63),
    eul(D1),          % As in "measure"
    cwp(D2),          % Count white pixels in left-most quarter of window
    swi.
```

Both **measure** and **measure1** may be regarded as performing a pair of orthogonal scans (one horizontal and the other vertical) across the character being read. A set of binary measurements is derived for each scan. The process can be likened to bar-code reading, as is evident from Figure 7.30.

Moments

Moments form a family of easily computed measurements, which can be derived from either binary or grey-scale images. The $(p,q)^{th}$ moment, $M_{p,q}$, of an image $\{F(x,y)\}$ is defined thus

$$M_{p,q} = \int_{-\infty}^{\infty} \int_{-\infty}^{\infty} F(x,y).x^p.y^q.dx.dy$$

F(x,y) is a continuous intensity function, defined in the (x,y)-plane. In practice, of course, the space is quantised and we replace integration with summation.

A finite set of moments, for example

$\{M_{0,0}, M_{0,1}, M_{1,0}, M_{1,1}, M_{2,0}, M_{2,1}, M_{2,2}, M_{0,2}, M_{1,2}, \ldots\}$

might be used as the basis for recognition in a variety of applications, including OCR. Moments may be computed from either grey-scale or binary images. The low-order moments are related to concepts which are already familiar from Chapter 2. For example, $M_{0,0}$ is (proportional to) the average intensity (grey-scale image), or the number of white pixels (binary image). The ratios ($M_{0,1}/M_{0,0}$) and ($M_{1,0}/M_{0,0}$) indicate the position of the centroid (binary image) or the centre of gravity of the intensity function (grey-scale image). The ratios ($M_{0,2}/M_{0,0}$) and ($M_{2,0}/M_{0,0}$) indicate the variance of the intensity function. Combinations of moments can be used to good effect. For example, the so-called normalised moments

$$\{ [(M_{0,j} - M_{0,1})/M_{0,0}], [(M_{i,0} - M_{1,0})/M_{0,0}] \}$$

are independent of position. It is also possible to obtain measurements that are independent of skew. This has been used to good effect, when trying to recognise sloping hand-printing.

Low-order moments can be calculated using Prolog+:

```
moment(P,Q,Z) :-
    wri(image),          % Save input image, usually binary
    wgx,                 % Calculate X^P
    tra(p_power),        % Use LUT to calculate X^P
    wri(temp),           % Save result
    wgx,                 % Calculate X^Q
    tra(q_power),        % Use LUT to calculate X^Q
    yxt,                 % Rotate image by 90°, to give Y^Q
    rea(temp),           % Restore image of X^P
    mul,                 % Multiply X^P and Y^Q
    rea(image),          % Read input image
    mul,                 % Calculate F(X,Y).X^P.Y^Q
    rin,                 % Integrate along X
    yxt,                 % Rotate by 90°
    int,                 % Effectively integrating along Y
    ppi(512,512,Z).      % Get result
```

This procedure is not very accurate for large values of P and Q, since it relies on 8-bit integer arithmetic throughout.

A set of moments may be used to provide alternative image measurements to those computed by **measure** and **measure1**. Of course, there is no reason why we cannot combine these measurements together, or with others that we might devise.

Walsh/Haar Transforms

The Walsh and Haar transforms offer computationally convenient alternatives to the Fourier transform. All three of these integral transforms are used extensively in (one-dimensional) signal processing and have been used for image analysis. The two-dimensional Walsh transform may be defined thus

$$\bar{F}(p,q) = \frac{1}{M \cdot N} \left(\sum_{j=1}^{M} \sum_{i=1}^{N} F_{i,j} \cdot W_{p,q} \left(\frac{i}{N}, \frac{j}{M} \right) \right)$$

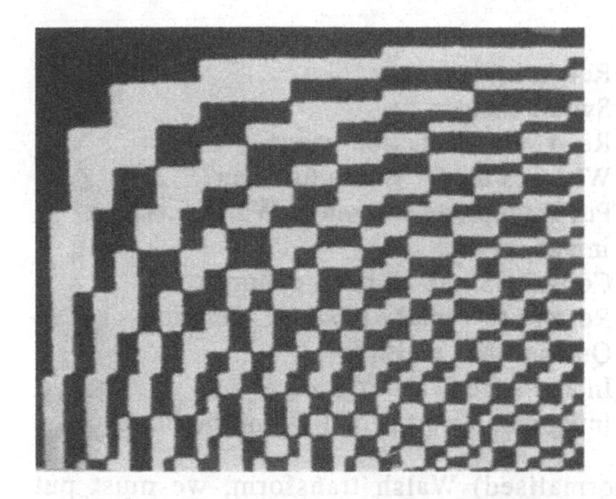

Figure 7.31. A number of Walsh functions can be stored as column (or row) vectors in an image. An image of resolution 512 × 512 pixels can obviously store 512 functions, each defined at 512 sample points. This is the top-left corner of such an image.

where the image is sampled at M.N equally spaced points. The $W_{p,q}(\_,\_)$ are termed Walsh functions and can assume one of only two values: ±1 (Figure 7.31).

The Walsh functions can be defined recursively. For example, the following formula relates a one-dimensional, N-point Walsh function of order (2j + q) in terms of a function of order j

$$W_{2j+q}(n) = (-1)^{[j/2]+q} \cdot [W_j(2n)+(-1)^{j+q} \cdot W_j(2(n-N/2))]$$

The Haar transform is defined in a very similar way in terms of the of so-called *Haar functions*, but can adopt three values: 0 and $\pm\sqrt{2}$. Walsh (and Haar) functions can be defined recursively, or expressed in terms of even simpler functions, called *Rademacher functions*. Of course, the word "recursion" immediately suggests the use of Prolog and it is indeed possible to write a generator for Walsh functions using the language. However, these details are not important. The reader should simply take note of the fact that we can either compute the Walsh functions using Prolog+, or simply define them using the computer keyboard. (The task only has to be done once.)

Let us assume that a set of Walsh functions has already been computed and that the results have been stored in an image; each row stores a different Walsh function.[10] The $(p,q)^{th}$ Walsh function $W_{p,q}$ is assumed to be held in column R of this image, where R is derived by referring to a database of facts of the following form:

```
sequency_order(P,Q,R).
```

To simplify the program for the purposes of illustration, we shall limit our attention to one-dimensional transforms. We begin by presenting a program for computing the Pth Walsh coefficient of each column in an image. Notice that a large number of different Walsh functions can be held in a single image, stored in the disc file named **walsh_functions.** In addition, observe that these stored functions assume the values 0 and 255, and not ±1 as stated earlier. The result is a constant offset which can be subtracted later, although this is often not necessary.

```
walsh_coefficient(P) :-
    sof,                        % Retain input image
    swi,                        % Switch images
    rea(walsh_functions),       % Read Walsh function image
    sequency_order(P,Y),        % Where is the Pth Walsh function?
    psh(Y,0),                   % Put appropriate column of Walsh function
                                  image on RHS
    csh,                        % Copy RHS into all other columns
    son,                        % Restore two-image mode of operation
    min,                        % Quicker than multiply
    yxt,                        % Interchange X and Y axes
    rin.                        % Integrate. Result is on RHS of the image
```

To calculate the complete (non-normalised) Walsh transform, we must put **walsh_coefficient** into a loop:

```
walsh_transform :-
    wri(temp),                  % Retain input image on disc
    walsh_transform(0).         % Enter main program

walsh_transform(512) :-         % Truncate Walsh series at appropriate order
    rea(result).                % Recover result from disc

walsh_transform(P) :-
    rea(temp),                  % Read input image again
    walsh_coefficient(P),       % Calculate Pth Walsh coefficient
    vgt(512,1,512,512,X),       % Copy RHS into buffer X
    rea(result),                % Read results image
    vfx(P,1,P,512,X),           % Overwrite its Pth column
    wri(result),                % Save intermediate result on disc
    Q is P + 1,                 % Increment counter
    !,                          % Makes tail-end recursion more efficient
    walsh_transform(Q).         % Repeat
```

The reader should be aware that it is likely to take a very long time to satisfy the goal **walsh_transform**. A similar technique could be used to compute the Fourier transform, although this would take even longer to perform!

7.9 Learning

It is agreed by most workers that, when new products are to be introduced to factory-floor inspection systems, only the very simplest type of programming should be needed. In fact it is often argued that, ideally, the machine should be able to adjust its internal parameters, by learning from a few well chosen samples presented to it. This is known as *Teaching by Showing*. After it has finished learning, the machine should be able to *recognise* the most appropriate class or type of object.

There are some philosophical arguments against this approach, which cause concern to some research workers, including the author. These objections are centred on the difficulty of obtaining a truly representative sample of objects which properly reflect the full range of variation of each of the parameters upon which the recognition is to be based. It is impossible, in almost all cases, to manufacture products with more than a very small subset of the important parameters set to predefined values; most tolerance parameters cannot be directly adjusted at will by a production engineer. For example, an injection moulding might have 250 important parameters, but the machine which makes it might have only six adjustable controls. Thus, there are at least 244 (= 250 – 6) parameters which cannot be modified merely by the machine operator twiddling a few knobs and can only be adjusted with some difficulty. Many parameters are determined by the mould geometry, and obviously cannot be adjusted easily. It is clearly impractical to expect hundreds of different moulds to be made, just to provide a teachable inspection system with examples of products covering the complete tolerance range of each parameter.

Notwithstanding these objections, the concept of Teaching by Showing remains very popular, because it claims to be able to make use of very low skill levels. As a consequence, it certainly has a role to play in some situations, although the author warns against the over-enthusiastic use and subsequent over-reliance upon this apparently seductive idea.

We cannot do full justice to the fascinating topic of machine learning in the limited space available here. This has been a rich source of material for researchers and a great many different learning programs, written in Prolog, have been published [COE-88]. Of course, any of these could be incorporated into Prolog+, once again demonstrating that Prolog+ provides a bridge, enabling standard, proven AI techniques to be applied to image data. We shall merely describe a few simple ideas for machine learning, without attempting to describe the latest, most sophisticated procedures.

Recognition Program

First, we present a very simple recognition program for recognising flat industrial piece-parts, such as metal stampings, leather cuttings, etc. Imagine that flat, opaque, back-lit blob-like objects (stampings) are presented for inspection. The implication of this assumption is that it is possible to use fixed-parameter thresholding to obtain a binary image.[11] The predicate **recognise_object** relies upon three simple measurements:

(a) Area of the stamping.
(b) The number of holes.
(c) The distance of the furthest point on the object boundary from its centroid. (Called its "radius".)

These three parameters will be used merely to demonstrate our first, very naive approach to learning. Here is the recognition program, which simply digitises an image, derives these three measurements and calculates the classification

```
% Top-level predicate for the recognition phase
recognise_object :-
    recognise(A,H,D,C),             % Object is from class C. Measurements:
                                    [A,H,D]
    message(['Parameters:',[A,H,D],' Class:',C]).
                                    % Standard MacProlog predicate
    !,
    recognise_object.               % Repeat
recognise_object :-
    message(['Sorry! The object was not recognised']),
    !,
    recognise_object.

% Second level predicate. Used in both recognition and learning phases

recognise(A,H,D,C) :-
    frz,                            % Digitise image from the camera
    thr(128),                       % Convert it to binary form
    get_features(A,H,D),            % Measure features
    !,                              % Prevent back-tracking past here
    object(A1,A2,H,D1,D2,C),        % Consult db for tolerance parameters
    A ≥ A1,                         % Is area greater than minimum?
    A ≤ A1,                         % Is area less than maximum?
    D ≥ D1,                         % Is "radius" greater than minimum?
    D ≤ D2.                         % Is "radius" less than maximum?

recognise(_,_,_,_) :-
    !,
    fail.                           % Give up the search. Do not back-track

% Feature measurement

get_features(A,H,D) :-
    lifo_push,                      % Save the image on the stack
    cwp(A),                         % Count white points
    cgr(X,Y),                       % Find centroid
    lic(X,Y),                       % Generate intensity cone
    min,                            % Mask cone with the silhouette
    hil(0,0,255),                   % Remove background from consideration
    gli(D1,_),                      % Find darkest point, i.e. furthest from centroid
    D is 255 – D1,                  % Convert to distance
    lifo_read.                      % Read top element of image stack. Do not pop
    blb,                            % Fill holes
    xor,                            % Isolate the holes
    ndo(3,H),                       % Count holes
    lifo_pop.                       % Restore image
```

```
% Typical database entries

object(1832,1897,3,319,325,widget).
object(1537,1624,5,573,545,wangler).
......
object(1535,1607,4,489,535,briddle).
object(_,_,_,_,_,'unknown object').      % Catch-all rule for non-recognised objects
```

This naive program has a number of deficiencies:

(a) The output is in the form of a written message (**message**) presented to the user. In practice, a program would need to operate some accept/reject mechanism, such as an *accept/reject* solenoid. Of course, this is easy to achieve using Prolog+.
(b) The number and type of parameters measured is fixed. Reprogramming is necessary, if new feature measurements are to incorporated.
(c) The program consists of an unending loop; there is no means of escaping, to enter the learning program. Once again, this is readily programmed in Prolog+.
(d) An exact match is necessary before the object can be recognised. Later, we shall remove this requirement by using a different type of recognition rule which uses the concept of *similarity*.

Learning Program

Now, we can consider how the database of **object** facts might be obtained. The top-level learning predicate **learn_object** puts the familiar built-in predicate **assert** to good use.

```
% Top-level learning predicate

learn_object :-
      prompt_read(['Specify the name of the object'],X),
                                                  % Ask user for class name
      (object(_,_,_,_,_,X); yesno(['This is a new class of objects. Is this correct?])),
      learn(X),                                   % Learn new parameter values
      learn_object .                              % Repeat process

% Ask user whether learning has finished

learn_object :-
      not(yesno(['Do you want to finish?']),
      !,
      Learn_object.

% Finish nicely, by succeeding

learn_object.
```

```
% Do nothing if the object is correctly classified

learn(X) :-
    message(['Please present an object for viewing']),
    recognise(_,_,_,,X),
    message(['Object correctly classified, so no parameters were changed']),
    !.

% The object was not correctly classified, so modify the stored parameters

learn(X) :-
    get_features(A,H,D),                          % Recalculate the parameters
    yesno(['Parameters: ',[A,H,D],'Do you really want to modify the database
        relating to class' ,X,'?']),
    modify_database(A,H,D,X),
    !,

learn(_) :- !.
% A is less than lower acceptable limit for the area, A1

modify_database(A,H,D,X) :-
    object(A1,A2,H,D1,D2,X),                      % Consult database
    A < A1,                                       % Is A the new minimum area?
    assertz(object(A,A2,H,D1,D2,X)),              % Add new clause to db
    retract(object(A1,A2,H,D1,D2,X)),             % Remove old clause
    modify_database(A,H,D,X).                     % Modify other parameters

% A is greater than upper acceptable limit for the area, A2

modify_database(A,H,D,X) :-
    object(A1,A2,H,D1,D2,X),
    A > A2,                                       % Is A the new maximum area?
    assertz(object(A1,A,H,D1,D2,X)),
    retract(object(A1,A2,H,D1,D2,X)),
    modify_database(A,H,D,X).

% Add new clause to allow for different numbers of holes for same object class

modify_database(A,H,D,X) :-
    not(clause_exists(H,X)),                      % Do not duplicate clauses
    assertz(object(A,A,H,D,D,X)).

% D is less than lower acceptable limit for the "radius", D1

modify_database(A,H,D,X) :-
    object(A1,A2,H,D1,D2,C),
    D < D1,                                       % Is D new minimum "radius"?
    assertz(object(A1,A2,H,D,D2,C)),
    retract(object(A1,A2,H,D1,D2,C)),
    modify_database(A,H,D,X).
```

```
% D is greater than upper acceptable limit for the "radius", D2

modify_database(A,H,D,X) :-
        object(A1,A2,H,D1,D2,C),
        D > D2,                              % Is D new maximum "radius"?
        assertz(object(A1,A2,H,D1,D,C)),
        retract(object(A1,A2,H,D1,D2,C)),
        modify_database(A,H,D,X).

% Force successful termination of this goal. End recursion.

modify_database(_,_,_,_).

% Check to see whether clause "object(_,_,A,_,_,B)" exists for given A and B

% It does, so fail

clause_exists(A,B) :-
        object(_,_,A,_,_,B),
        !,
        fail.

% It does not, so succeed

clause_exists(_,_,_,_,_,_).
```

Practical Applications of Teaching by Showing

In order to illustrate the practical use of teaching by showing, first consider *lathe-turned* components which might have several different critical diameters (see Figure 7.32).

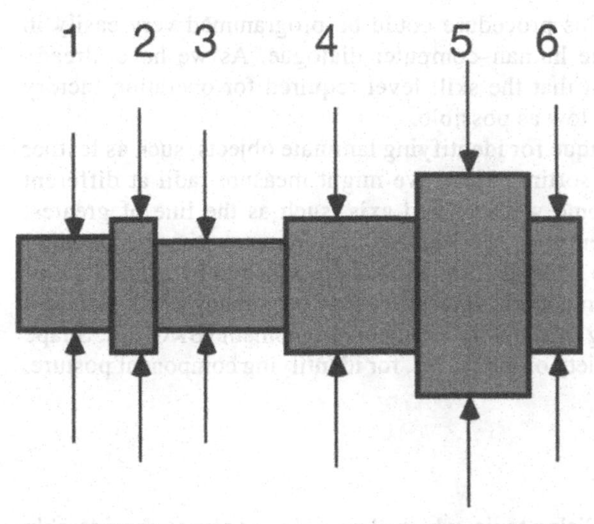

Figure 7.32. Silhouette of a hypothetical lathe-turned object. Critical diameters are indicated by the arrows.

It is important to the manufacturer and to his customer that the various diameters be measured and checked against *unknown* tolerance values. Rather than specifying these tolerance values by entering numeric data, the user of the inspection machine will select a set of known "good" components. These will then be shown to the machine, which will adjust its internal parameters in order to refine its recognition rules. The components are assumed to be presented for inspection in standard orientation and position. A simple and obvious *modus operandi* for such an inspection machine is as follows:

Learning Phase:

(i) The machine displays a silhouette of the *reference* component, with vertical bars superimposed on it.
(ii) The user adjusts the horizontal positions of these bars, with a keyboard or mouse-controlled cursor. These bars indicate the positions where the component diameters are to be measured. (Notice that the horizontal position of a lathe-turned object could vary by a small amount without upsetting the measurements. However, component skew is not allowable, in this simple scheme.)
(iii) The user initiates the learning procedure by simply pressing a single button, or using a pull-down menu.
(iv) The system measures the various diameters of the reference components and adjusts its database accordingly, as described above.
(v) Steps (i) – (iv) are repeated many times over.

Recognition Phase:

The user initiates the recognition phase, by pressing another button, or using a pull-down menu.

It is clear that both phases of this procedure could be programmed very easily in Prolog+. Notice the very simple human–computer dialogue. As we have already emphasised, it is very important that the skill level required for operating factory inspection systems should be as low as possible.

We might use a similar technique for identifying laminate objects, such as leather components for shoes, prior to sorting. Here, we might measure radii at different angular positions, relative to some well-defined axis, such as the line of greatest "radius" (measured from the centroid; see Figure 7.33). Once again, the learning procedure would have the same general form as that described above. The task of recognising shoe components is important, since there may be as many as 10^4 different types of leather shape circulating in a factory at any given moment [BRO-86]. Shape recognition is important, in a variety of industries, for identifying component posture, prior to robot manipulation.

Analysing Surface Texture

Texture is an interesting and tantalising topic, which almost always causes considerable vexation to industrial inspection staff, since it cannot be defined in a completely

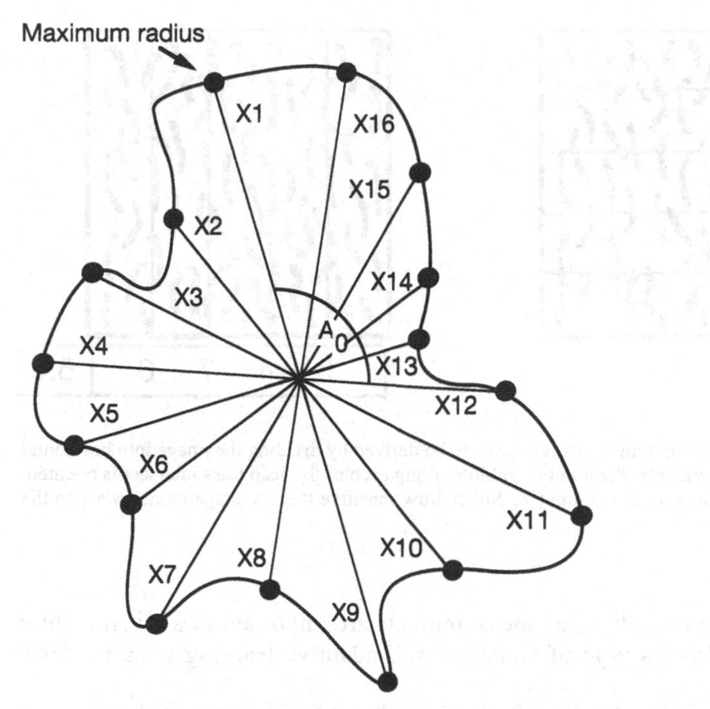

Figure 7.33. Radial coding of shape. A 16-element measurement vector [X1,X2,...,X16] might be derived, where Xj is the distance from the centroid to the edge of the blob, measured at an angle of $(22.5j + A_0)$ degrees, where A_0 is the orientation derived from the greatest radius (X1).

objective way. As a result, definitions of what constitutes "good" or "bad" texture tend to be based upon pictorial examples and loosely defined recognition rules. We all know that the texture of bread differs from loaf to loaf and that to some extent at least, texture indicates the degree of difficulty we will have when chewing it. It has been realised that the "crunchiness" of vegetables, such as carrots, can be predicted from the inter-cellular wall structure, which is visible under a microscope. Decorative woodwork is attractive because it has a certain type of grain, which is, of course, another term for texture. Microscopic texture variations affect the appearance of the paintwork on an automobile, or refrigerator body. Surface texture is of great importance when trying to judge the quality of finish on machine bearings, or other surfaces which are going to rub together. Texture is also of great significance in a range of other industrial products and processes, including a wide variety of domestic goods, processed foodstuffs, paper, knitwear, non-woven fabrics, foam, steel sheet, magnetic materials, materials/objects whose optical properties are important, sand-blasting, grinding, polishing, milling, etc.

Despite the lack of objective criteria for judging texture, it is often quite easy for an experienced inspector to look at a surface and qualitatively classify it as either "good" or "bad". Very often the inspector is able to discriminate several different classes of texture. It is relatively easy to define a multitude of different types of measurements which might be used collectively to characterise texture. Fourier and Walsh transforms offer two obviously attractive approaches to texture analysis. Some of the alternatives are listed below. It should be understood that it is often very

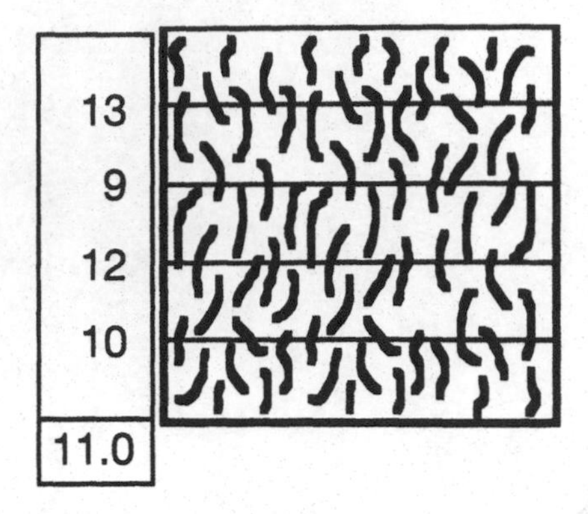

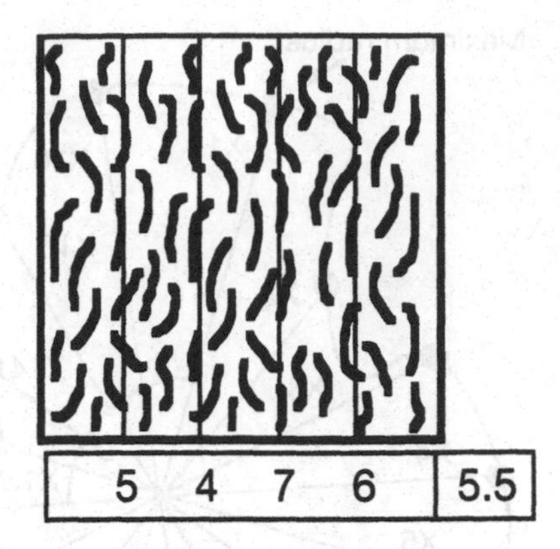

Figure 7.34. A set of simple texture measurements can be derived by dividing the image into horizontal (or vertical) strips. In this example, the number of blobs along each of the scan lines intersect is counted. The average of these numbers is also computed. Notice how sensitive these average numbers are to the direction of the texture.

difficult to be sure which individual measurements are important in a given texture discrimination task. In this type of situation, self-adaptive learning is particularly valuable.

The first task in most applications is to generate a binary image of the textured surface, given a grey-scale image digitised from the camera. This can be done in a variety of ways: high-pass filtering, followed by thresholding is particularly useful. An even better possibility in many applications is *local area histogram equalisation*, followed once again by thresholding (Figure 2.10). Once a binary image has been created, it is necessary to characterise it using some numeric, or even symbolic, descriptors such as those listed below (Figure 7.34).

(i) Mean of the lengths of horizontal runs of white pixels (run lengths)
(ii) Mean of the vertical run lengths
(iii) Ratio of (i) and (ii)
(iv) Mean blob area
(v) Mean blob perimeter
(vi) Mean of the ratio of the blob area, divided by the square of the blob perimeter
(vii) As (i) to (vi), after first applying expansion and shrinking operators
(viii) The mean number of joints in a unit area, after applying the skeletonisation operator
(ix) Intensity histogram of the image obtained by applying the grass-fire transform (this yields many descriptors)

It is not our intention to try and list all of the measurements that might prove to be important for texture analysis; we have merely listed a small representative proportion from amongst numerous other possibilities. All of the above and many more texture descriptors can be coded in Prolog+. However, the main point that we wish to emphasise here is that, whatever measurements are finally selected to characterise texture, there are likely to be many of them. A simple computational

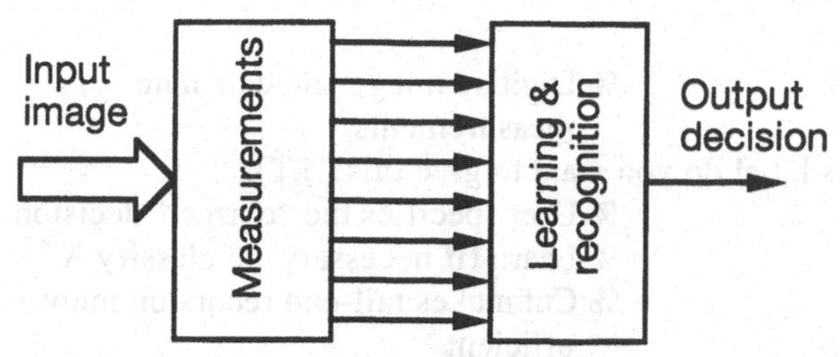

Figure 7.35. Archetypal pattern recognition system. Many measurements may be obtained and the output decision may be a binary or multi-state signal.

model may be envisaged, in which a large number of "low-cost" texture measurements are calculated and a learning program is used to analyse them (Figure 7.35). This *systematic* approach to texture analysis contrasts strongly with the rather haphazard approach to machine learning which **learn_object** embodies. This leads us into our next topic.

Another Classification and Learning Scheme

In this section, we shall touch on the much misunderstood subject called *Statistical Pattern Recognition*. This offers a number of well developed techniques for feature-set evaluation, classification, self-adaptive learning, cluster analysis and the estimation of probability density functions. The reader is referred elsewhere for a more complete description of both the particular techniques which we shall describe, and of the subject in general [BAT-78].

Figure 7.35 is an archetypal pattern recognition system and is applicable to many tasks in addition to texture analysis. We may assume that a large number of measurements have already been calculated (**measure**) and that what remains is to derive an N-ary decision that the image is from one of N classes. Many examples of these N classes are presented for analysis in random order, in the form of a training set of data, using which our learning program must design an appropriate decision strategy. In our next program, the user is asked to indicate the desired output for each image.

The following program implements the so-called *Store-when-wrong* rule for training a *Nearest Neighbour Classifier*.

```
% Top-level predicate

start_learning :-
    measure(X),                         % Digitise image and calculate measurements
    prompt_read(['What class label do you want to give this?'],T),
                                        % User specifies the "correct" decision
    assert(exemplar(X,T)),              % Remember the details prior to learning
    learning_cycle.                     % Enter learning cycle

restart_learning :-
    learning_cycle.                     % Continue learning after a coffee break
```

```
learning_cycle :-
    measure(X),                          % Digitise image and calculate
                                           measurements
    prompt_read(['What class label do you want to give this?'],T),
                                         % User specifies the "correct" decision
    learn(X,T),                          % Learn (if necessary) to classify X
    !,                                   % Cut makes tail-end recursion more
                                           efficient
    learning_cycle.                      % Repeat

learn(X,T) :-
    remember(nnc,[1000000.0, rubbish]),  % Initialise variables
    nnc(X,M),                            % Nearest neighbour classifier
    not(M = T),                          % Is decision by NNC wrong?
    assert(exemplar(X,T)),               % Add new exemplar to the database
    nl,
    write('Incorrectly classified. Exemplar added to database. [X,M,T] = '),
    write([X,M,T]).

learn(_,_) :-
    nl,
    write('Correctly classified – no exemplar added to the database').

nnc(X,_) :-
    exemplar(Y,M),                       % Consult database for a stored vector
    euclidean_distance(X,Y,0,D),         % D is distance between X and Y
    recall(nnc,[E,_]),                   % Find smallest distance so far
    E > D,                               % Is new exemplar closer?
    remember(nnc,[D,M]),                 % Store details of new closest neighbour
    fail.                                % Back-track to top of this clause

nnc(_,M) :-
    recall(nnc,[_,M]).                   % Find classification of nearest
                                           neighbour

start_recognition :-
    measure(X),                          % Digitise image and calculate
                                           measurements
    nnc(X,M),                            % Nearest neighbour classifier. Decision
                                           M
    message(['This is class',M]),
    !,                                   % Cut makes tail-end recursion more
                                           efficient
    start_recognition.                   % Repeat process

euclidean_distance([],_,A,B) :-
    sqrt(A,B).                           % Square root. Redundant calculation

euclidean_distance(_,[],A,B) :-
    sqrt(A,B).                           % Square root. Redundant calculation
```

```
euclidean_distance([A|B],[C|D],E,F) :-
        G is (A - C)*(A - C) + E,              % Sum of squares
        euclidean_distance(B,D,G,F).           % Repeat for all elements

% Simple image blob measurement procedure, appropriate for recognising letters

measure([A,B,C,D]) :-
        ctm,                                   % Let user see image from camera
        message(['Click on "OK" when you are ready to continue']),
        frz,                                   % Digitise image
        thr(128),                              % Threshold
        blb,                                   % Fill holes in white regions
        xor,                                   % Show holes as white blobs
        ndo,                                   % Shade blobs according to area
        thr(255),                              % Select biggest blob
        exw,skw,skw,exw,                       % Filter to reduce the effects of noise
        edg(0,4),                              % Mask edge of window
        dim(X1,X2,Y1,Y2),                      % Find dimensions of box around blob
        swi,                                   % Switch images
        DX is X2 - X1 + 1,
        DY is Y2 - Y1 +1,
        swc(8,X1,Y1,DX,DY),                    % Set window coordinates
        wgx,                                   % Wedge
        min,                                   % Mask it using the blob
        thr(192),                              % Select right-most quarter of blob
        cwp(A),                                % Count white pixels in right-most quarter
        swi,                                   % Switch images
        thr(128,191),                          % Select second quarter of blob
        cwp(B),                                % Count white pixels
        swi,                                   % Switch images
        thr(64,127),                           % Select third quarter of blob
        cwp(C),                                % Count white pixels
        swi,                                   % Switch images
        thr(1,63),                             % Select fourth quarter of blob
        cwp(D).                                % Count white pixels

/* Database format

exemplar([2199, 1853, 1235, 777], 'A').
exemplar([1914, 1382, 1292, 1664], 'K').
exemplar([1815, 2006, 1178, 1482], 'Q').  */
```

Notice that the concept of distance has been extended here into high-dimensional (actually four-dimensional) space. Moreover, the distance between two points in such a space is assumed to have an inverse relationship to the *similarity* between the events or patterns which those points represent. Hence, the Nearest Neighbour Classifier can be regarded as being equivalent to a *maximum similarity* classification rule. Since we cannot draw four-dimensional spaces, we must be content to explain the concept of a Nearest Neighbour Classifier using just two dimensions (see Figure 7.36).

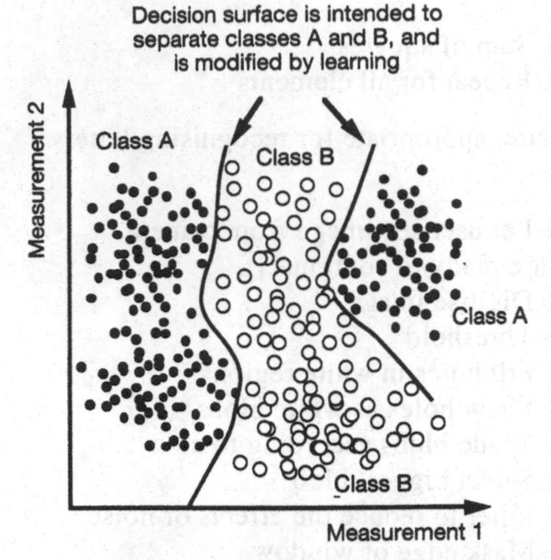

Figure 7.36. A pattern classifier consists of an analysis unit which calculates several/many measurements in parallel. In this simple case there are two. A two-way discrimination can be made by finding on which side of the Decision Surface a point falls. During learning, the position of the surface is iteratively adjusted, by altering the values of the parameters which define it.

Figure 7.37 shows a set of 48 handwritten characters, drawn without special care by the author and which were used to train a nearest neighbour classifier by the store-when-wrong rule. Although only seven samples of each of the eight character classes were included in the training set, the results were encouraging: a total of 12 exemplars were stored by the learning rule. Each class was represented by just one exemplar, except that class "6" required two exemplars and "A" required three. This represents an error rate of 8.3%, although further training is likely to yield even better results.

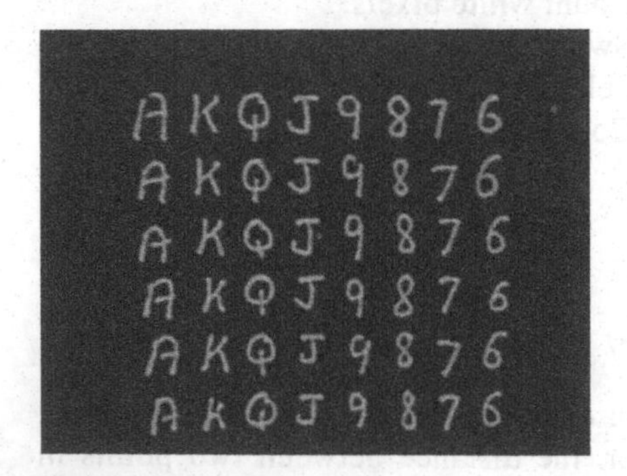

Figure 7.37. Forty-eight samples of letters, produced by the author in his normal handwriting (shown here about twice full size). A total of 16 measurements was used: the predicates **measure** and **measure1** in Section 7.8 were combined, with eight horizontal stripes being considered. The eight characters in the bottom row were used to initialise the nearest neighbour classifier. The remaining characters were shown to the classifier while it was learning by the store-when-wrong rule. When the resulting classifier was tested on a set of 48 characters, there were four misclassifications: one "6" and three examples of "A". Further learning would certainly improve this performance.

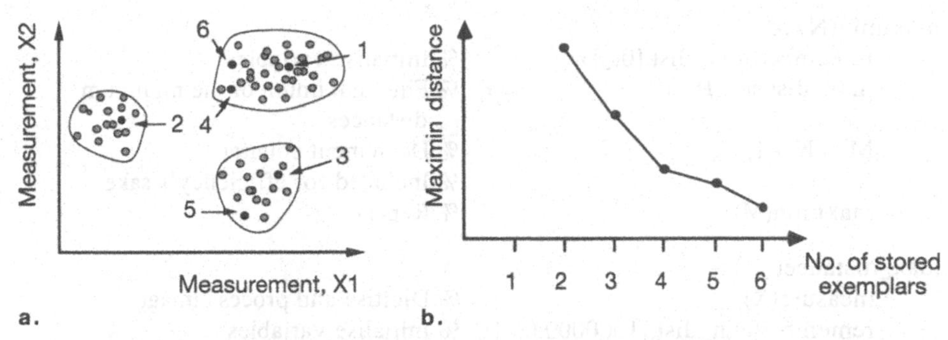

Figure 7.38. Cluster analysis and the **maximindist** procedure. (**a**) The measurement space. Clearly, there are three clusters. The points labelled 1 to 6 were found by **maximindist**, in that order. Notice that the first three points found (1, 2 and 3) are from different classes. (**b**) The maximin distance plotted against the number of points stored by **maximindist**.

The store-when-wrong learning rule is conceptually attractive and works best in those situations in which the pattern classes are well separated. It does not handle inter-class overlap at all well; more refined techniques are required then. This situation can be detected easily; the number of **exemplar** clauses in the database becomes large, while the classification accuracy does not improve significantly. The reader may wish to consider how this could be represented using Prolog+.

It is particularly easy to express the store-when-wrong learning procedure in Prolog+; the use of **assert** makes the extension of the database trivial. However, some of the other standard pattern recognition procedures use iterated parameter adjustment and hence are likely to be implemented more efficiently in other languages, especially those such as Occam, APL and Concurrent C which are intended to represent parallel processes.

Cluster Analysis

Let us briefly return to the task of identifying texture. Suppose that we are given a set of images showing texture which can be analysed, as mentioned above, by some unspecified procedure **analyse_image**. Our task now is to understand the nature of the variations in the texture. How many distinct types of texture are there? In other words, can we find clusters of points in the measurement space (Figure 7.38)? Can we find representatives of each cluster? One procedure which simultaneously answers both of these questions is called **maximindist**. The core of this procedure is coded below in Prolog+.

```
% Top-level predicate. N specifies the limit on the number of exemplars to be found

maximindist(N) :-
        measure(X),                  % Process image to get measurement vector X
        assert(exemplar(X)),         % Save as first exemplar
        maximin(N).                  % Obtain N exemplars

maximin(0).                          % Terminate recursion
```

```
maximin(N) :-
        remember(max_dist,[0,_]),            % Initialise variables
        max_distance(P,_),                   % Find maximum of the minimum
                                               distances
        M is N - 1,                          % Decrement counter
        !,                                   % Included for efficiency's sake
        maximin(M).                          % Repeat

max_distance(_,_) :-
        measure(X),                          % Digitise and process image
        remember(min_dist,[1000000.0,_]),    % Initialise variables
        min_dist(X,E,Z),                     % Find minimum distance, E, from X.
                                               Closest point is Z.
        recall(max_dist,[A,Y]),              % Find previous estimate of maximum
        A < E,                               % Is new value bigger?
        remember(max_dist,[E,X]),            % Retain new value
        fail.                                % Back-track until "measure" fails12

max_distance(P,Q) :-
        recall(max_dist,[P,Q]),              % Recover maximum distance
        nl,                                  % Information for the user
        write([P,Q]),
        assert(exemplar(Q)),                 % Store vector as an exemplar
        nl,
        write('Maximindistance ='),
        write(P).

min_dist(X,_,_) :-
        measure(Y),                          % Digitise and process image
        euclidean_distance(X,Y,0,D),         % Calculate distance (D) from X to Y
        recall(min_dist,[E,_]),              % Find previous estimate of minimum
                                               distance
        E > D,                               % Is this smaller than before?
        remember(min_dist,[D,Y]),            % Remember value of distance and vector
        fail.                                % Back-track until "measure" fails

min_dist(_,A,B) :-
        recall(min_dist,[A,B]).              % Return minimum value and vector
```

The reader is referred to the literature for a description of the interpretation of the results of running **maximindist** [BAT-78]. Let it suffice to say that it can estimate the number of clusters and find a representative of each one.

Another task is that of deciding whether a given pattern is likely to belong to the normal (i.e. familiar, not Gaussian) distribution that an inspection machine encounters. This is teaching by showing in a different guise. To illustrate the point, consider on-line monitoring of the texture of some sheet-like product. A human inspector may be able to stand beside the inspection machine for a short while, as it learns. To indicate that the surface texture on the sheet is "good", he holds a button down. (He releases it if he detects any defects, but these are not used at all by the learning machine.) After a few minutes he releases the button, the machine stops learning and takes over the inspection process. The human operator then goes onto other work. During the teaching-by-showing process, the machine analyses the surface texture on the web and uses the very simple learning rule explained below. The recognition procedure is a slight variant of the nearest neighbour classifier.

```
learning_normality :-
    maximindist(15),                          % Initialise database with 15 exemplars
    remember(dist,0),                         % Initialise variable
    1000•(measure(X,D),                       % Cycle 1000 times...
    recall(dist,E),                           % ... and find the ...
    ((D > E, remember(dist,D))) ; true).      % ... maximum distance

normal :-
    measure(X),                               % Digitise and analyse an image
    min_distance(X,E),                        % Find minimum distance from X to
                                                exemplars
    recall(dist,F),                           % Look up maximum permissible
                                                distance
    E < F,                                    % Is F within prevously learned limits?
    !,                                        % Makes tail end relationship more
                                                efficient
    normal.                                   % Repeat until non-normal pattern is
                                                found
```

This procedure was applied to optical character recognition (see Table 7.1 and Figure 7.39).

Table 7.1. Learning on a single class. **maximindist** was used to select 15 exemplars from a training set of 64 samples of the character "8" (Figure 7.39). A nearest neighbour technique was then used to calculate a score for six examples of each of the characters {6, 7, 8, 9, J, Q, K, A}. The maximum-minimum ranges of this score were then calculated. The maximum value of the score for the class "8" was found (72.02). Only one character (a "6") produced a score lower than this value. In other words, we have devised a classifier which can discriminate class "8" from class "not-8".

Character	Range of distances	Number with score < min. score for "8" (6 maximum)
6	67.30 – 109.65	1
7	95.31 – 121.50	0
8	28.96 – 72.07	–
9	73.08 – 99.81	0
J	99.88 – 128.14	0
Q	159.87 – 196.07	0
K	79.26 – 105.21	0
A	99.27 – 153.76	0

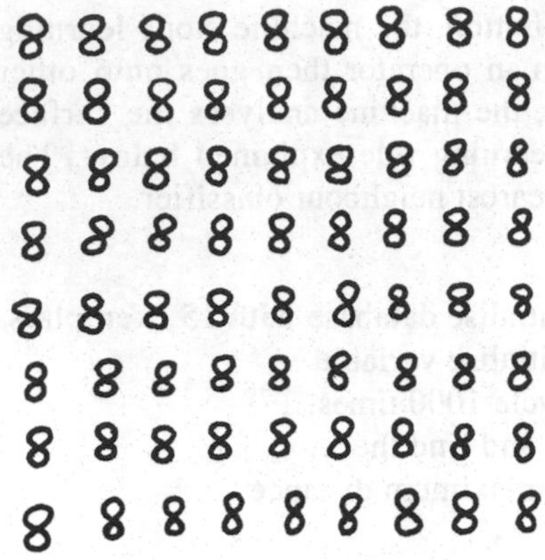

8 8 8 8 8 8 8 8 8
8 8 8 8 8 8 8 8 8
8 8 8 8 8 8 8 8 8
8 8 8 8 8 8 8 8 8
8 8 8 8 8 8 8 8 8
8 8 8 8 8 8 8 8 8
8 8 8 8 8 8 8 8 8
8 8 8 8 8 8 8 8 8

Figure 7.39. Examples of hand-printed numerals used in the experiment to learn the characteristics of the class "8".

Alternative Methods of Decision Making and Learning

Another Prolog program for learning is described and listed in full by Ross [ROS-89]. It is not necessary to discuss this program's finer points in detail, in order to understand what it does and how it might be applied to industrial image processing. The program is able to learn a logical expression (i.e. a concept) that can be written in the form of a *Disjunct of Conjuncts*. (Electrical engineers are more familiar with the equivalent term *Sum of Products*, which is commonly used in digital circuit design.) We shall explain what this means later, but first we need to develop some suitable mathematical notation.

A set, S, of patterns (i.e. images) is used to train the learning program. S is subdivided into two mutually exclusive subsets, which we shall call S_Y and S_N, where

$S = S_Y + S_N$

and

$S_Y \cdot S_N = \emptyset$

($\emptyset$ is the empty set, "+" denotes logical **OR** and "·" denotes **AND**.) The members of S_Y are associated with the decision **YES**, while those in S_N are associated with the decision **NO**. The ith member of S will be denoted by X_i and is represented by a set of descriptors:

$X_i = \{X_{i,1}, X_{i,2}, X_{i,3}, \ldots, X_{i,N(i)}\}$

Notice that N(i), the number of descriptors used to represent X_i is a function of i. Furthermore, each of the $X_{i,j}$ is a member of one of several sets, each of which is organised in an hierarchical manner. For example one of the $X_{i,j}$ might take the value associated with any one of the nodes in the tree shown in Figure 7.40. Another descriptor, say $X_{i',j'}$, might assume one of the values associated with one of the nodes

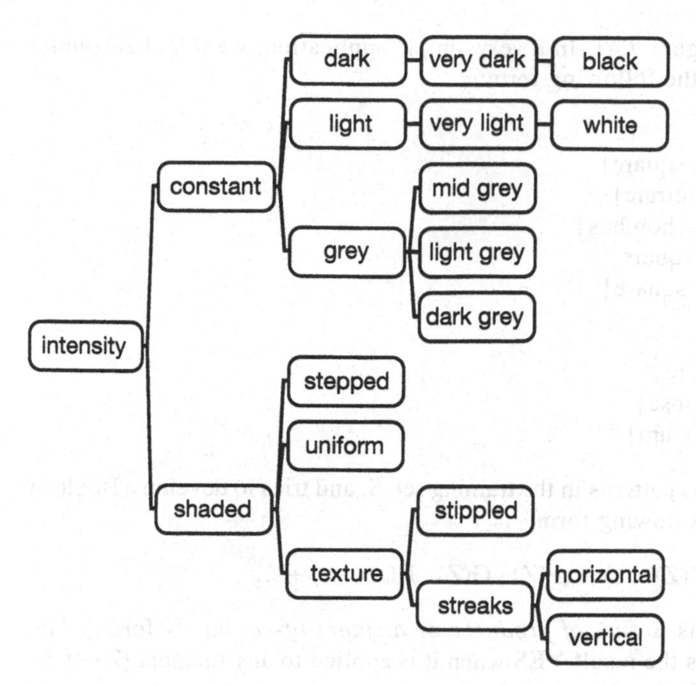

Figure 7.40. Hierarchical structuring of the concept *"intensity"*.

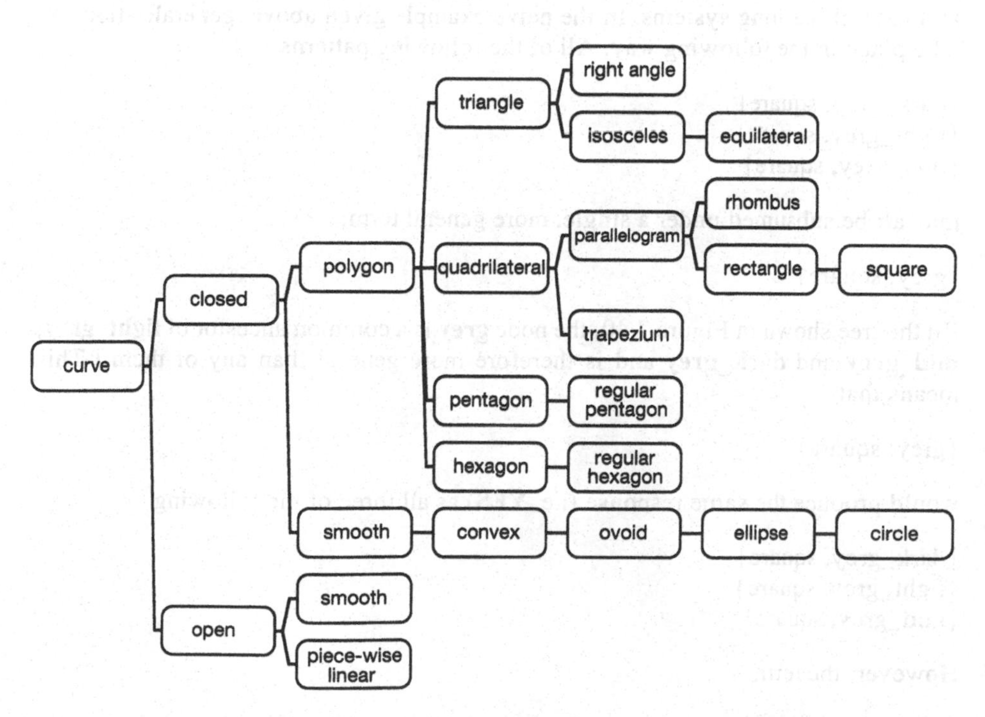

Figure 7.41. Describing a binary image in a format suitable for Ross's learning program.

in the tree shown in Figure 7.41. In a very simple application, we might encounter pattern descriptions of the following form:

Subset S_Y

{light_grey, square}
{dark_grey, circle}
{light_grey, rhombus}
{mid_grey, square}
{dark_grey, square}

Subset S_N

{white, square}
{texture, ellipse}
{dark, trapezium}

The program analyses patterns in the training set, S, and tries to develop a Boolean function, Q(Z), of the following form

$$Q(Z) = (A(Z) \cdot B(Z) \cdot C(Z) \cdot \ldots) + (F(Z) \cdot G(Z) \cdot H(Z) \cdot \ldots) + \ldots$$

(This expression has a *sum-of-products* or *disjunct-of-conjuncts* form.) The function Q(Z) generates the result **YES** when it is applied to any member (Z) of S_Y or **NO** for members of S_N.

The program described and listed by Ross can *generalise*. As the reader will no doubt appreciate, the ability to generalise is an important function of both machine and natural learning systems. In the naive example given above, generalisation can take place in the following way. All of the following patterns

{dark_grey, square}
{light_grey, square}
{mid_grey, square}

may all be subsumed under a single, more general term:

{grey, square}

(In the tree shown in Figure 7.40, the node **grey** is a common ancestor of **light_grey**, **mid_grey** and **dark_grey** and is therefore more general than any of them.) This means that

{grey, square}

would produce the same response (i.e. **YES**) as all three of the following

{dark_grey, square}
{light_grey, square}
{mid_grey, square}

However, the term

{constant, square}

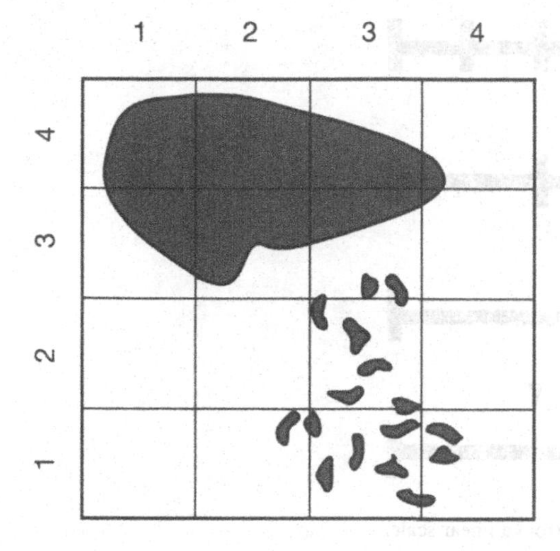

Format :

[[X1, Y1], No. of blobs, Area of white], ...]

Description of this image :

[[[1, 3], 1, 35], [[1, 4], 1, 70], ..., [[4, 1], 3, 5]]

Figure 7.42. Creating an artificial hierarchy from a linear scale.

is too general and would not produce the correct response for
{white, square}

The learning procedure implemented in the program listed by Ross has two important features:

(a) It can generalise.

(b) Generalisation takes place as a direct result of the hierarchical organisation of the descriptor values. Without that hierarchical organisation, learning would be no more efficient than that achieved by the simpler program, **learn_object.**

Hence, to make good use of the program which Ross has developed, we need to find applications in which there is some natural hierarchical structuring of the descriptor values. Finding appropriate industrial image processing applications which have these properties has not been easy. It is possible to *impose* a hierarchical structuring in certain applications. For example, we may create an *artificial* tree-like structuring by associating together spatial or other numeric measurements, as we shall show in the following example.

Let us now assume that X_i is the ith picture in a series and let $X_{i,j}$ be a measurement derived from the jth region of X_i. (See Figure 7.42, where j is a two-element list.) Notice, however, that these regions can be chosen quite arbitrarily and may, or

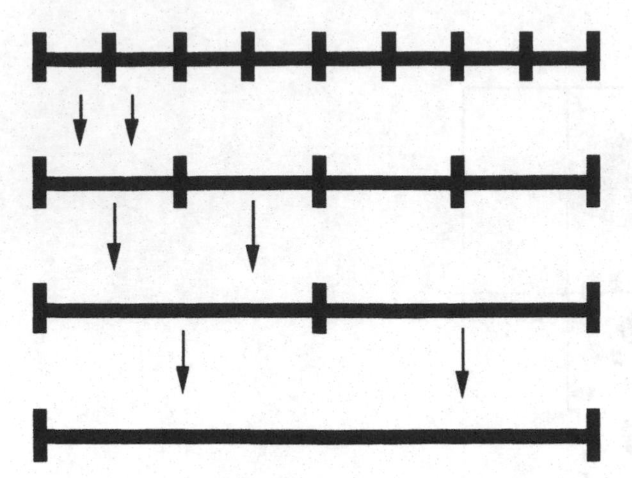

Figure 7.43. Creating an artificial hierarchy from a linear scale.

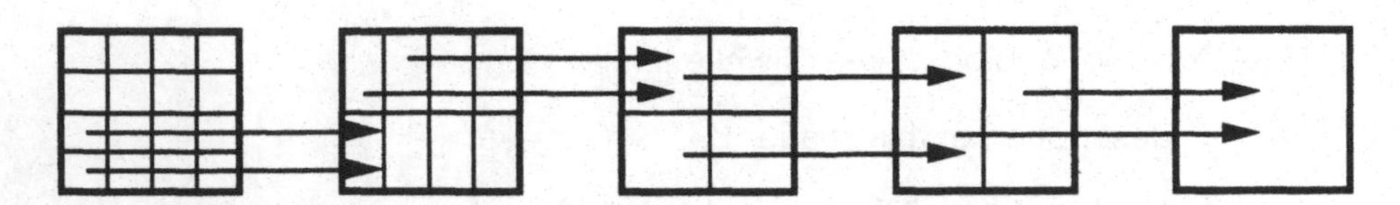

Figure 7.44. Creating an artificial hierarchy from a two-dimensional field, i.e. an image.

may not, overlap. In order to fix our thoughts on something tangible, we might consider computing $X_{i,j}$ in the following way:

```
compute_descriptor(I,[P,Q],[[P,Q],M,N]) :-
        rea(image(I)),          % Read the Ith image from disc
        raf(11,11),             % Low-pass filter
        sub,                    % Subtract images
        thr(135),               % Threshold
        cwp(M),                 % Count white points
        eul(N).                 % Count distinct blobs
```

Notice that this computes descriptors in the form of a list, [[P,Q],M,N], which implicitly indicates the region identity (i.e. its "address" [P,Q]), the number of blobs in that region (N) and their total area (M). The reason for including J is that we create a descriptor whose range of possible values is different from that which can be derived from all other regions. (This is required by the program which Ross lists.) Figure 7.43 explains a naive method for constructing a hierarchy from a continuous linear scale. (The region identity is ignored simply to make understanding easier.) It is also possible to create a hierarchy from an image by grouping together neighbouring regions (Figure 7.44).

7.10 In Conclusion

There has been a deliberate attempt to cover as wide a range of topics as possible in this chapter. The reason is that the author has tried to demonstrate that Prolog+ is readily applicable and natural to use on a broad spectrum of tasks requiring a combination of intelligent, inferential decision-making and image processing. The study of any topic is made infinitely easier if there is a suitable language in which ideas can be expressed to communicate ideas both to other people and to a computer. It is mooted that Prolog+ is a language suitable for both of these functions, when tasks involving intelligent image processing are concerned.

There has been a great deal of research work applying (standard) Prolog to a number of classical AI tasks, such as planning and learning. This has led to some ingenious programs, exemplified by Warplan and Ross's program [ROS-89]. A major role for Prolog+ is in allowing programs such as these to be applied to image processing; Prolog+ forms a bridge between what has hitherto been abstract research and the practical needs of industrial vision systems.

Notes

1. This procedure is rather inefficient, since it normalises the position and orientation of the card. It would require far less computational work to sample the intensity along a vector whose parameters are calculated using **lmi**. However, we have already devised a means for normalising the card position and orientation, so it is easier to explain the concept in these terms.
2. In order for this method to work effectively, the picture cards must be quite different from one another; traditional playing cards may not be sufficiently different. (Recall that they show a series of very similar stylised pictures of 16th century European noble-men and women.) However, a more modern picture card might simply contain large bold icons.
3. Although the list [A,B] has two items, there is only one remaining block to be moved. The reason is that the identity of the previous block taken from the list (A) is always kept for reference purposes.
4. This is a very much more complicated task in practice, since there are many different constraints upon which advertisements can/cannot appear at a given time and within a given television programme.
5. A parallel situation arose with the computer language APL, whose development was prompted by K.E. Iverson's dissatisfaction with conventional modes of expressing ideas using mathematical notation.
6. An ad hoc style of programming is clearly evident in this section. No apology is made about this, since it is very often found that rigour leads to excessive complication in the program structure. A good working rule is that, if a simple technique works, use it. Always avoid over-complication. Testing becomes even more important when programming heuristics, rather than algorithms.
7. Such a limitation is imposed by the camera/optical system, not by the analysis procedure.
8. Notice that **search_maze** asserts a set of facts in the database, while the simplification of blind alleys requires list notation. Clearly, it is a trivial matter to modify **search_maze**, or write a simple conversion program.
9. The robot and cutting techniques are being developed by The Institute of Engineering Research, Agriculture & Food Research Council, Wrest Park, Silsoe, Bedford, U.K.
10. A similar technique of storing a set of waveforms (sinusoids) in a single image was used in the Autoview system for calculating the spectral response of image filters.
11. It is a straightforward matter to accommodate a less well controlled lighting situation in which the background intensity is not constant. However, this is a distraction from the theme of this section.
12. This is a slightly different function from that defined earlier.

CHAPTER 8

ADVANCED INSPECTION TECHNIQUES

"Gweledydd yw sail pob gwybodaeth"
(Vision is the basis of all knowledge.)
"Nid byd byd heb wybodaeth"
(The world is no world without knowledge.)
Welsh Proverbs

8.1 What is "Advanced Inspection"?

To date, Automated Visual Inspection has concentrated almost exclusively on mass-produced objects, because the high cost of designing an inspection system has made it uneconomic to apply these techniques to any other type of artifact. However, there are several possible applications areas for industrial vision systems that have been almost totally neglected to date and which we shall address in this chapter:

(i) Objects made in small quantities. (It has been estimated that 75% of manufactured goods are made in batches of 50, or fewer, items.)

(ii) Complex (monolithic) objects. Examples are car engine blocks, castings, mouldings and currency notes.

(ii) Inspecting assemblies of components (e.g. pianos, printed circuit boards, electric motors, hair-dryers, in-flight meal trays, computer keyboards, etc.).

(iv) Mass-produced goods which are deliberately made with a high degree of individual variation, to simulate their being hand made. (There has always been a high premium paid for hand-made objects, such as wood carvings, hand-thrown pottery, embroidery, painted china, etc. While it is a relatively straightforward matter to manufacture goods with the same level of variability that hand-made goods possess[1], it is not yet feasible economically to inspect

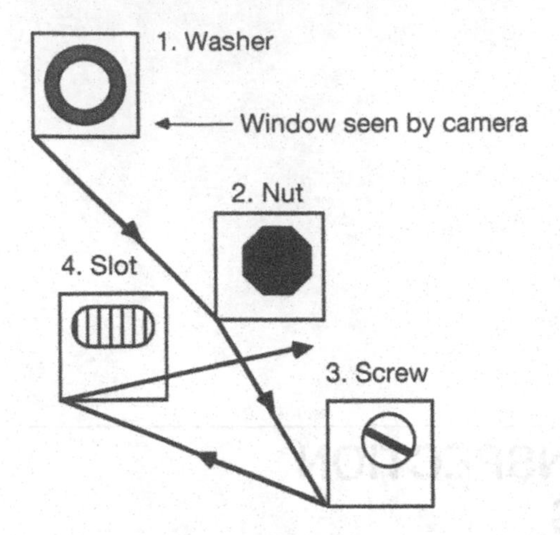

Figure 8.1. Step-by-step inspection of a complex object. Such a procedure might be followed using a Flexible Inspection Cell.

them automatically. If this did come about, then "individual character", similar to that so prized in hand-made goods, could be added to many more mass-produced artifacts.)

(v) Objects which have, by their very nature, a high degree of variability. Good examples of this group are to be found in the food manufacturing industry. (Pizzas are like fingerprints – no two look exactly alike!)

(vi) Objects which are either flexible or consist of an assembly of jointed parts.

(vii) Natural objects which also exhibit a high degree of variability and which have ill-defined quality criteria.

So far, Automated Visual Inspection has had virtually no impact on the manufacture/processing of these products, for the simple reason that it has just not been cost-effective. We hope to change this, using AI techniques. These can be used to assist in the analysis of the images, controlling feature search patterns and in a number of other ways. We shall also discuss the programming of vision systems using natural language.

The Flexible Inspection Cell (FIC, Figure 3.1) was devised especially for inspecting low-volume products and complex objects. Indeed, the motivation for developing Prolog+ came, in large part, from the need that exists for controlling such a cell and for inspecting complex and small batch products. One possible approach to inspecting the various categories of artifacts listed above is to employ a robot to move either the camera or the object to a sequence of positions/postures. At each one, a single object feature is examined. The sequence of moves may be either pre-defined, or calculated as the inspection operation takes place. We shall consider both possibilities in this chapter. Figure 8.1 represents this process of step-by-step inspection in diagrammatic form. Notice that each view requires its own special lighting conditions, optical and camera set-up etc., as well as a specially designed/selected image processing algorithm. As we have explained in several places in this monograph, this is exactly what Prolog+ was designed to do.

We shall consider three major topics under the general heading which forms the title of this chapter. These may, at first sight, seem to be unrelated, but they are, in fact, closely inter-linked. The major topics that we shall discuss are:

(a) *Model-based inspection*, for use when there exists a model, drawing or other specification of what an object or scene should look like.
(b) *Rule-based inspection*, in which the experience of a human inspector can be encapsulated in a set of simple rules.
(c) *Declarative programming*, in which the programmer specifies what the vision system can expect to see. The user specifies its program using standard English.

The fact that these are related and share many of the same ideas will become apparent as we progress in our discussion.

Model-Based Inspection

The term model-based inspection could be interpreted in a number of different ways. The intended use in this chapter is rather specific. To understand the concept, consider the emblem associated with the Olympic Games:

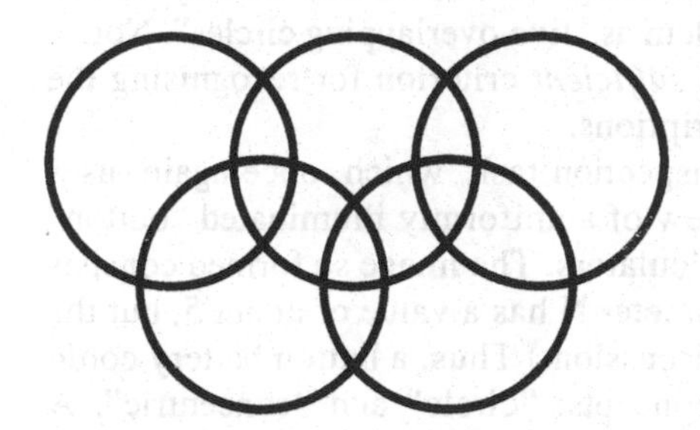

The reader will, no doubt, be aware that this consists of

"five overlapping circles".

By using this description of the Olympic emblem, the author has made use of three of the reader's pre-existing mental models:

(a) The number "5"
(b) The idea of two objects overlapping
(c) The word "circle"

The compound term "five overlapping circles" clearly permits certain patterns but does not allow others. Hence, even if the reader had no prior knowledge about the form of the Olympic emblem, he/she could use the description given above to classify a given picture into one of two classes:

"Possibly Olympic emblem"
or
"Not Olympic emblem".

Figure 8.2. Grey-scale image of Bakewell tarts. Compare this with Figure 2.21.

Of course, the description "five overlapping circles" is imprecise but it could be made progressively more exact, by adding further constraints on the relative sizes and positions of the circles. We have a mental image of a circle and we rely on this knowledge when we define the Olympic emblem as "five overlapping circles". Notice that the description is a *necessary* but not a *sufficient* criterion for recognising the Olympic emblem. This is typical of AI descriptions.

Let us turn our attention now to a real inspection task, which, once again, uses the concept of a circle. Consider the plan view of a uniformly illuminated "button" battery, such as those used in watches and calculators. The image so formed consists of a series of N concentric circles. (The parameter N has a value of about 5, but the exact value is unimportant for our present discussion.) Thus, a button battery could be inspected by making use of only two concepts: "circle" and "concentric". A Bakewell tart (Figure 8.2; also see Figure 2.21), the top of a can of baked beans, a countersunk hole and the mouth of a bottle can also be defined in the same way. We shall return to these ideas again in Section 8.6, where we shall explain how we can use models like circle, rectangle, etc. to describe images.

Let us consider another application, namely that of identifying resistors on a PCB. A (low-power) resistor may be defined as consisting of

(i) a rounded, approximately rectangular body, whose length and width are within defined limits.

(ii) two leads, one at each end of the body.

A lead may, in turn, be defined as:

(iii) a grey, approximately linear feature, whose width and length are within a defined range.

Notice the hierarchical structure in the model of a resistor. Clearly, ICs, capacitors etc. can be defined in a similar way. Hence, a PCB appears as a rectangular grey region (R) in an image, which contains an unknown number of ICs, resistors, capacitors, etc. This definition of a PCB nicely illustrates the concept of model-based inspection.

Model-based inspection, in a completely different guise, is already in use by certain machines that have been designed for examining non-populated printed circuit boards. Some machines, such as that developed by Applied Vision Systems, Inc.[2] use the board design data to adjust the inspection algorithm. At the moment, this is limited to "fine tuning" the inspection algorithm; the number of iterations of a thinning procedure is adjusted to take into account the varying widths of conductors on different boards. What we mean to imply in this chapter by the term *model-based inspection*, normally operates at a higher conceptual level, than merely adjusting the value of one or a few parameters.

Rule-Based Inspection

Rule-based inspection techniques may be used when a human inspector is able to express his/her knowledge in the form of a number of IF...THEN... rules. Representing knowledge in this way can be useful in a number of application areas. For example, grading fruit and vegetables has traditionally been based upon a set of rules, which relate a quality index to colour, shape, splitting, insect damage, etc. Later, we shall present rules for recognising carrots, bananas and Bakewell tarts.

Rule-based methods have also been studied in relation to inspection of PCB solder joints and multi-layer hybrid circuits. It is not unreasonable to expect rule-based inspection to be used in a host of other applications where there is a high degree of variability of the objects to be inspected. The list of possible applications areas for rule-based inspection includes:

(i) Food products (cakes, pizzas, loaves, etc.[3])
(ii) Timber (grading for different uses)
(iii) Leather (grading for use on different parts of a shoe)
(iv) Decorative plants (potted rhododendrons, cyclamen, etc.)
(v) Mineral products (gems, decorative facing stone, marble, etc.)

8.2 Performing a Pre-Defined Sequence of Inspection Operations

There is, in most instances, far too much detail for a whole printed circuit board to be represented in a single stored digital image. The same is true of a panel for a car door, refrigerator, washing machine, etc. Furthermore, it may be necessary to use quite different lighting arrangements for the various parts of a complex object. A lighting arrangement that is optimal for one feature of a complex object may produce severe shadows or glinting for another. This is typical of most complex objects; a large number of digital images have to be acquired and processed, in order to examine

Table 8.1. Arguments for the view predicate.

Variable	Function
N	View number
F	Type of feature being viewed
L	List specifying the lighting configuration
X	X-position of the (X,Y,θ)-table
Y	Y-position of the (X,Y,θ)-table
T	θ-position of the (X,Y,θ)-table
C	Camera identification
P	Image processing predicate to be performed

all relevant aspects of it. Indeed, this may be taken to be an operational definition of a complex object.

It is very easy to write a Prolog+ program which can operate as a sequence controller for an FIC. In the program which we shall present in a moment, the moves are assumed to have been stored beforehand in the database, in the following form:

```
view(N,F,L,X,Y,T,C,P).
```

where the variables have the meanings indicated in Table 8.1.

The program to interpret and operate upon a sequence of facts of this form can, of course, be very simple:

```
inspect :-
    get_next_object,                          % Deliver next object to be inspected
    view(N,F,L,X,Y,T,C,P),                    % Consult database
    move_robot_to(X,Y,T),                     % Move the (X,Y,θ)-table
    information(N,F,L,X,Y,T,C,P),             % Information for the user
    csl(C),                                   % Select appropriate camera
    (call(P),write('PASS')); write('FAIL')),  % Perform test report result
    nl,
    fail.                                     % Back-track to "view"

inspect :- inspect.                           % Repeat inspection on next object

% Information for the user

information(N,F,L,X,Y,T,C,P) :-
    nl,
    write('View no:'),
    write(N),
    nl,
    write('This is a'),
    write(F),
    write('and is located at'),
    write([X,Y,T]),
    nl,
    write('Camera'),
    write(C),
```

```
    write('was used and the lighting vector was'),
    write(L),
    nl,
    write('The inspection procedure used was'),
    write(P),
    nl,
    write('Result of the inspection process:').
```

Setting up the Database

The database of **view** facts required by **inspect** could be set up quite conveniently using appropriate pull-down menus, or by a specially programmed dialogue, written in Prolog+. The ideas are very simple, as will soon become apparent.

As we saw in Chapter 3, a set of lights, an (X,Y,θ)-table and an input video multiplexer can all be operated using pull-down menus. This facility makes it possible to make remote adjustments to the lighting and viewing conditions, to obtain a good view of a given feature.[4] Once the image acquisition conditions have been optimised, it is necessary to record the complete state of the FIC. This can be determined using **state** which relies upon three lower-level predicates:

```
state(X,Y,T,L,C) :-
    where_is_robot(X,Y,T),          % An obvious extension to Chapter 6
    light_state(L),                 % Instantiate L to lighting vector
    camera(C).                      % Find currently selected camera
```

Although we have not described them in detail, the predicates **where_is_robot**, **light_state** and **camera** operate in a fairly obvious way. Using **state**, an *interactive* method for building up the database required by **inspect** can be developed:

```
build_database :-
    yesno(['To set up another inspection task, adjust the table, camera and
    lights before clicking on "YES", otherwise select "NO".']), 5
    genint(step_number,N),          % Increment integer by 1
    state(X,Y,T,L,C),               % Find present state of the FIC
    prompt_read(['What type of feature is this?'],F),
    prompt_read([Specify test needed for this feature'],P),
    assertz(view(N,F,L,X,Y,T,C,P)), % Add fact to database
    build_database.                 % Generate rest of sequence

build_database :-
    yesno(['Do you want to enter the inspection cycle now?']),
    inspect.

build_database.
```

Of course, it would be sensible to incorporate facilities like **build_database** and **inspect** into the pull-down menus. (The menus "Utility" or "Macro" appear to be most appropriate.)

The image processing predicates are assumed to have been programmed beforehand. (They can of course, be programmed afterwards, provided that **inspect** is not invoked first.)

8.3 Scan and Search Patterns

So far, we have considered only fixed inspection sequences. Let us now turn our attention to consider inspection sequences which cannot be predicted beforehand. An obvious requirement is for some mechanism for searching for named features. For example, it may be necessary to look for an edge, a hole, bolt, or some other nominated feature on the object that is to be examined. We shall therefore discuss search patterns next.

There are several important feature-search methods which are useful when searching for features in a scene or an image:

(a) Extreme-feature location
(b) Raster scan
(c) Random scan
(d) Space-filling scan (e.g. Peano scan)
(e) Coarse–fine scan search
(f) Edge/contour following
(g) Navigation by landmarks
(h) Inspecting all … on every … of a …
(i) Hill climbing

We shall now briefly discuss each of these in turn.

Extreme Feature Location

Extreme-feature location is one of the most common, and conceptually simplest, types of search used in model-based inspection. A typical task is that of finding the right-most edge point of an object silhouette. Once a binary image has been created, the primitive operator **dim** will locate all four extremal points.

Another example of a search pattern of this type is that of finding the position of the brightest/darkest point in an image. This is straightforward in Prolog+:

```
brightest_point(X,Y) :-
        gli(_,Z),               % Find maximum intensity
        thr(Z),                 % Threshold at this level
        ndo(1,_),               % May be more than one blob, so separate them …
        thr(255),               % … and find the top-left-most blob
        cgr(X,Y).               % Find its centroid
```

A third example of an extreme-feature search pattern is that of finding the position of the largest hole in the silhouette of an object. Here is the Prolog+ program:

```
biggest_hole(X,Y) :-
    blb,                 % Fill holes
    xor,                 % Exclusive OR identifies the holes
    ndo(3,_),            % Shade according to area
    thr(255,255),        % Select the largest one
    cgr(X,Y).            % Find its centroid
```

Yet another example is provided by the following program which tries to find the blob that is closest to a given point, [X1,Y1].[6] The coordinates [X2,Y2] of the centroid of the blob so identified are then computed.

```
closest_blob(X1,Y1,X2,Y2) :-
    hic(X1,Y1),          % Draw intensity cone centred at (X1, Y1)
    min,                 % Use input image as a mask
    gli(_,Z),            % Find intensity of brightest pixel
    thr(Z),              % Threshold at this level
    ndo(1),              % Shade blobs
    thr(1,1),            % Select one of them
    cgr(X2,Y2).          % Find its centroid
```

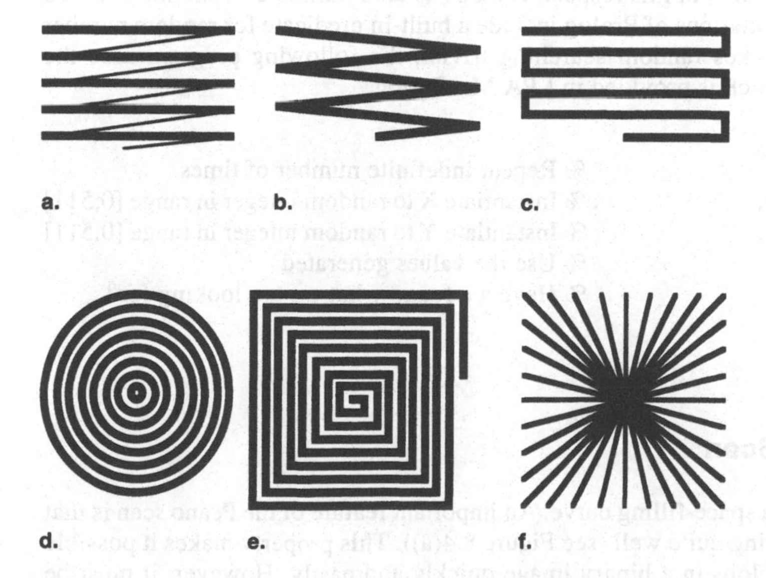

Figure 8.3. Common scan patterns. (**a**) Left-to-right, top-to-bottom with rapid "flyback". This is suitable for electronic scanning, where rapid flyback is possible. (**b**) Backwards and forwards, with steady progress in the orthogonal direction. This is suitable for scanning by a slow electro-mechanical device, such as a robot or (X,Y)-table. (**c**) As (**b**), but with incremental motion along the "slow" axis. (**d**) Concentric circles. (**e**) Spiral. The spiral may be square or circular, linear or logarithmic. (**f**) Wheel-spoke.

Raster Scan

Raster scanning is important because it provides the basis of the majority of systematic search procedures, and clearly is the most useful in those situations in which there is no, or very little, prior knowledge about the location of a feature of interest in an image. There are six important raster scan patterns, each of which has different merits and uses (see Figure 8.3). Generating raster scan patterns like these is trivial in Prolog+, as it is in nearly all other computer languages. For this reason we shall not dwell on this topic further.

Random Scan

Random scanning is, of course, usually sub-optimal as a search strategy but if there is no prior knowledge, it may well be very effective at locating large objects in an image or scene. Random scanning is, of course, the basis of Monte Carlo estimation techniques and it is possible to perform mathematical integration of numerically defined functions, using random search patterns. Remember that the result of the integration becomes more accurate as time progresses. This is useful in many heuristic methods which require a quick, if crude, preliminary result and a more refined one later.

It is, of course, possible to write pseudo-random number generators in almost any computer language and, in this respect, Prolog+ is unremarkable. It should be noted that some implementations of Prolog include a built-in predicate for random number generation This makes random searching trivial; the following program uses the predicate **irand** which is provided in LPA MacProlog.

```
random_scan :-
    repeat,             % Repeat indefinite number of times
    irand(512,X),       % Instantiate X to random integer in range [0,511]
    irand(512,Y).       % Instantiate Y to random integer in range [0,511]
    use(X,Y),           % Use the values generated
    test.               % Have we found what we are looking for?
```

Space-Filling Scan

The Peano scan is a space-filling curve. An important feature of the Peano scan is that it preserves clustering quite well (see Figure 8.4(a)). This property makes it possible to find the larger blobs in a binary image quickly and easily. However, it must be appreciated that this procedure is non-algorithmic. An equivalent search pattern can be defined which performs a three-dimensional scan (Figure 8.4(b)).

The two-dimensional Peano scan is most easily defined using recursion and hence can be expressed very conveniently in Prolog+.

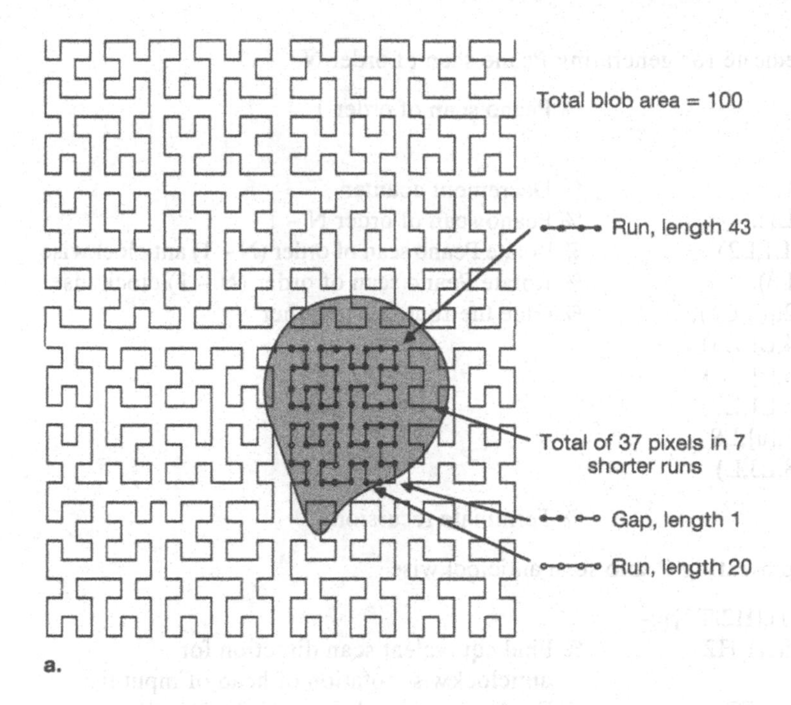

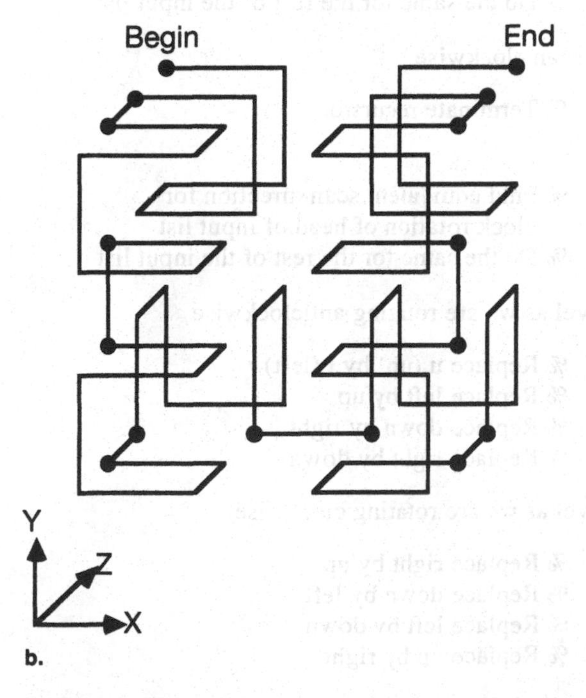

Figure 8.4. Peano scan. (**a**) two-dimensional scan. Notice that the Peano scan preserves clustering; once the scan curve enters a compact blob, it usually stays there for some considerable time. (This is called a *run*.) In this instance, there are 100 pixels within the blob. The two longest runs, of lengths 20 and 43 pixels, are separated by a gap of only one pixel. Shorter runs have a total length of 37 pixels. (**b**) three-dimensional scan.

```
% Top-level predicate for generating Peano scan of order N

peano(1,[d,r,u]).               % Peano scan of order 1

peano(N,L) :-
     M is N – 1,                % Decrement counter
     peano(M,L1),               % Peano scan of order N – 1
     anticlock(L1,L2),          % Rotate Peano scan of order (N – 1) anticlockwise
     clock(L1,L3),              % Rotate Peano scan of order (N – 1) clockwise
     append(L2,[d],L4),         % Glue the four bits together
     append(L4,L1,L5),
     append(L5,[r],L6),
     append(L6,L1,L7),
     append(L7,[u],L8),
     append(L8,L3,L).

anticlock([],[]).               % Terminate recursion

% Rotate list representing Peano scan anticlockwise

anticlock([H1|T1],[H2|T2]) :-
     rotate_anti(H1,H2),        % Find equivalent scan direction for
                                  anticlockwise rotation of head of input list
     anticlock(T1,T2).          % Do the same for the rest of the input list

% Rotate list representing Peano scan clockwise

clock([],[]).                   % Terminate recursion

clock([H1|T1],[H2|T2]) :-
     rotate_clock(H1,H2),       % Find equivalent scan direction for
                                  clock rotation of head of input list
     clock(T1,T2).              % Do the same for the rest of the input list

% What to do at the elemental level as we are rotating anticlockwise

rotate_anti(u,l).               % Replace u (up) by l (left)
rotate_anti(l,u).               % Replace left by up
rotate_anti(d,r).               % Replace down by right
rotate_anti(r,d).               % Replace right by down

% What to do at the elemental level as we are rotating clockwise

rotate_clock(r,u).              % Replace right by up
rotate_clock(d,l).              % Replace down by left
rotate_clock(l,d).              % Replace left by down
rotate_clock(u,r).              % Replace up by right
```

The Peano scan is only one of the many space-filling curves that could be used for scanning. It is a particularly convenient one, because it fills a *rectangular* space and uses movements that are always parallel to the coordinate axes of the space. Moreover, the resolution of the scanning pattern is controlled by a single parameter.

Coarse/Fine Scan Search

Suppose that we know that small objects are likely to be clustered together in the space that is to be searched. This is often the case, for example, with defects on a web product (e.g. sheet steel, plastic, cloth, etc.), although the general principle of clustering is much more common than this.

A sensible search heuristic when clustering is likely to occur is to use a coarse scan to find one object and then hunt nearby for others. This is the basis of the coarse/fine scan search method. An alternative reason for using coarse/fine scan is, of course, to examine objects in varying detail. For example, a car engine block might be located by using a low-resolution view of a wide scene. Then, a high-resolution view might be employed to look for debris in its various holes, exhaust ports etc.

Coarse/fine scanning can be achieved in several ways:

(i) *Mechanical* – a coarse initial search by a robot is followed by a finer one.

(ii) *Optical* – the camera zoom lens is set to "out" initially. When an object has been located, the lens is adjusted to zoom in, to examine the area of interest in greater detail.

(iii) *Camera* – Certain types of camera, notably the image dissector and certain types of solid-state sensor, offer random access. Such a camera is supplied with an (X,Y)-address and returns a number representing the light intensity at that point. With suitable scanning software, a random access camera can perform coarse/fine scanning, or edge/streak following, which is discussed below.

(iv) *Software* – a high-resolution stored image is sampled initially to give a low-resolution image which is then analysed. After an object has been found, a small high-resolution window is selected from the main image for more detailed scrutiny.

Here is just the top level of a program for searching by zooming in and out.

```
zoom_search(U,V) :-
    pan_and _tilt(middle _of_scene),  % Set camera pan and tilt position
    zoom(out),                        % Obtain low resolution view of the scene
    frz,                              % Digitise image
    analyse_image1,                   % Analyse image, to obtain single blob
    dim(A,B,C,D),                     % Find blob size and position
    P is 0.5*(A + C),                 % Find centre of rectangle enclosing blob
    Q is 0.5*(B + D),                 % Find centre of rectangle enclosing blob
    pan_and _tilt(P,Q),               % Shift camera to [P,Q]
    R is C – A,                       % Find size of rectangle
    S is D – B,                       % Find size of rectangle
    ((R > S, Z is R); (Z is S)),      % Instantiate Z to larger of R and S
    zoom(Z),                          % Set zoom to obtain higher resolution view
    frz,                              % Digitise again
    analyse_image2(U,V).              % More detailed analysis than before
```

A popular idea that is widely used in image processing is that of *quad-trees* [SCH-89]. A quad-tree is a series of images derived from the same initial image and of resolution:

$2^n \times 2^n$ pixels,
$2^{n-1} \times 2^{n-1}$ pixels,
$2^{n-2} \times 2^{n-2}$ pixels,
...,
$2^1 \times 2^1$ pixels,
$2^0 \times 2^0$ pixels.

The basic idea is that decisions obtained on a low-resolution image can guide the processing of higher resolution images. The objective is to avoid undertaking slow/costly image processing operations, unless we can be sure that they will be worthwhile. Clearly, this is an area in which Prolog+ should be able to find a niche, but so far the author has not had the opportunity to explore this exciting possibility.

Edge/Contour Following

In many industrial inspection problems, images are to be analysed which contain few small-area features, such as linear or gently curved arcs. These often arise from edges, cracks or scratches.[7] An image of a scratch, for example, may contain a total of say 512 × 512 pixels (262 144 pixels). If the background is of nearly uniform brightness, there may be only a few hundred pixels whose intensities differ significantly from the average value. Clearly, a great deal of processing effort could be saved if it were possible to concentrate upon such features as these, by a process of following edges, contours or streaks, as appropriate.

However, there may be other advantages, the most important of which is that certain (serial) processes can be implemented, which would not otherwise be possible. For example, an edge might appear "broken" after an edge detector and thresholding (**sed, thr**) have been applied to an image. In order to "join up" the disconnected segments, to form a complete edge contour, a serial process of the type described below could be used. This has no counterpart in parallel processing, although the Hough and Radon transforms [SCH-89] might arguably achieve a similar result. The technique uses a procedure called *Island Hopping*. There are two parts to this process:

(a) Following segments of arc which do not contain breaks
(b) Jumping over breaks, between the extant portions of a broken arc

The first of these makes use of a small window, called a Region Of Interest (or ROI). Processing is limited to the ROI and this may employ parallel algorithmic operations. Consider Figure 8.5, which shows a ROI "sitting astride" a skeletonised figure. The first part of the process analyses the intensities of those pixels within the ROI, to find where to place the next ROI, which will then be analysed. Here is the essence of a program to do this:

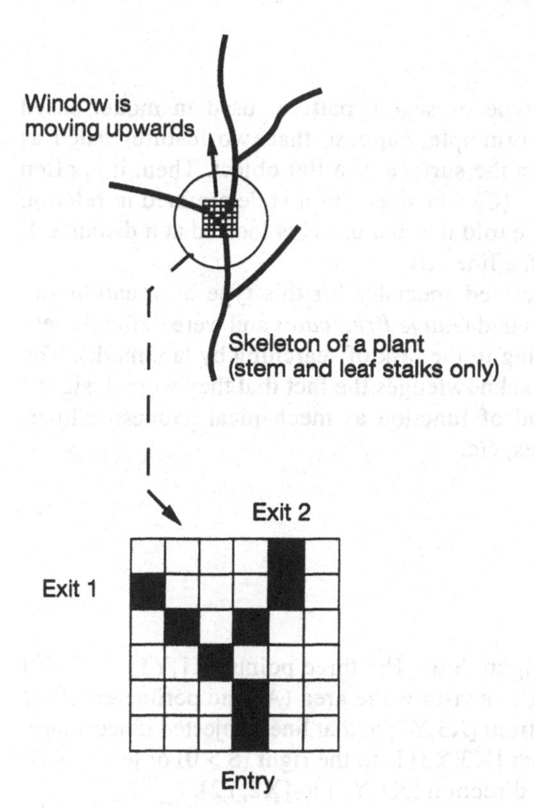

Figure 8.5. Using a region of interest (ROI), (i.e. a small processing window) to decide where to place the next ROI. In this way, it is possible to follow the stem of a plant and to locate joints quickly. Various rules can be formulated which allow the ROI to trace the main stem. (This idea was first suggested by the author's colleague, Mr I. P. Harris).

```
roi_mover(M,N,X1,Y1) :-
    swc(15,X1,Y1,M,N),                  % Window is at [X1,Y1] and has M × N
                                        pixels
    exits(L),                           % List of points where skeleton leaves ROI
    select_exit(L,X2,Y2),               % Select most appropriate member of L
    offset_adustment(X2,Y2,N,X3,Y3),    % Find position of next ROI
    !,                                  % Avoid back-tracking
    not(test_finished),                 % Has the process finished yet?
    roi_mover(M,N,X3,Y3),               % Repeat until done

roi_mover(_,_,_,_).                     % Terminate recursion on a note of
                                        success
```

A program which uses these principles and traces the stem of an open-structure plant has been written by one of the author's colleagues.[8] This program also locates the points where leaf stalks join the main stem (i.e. where the axial buds can be found) and automatically labels the points where the leaf stalks are attached to the leaves.

The remainder of the Island Hopping procedure will be explained later.

Navigating by Landmarks

This is perhaps the most common type of search pattern, used in model-based inspection. To illustrate the general principle, suppose that two features, such as holes (A and B), have been located on the surface of a flat object. Then, it is often found that there will be a third feature (C) whose position is determined in relation to A and B. We might, for example, be told that feature C is located at a distance X along a line inclined at Y degrees to the line AB.

A series of predicates has been defined specially for this type of situation and these are described below. These are called *Gauge Predicates* and were defined specifically to augment Prolog+ by assisting in the task of searching by landmarks. The choice of the name Gauge Predicates acknowledges the fact that they were designed specifically to perform the same kind of function as mechanical gauges: caliper, protractor, ruler, compass, plug gauges, etc.

8.4 Gauge Predicates

triangle(X1,Y1,X2,Y2,A,P,S) (see Figure 8.6). The three points [X1,Y1], [X2,Y2] and [X3,Y3] define a triangle. **triangle** measures the area (A) and perimeter (P) of the triangle and finds the distance (|S|) from [X3,Y3] to that line, projected if necessary. In addition, sign(S) determines whether [X3,Y3] is to the right (S > 0) or left (S < 0) of this line, as we look along it in the direction [X1,Y1] to [X2,Y2].

compass(X1,Y1,R,Theta,X2,Y2) (see Figure 8.7). A circle of radius R is centred at [X1,Y1]. [X2,Y2] is the address of the first white pixel encountered as we move anticlockwise from the point [X1+ R.cos(Theta), Y1+ R.sin(Theta)]. The four "input" argu-ments (X1,Y1,R,Theta) must all be instantiated when calling **compass.**

balloon(X1,Y1,R,X2,Y2) (see Figure 8.8). A circle of radius R is centred at [X1,Y1]. If there are no white points on this circle, then its radius is increased by a small amount and the process is repeated, recursively. The procedure stops when a white pixel is found and its address [X2,Y2] is returned. If there are two or more white pixels on the same circle, then the one forming the smallest angle with the horizontal is used to calculate these parameters.

Notice that **balloon** performs a very similar operation to that achieved by **closest_blob**, which was defined earlier.

caliper(X1,Y1,Theta,L,D) (see Figure 8.9). The simplest way to describe this gauge predicate is to imagine a robot picking up an isolated blob-like object. The centre of the gripper is placed at [X1,Y1]. The gripper has two parallel jaws of length L (L ≥ 0). The gripper opens by moving the jaws along a line inclined at angle Theta to the hori-zontal. The separation of the jaws as they close to grip the object is given by D. Clearly, there are two possible modes of operation of **caliper**, depending upon whether or not the object can slip over the table top as it is being picked up. (The robot arm is assumed to be absolutely rigid.) If the object slips as it is being picked

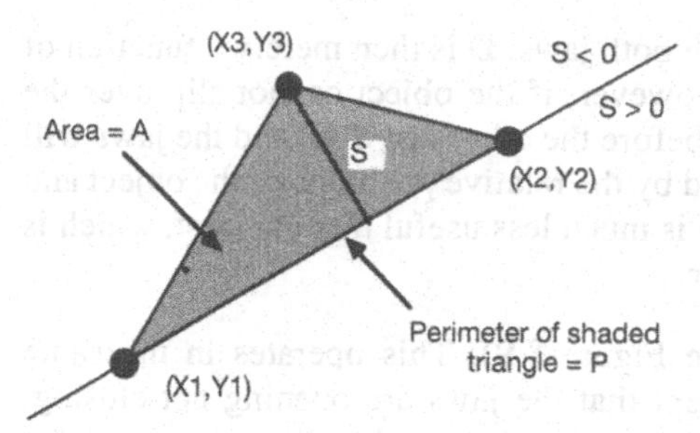

Figure 8.6. The **triangle** predicate.

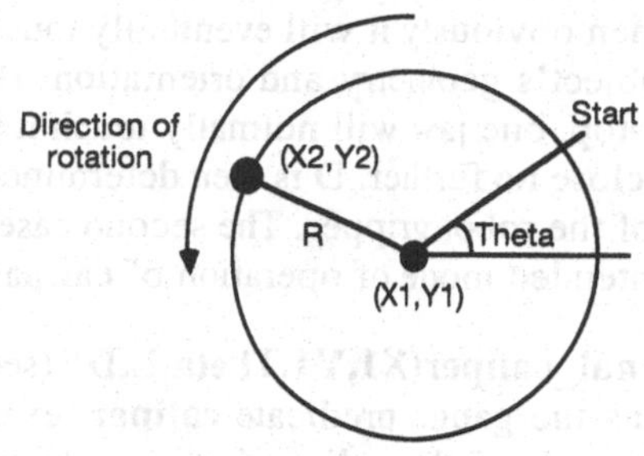

Figure 8.7. The **compass** predicate.

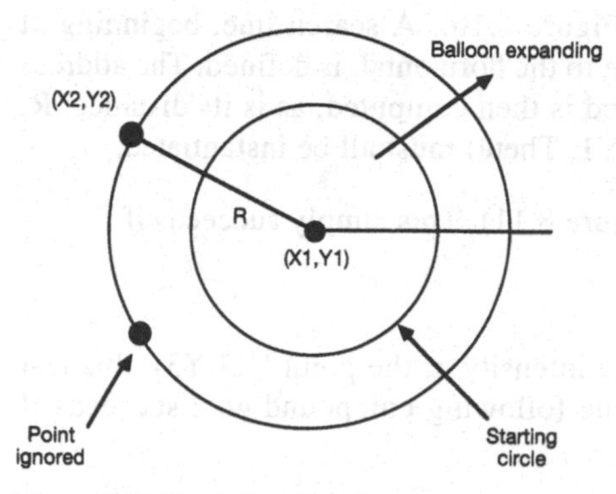

Figure 8.8. The **balloon** predicate.

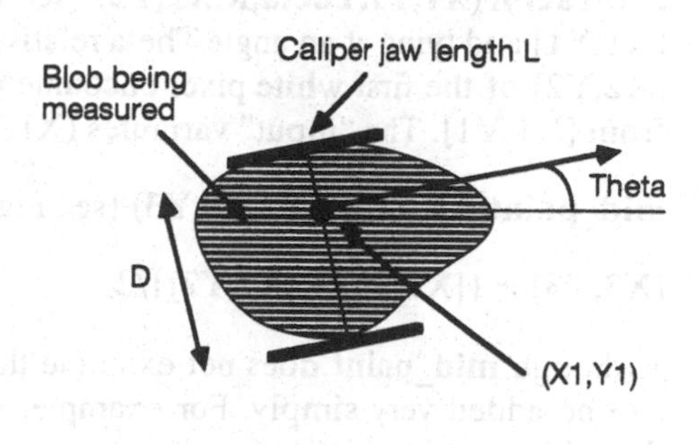

Figure 8.9. The **caliper** and **internal_caliper** predicates.

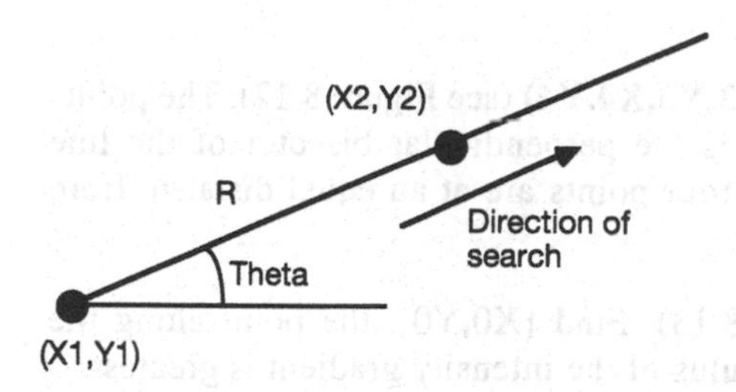

Figure 8.10. The **protractor** predicate.

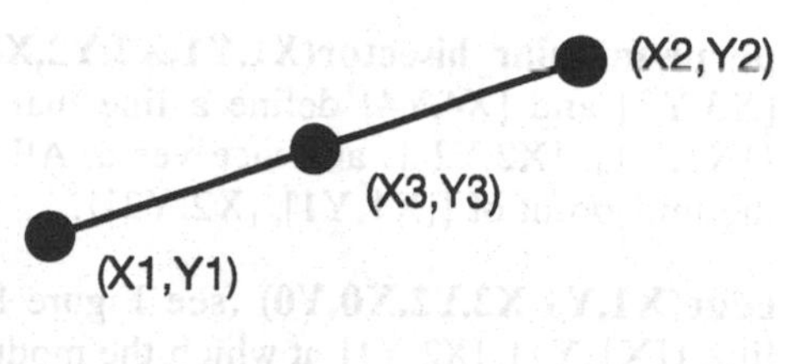

Figure 8.11. The **mid_point** predicate.

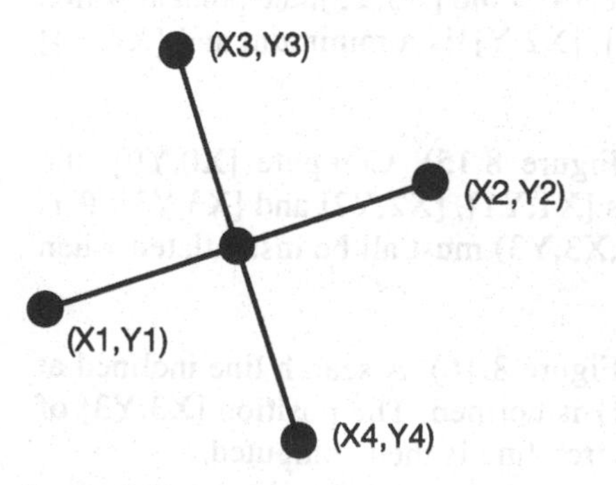

Figure 8.12. The **perpendicular_bisector** predicate.

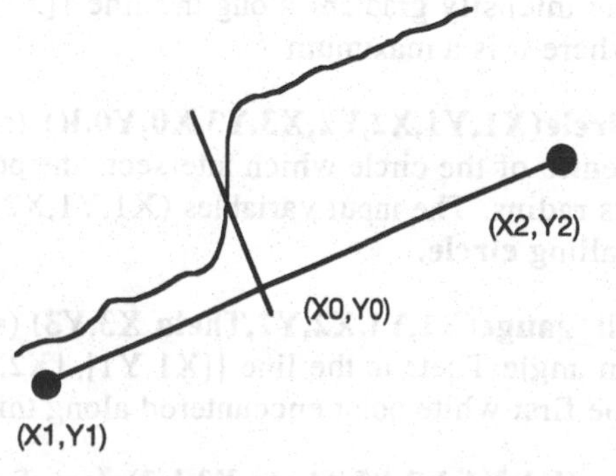

Figure 8.13. The **edge** predicate.

up, then obviously it will eventually touch both jaws; D is then merely a function of the object's geometry and orientation. However, if the object cannot slip over the table-top, one jaw will normally touch it before the other one does and the jaws will then close no further. D is then determined by the relative positions of the object and that of the robot gripper. The second case is much less useful than the first, which is the intended mode of operation of **caliper**.

internal_caliper(X1,Y1,Theta,L,D) (see Figure 8.9). This operates in the same way as the gauge predicate **caliper,** except that the jaws are opening not closing. The length of the caliper jaws can be set to zero. **internal_caliper** is useful for examining holes.

protractor(X1,Y1,Theta,R,X2,Y2) (see Figure 8.10). A search line, beginning at [X1,Y1] and lying at an angle Theta relative to the horizontal, is defined. The address [X2,Y2] of the first white pixel encountered is then computed, as is its distance, R, from [X1,Y1]. The "input" variables (X1,Y1, Theta) must all be instantiated.

mid_point(X1,Y1,X2,Y2,X3,Y3) (see Figure 8.11). This simply succeeds if

[X3,Y3] = {[X1,Y1] + [X2,Y2]}/2

Although **mid_point** does not examine the intensity at the point [X3,Y3], this test can be added very simply. For example, the following compound goal succeeds if the mid-point is white

```
mid_point(X1,Y1,X2,Y2,X3,Y3),
pgt(X3,Y3,255)
```

perpendicular_bisector(X1,Y1,X2,Y2,X3,Y3,X4,Y4) (see Figure 8.12). The points [X3,Y3] and [X4,Y4] define a line that is the perpendicular bisector of the line {[X1,Y1], [X2,Y2]}, and vice versa. All four points are at an equal distance from the mid-point of {[X1,Y1], [X2,Y2]}.

edge(X1,Y1,X2,Y2,X0,Y0) (see Figure 8.13). Find [X0,Y0], the point along the line {[X1,Y1], [X2,Y]} at which the modulus of the intensity gradient is greatest.

gap(X1,Y1,X2,Y2,X3,Y3,X4,Y4) (see Figure 8.14). Find [X3,Y3], the point at which the intensity gradient along the line {[X1,Y1], [X2,Y]}is a minimum and [X4,Y4] where it is a maximum.

circle(X1,Y1,X2,Y2,X3,Y3,X0,Y0,R) (see Figure 8.15). Compute [X0,Y0], the centre of the circle which intersects the points [X1,Y1], [X2,Y2] and [X3,Y3]. R is its radius. The input variables (X1,Y1,X2,Y2,X3,Y3) must all be instantiated when calling **circle**.

ell_gauge(X1,Y1,X2,Y2,Theta,X3,Y3) (see Figure 8.16). A search line inclined at an angle Theta to the line {[X1,Y1], [X2,Y2]} is defined. The position [X3,Y3] of the first white point encountered along this search line is then computed.

fan(X1,Y1,X2,Y2,Theta,X3,Y3) (see Figure 8.17). This is identical to balloon,

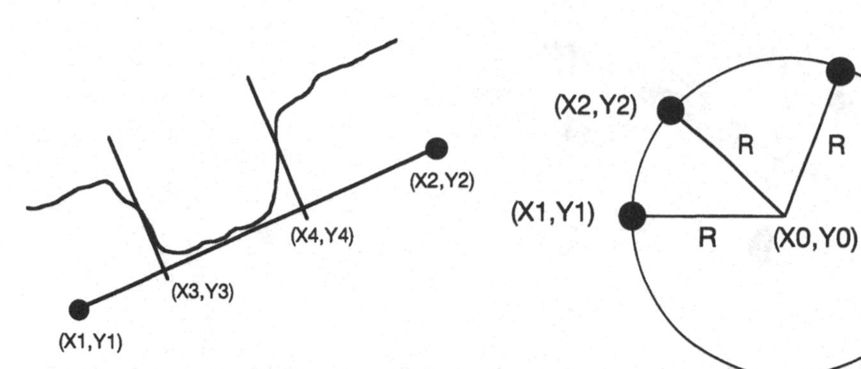

Figure 8.14. The **gap** predicate.

Figure 8.15. The **circle** predicate.

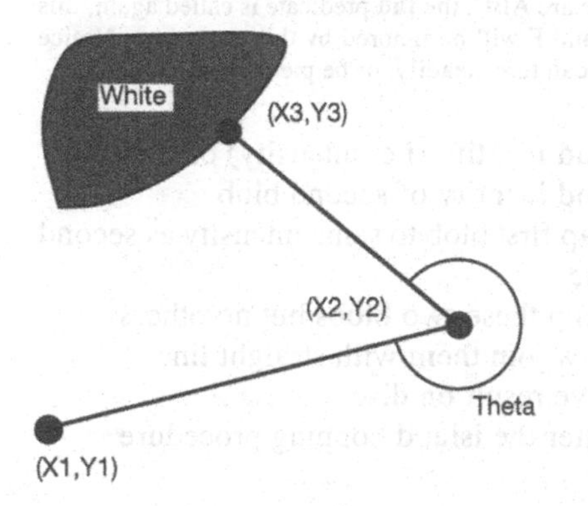

Figure 8.16. The **ell_gauge** predicate.

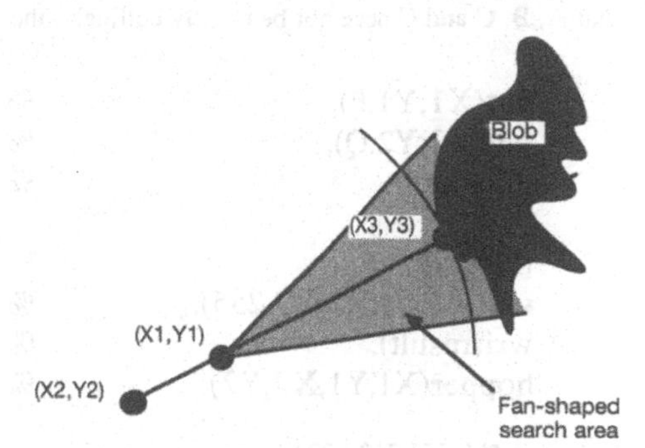

Figure 8.17. The **fan** predicate.

except that the search is restricted to a wedge-shaped region (the fan angle) of angular width Theta.

Applications of Gauge Predicates

We shall describe just two applications: the first one uses **fan** to reconstruct "broken" arcs, while the second employs **caliper** to measure the silhouette of a carrot.

The **fan** predicate may be used to good effect in joining together "broken" lines or arcs, provided that they do not change direction suddenly (Figure 8.18). Here is the kernel of the program:

```
island_hopping :-
    frz,                                 % Digitise an image
    convert_to_binary,                   % Obvious function but not defined here
    ndo,                                 % Shade blobs. ≤ 255 blobs allowed in image
    wri(temp),                           % Save on disc
    choose_first_island(X1,Y1),          % [X1,Y1] is its centroid
    get_nearby_island(X1,Y1,X2,Y2),      % [X2,Y2] is its centroid
    rea(temp),                           % Read shaded image back from disc
```

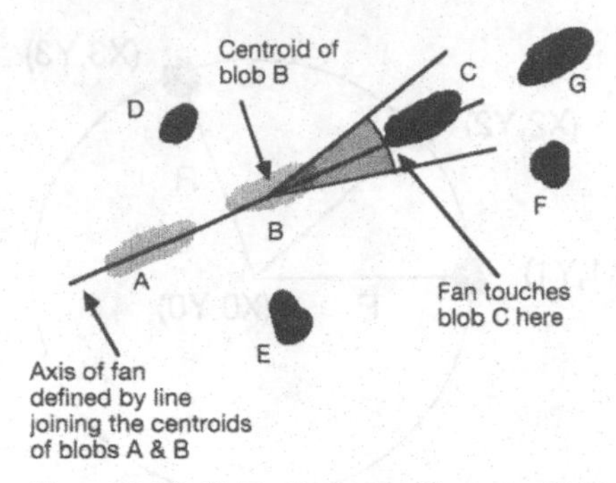

Figure 8.18. Using the **fan** predicate for "Island Hopping". Blobs A and B have already been identified as lying on a straight line. The **fan** predicate is then used to locate another blob, C, which is approximately collinear with A and B. To find G, the next blob along the arc ABC, the **fan** predicate is called again, this time using blobs B and C as "inputs". The blobs D, E, and F will be ignored by this procedure. Notice that A, B, C and G need not be exactly collinear; the arc can turn steadily, or be piece-wise linear.

```
        pgt(X1,Y1,P),                       % Find identity (i.e. intensity) of first blob
        pgt(X2,Y2,Q),                       % Find identity of second blob
        hil(P,P,Q),                         % Map first blob to same intensity as second
                                              one
        thr(Q,Q),                           % Keep these two blobs but no others
        vpl(X1,Y1,X2,Y2,255),               % Now join them with straight line
        wri(result),                        % Save result on disc
        hopper(X1,Y1,X2,Y2).                % Enter the island hopping procedure

hopper(X1,Y1,X2,Y2) :-
        fan_angle(F,D1),                    % Consult db for parameters F and D1
        fan(X1,Y1,X2,Y2,F,X3,Y3),           % Find another blob, if there is one
        distance([X2,Y2],[X3,Y3],D),        % Compute distance to new blob
        D ≤ D1,                             % Arbitrary distance threshold
        rea(temp),                          % Read shaded image
        pgt(X3,Y3,Z),                       % Discover identity of new blob
        thr(Z,Z),                           % Now isolate that blob
        cgr(X4,Y4),                         % Calculate its centroid
        rea(result),                        % Read intermediate result image
        max,                                % Add new blob to it
        vpl(X2,Y2,X4,Y4,255),               % Join new blob to string
        wri(result),                        % Save image on disc
        !,                                  % Included for efficient recursion
        hopper(X2,Y2,X4,Y4).                % Repeat

% "fan" failed or distance to new blob was too great

hopper(_,_,_,_) :-
        rea(result).                        % Read result image

fan_angle(20,17).                           % Fan angle is 20°. Distance threshold is
                                              17. Both parameters are chosen by
                                              experience.
```

This program, although essentially serial in nature, forms an attractive alternative to the Hough and Radon transforms, which are the most widely accepted techniques for detecting linear structures and collinear groups of blobs in binary images. The program **island_hopping** uses the heuristic that the blobs must be close to each other and collinear, within the limits set by **fan_angle**. Of course, it is possible that these parameters could both be changed while the program is running. For example, a simple heuristic might reduce the fan angle progressively as an arc is being constructed, whilst tolerating ever bigger gaps between blobs. **island_hopping** can reconstruct gently curved arcs of unknown form. Neither the Hough nor Radon transforms can do this.

If there were some information about the likely or expected form of the curve, such as might be obtained from a model, the fan-angle and distance-threshold parameters could be adjusted in an intelligent way. For example, if we were to trace a "broken" numeral "2", the fan angle should be small near the bottom of the figure, which is straight. The upper parts of the "2", where the arc is curved, would require a larger fan angle. (Of course, the abrupt change of direction at the lower-left corner of a "2" would cause problems for **island_hopping** in its present form.)

The following predicate uses **caliper** to verify that the "width" of a non-convoluted snake-like object of arbitrary length is within prescribed limits defined by **width_limits**.

```
snake :-
    silhouette,                          % Generate binary image of snake-like object
    wri(temp),                           % Save image on disc
    mdl,                                 % Skeleton
    sample_points(L),                    % List of point equally spaced along skeleton⁹
    rea(temp),                           % Read image of the snake from the disc
    check(L).                            % Perform measurements

check([]).                               % Verify that the snake is not too thin nor
                                         too fat

check([[X1,Y1],[X2,Y2]|T]) :-
    mid_point(X1,Y1,X2,Y2,X3,Y3),        % Mid-point of {[X1,Y1],[X2,Y2]}
    angle([X1,Y1],[X2,Y2],A),            % Orientation of {[X1,Y1], [X2,Y2]}
    B is A + 90,                         % Measure at 90° to this
    caliper(X3,Y3,B,0,C),                % Measure width
    width_limits(P,Q),                   % Consult db for tolerance limits
    !,                                   % Included for sake of efficiency
    C ≤ Q,                               % Is snake too fat?
    C ≥ P,                               % Is snake too thin?
    check([[X2,Y2]|T]).                  % Take more measurements

% Snake is too thin or too fat

check(_) :-
    !,
    fail.                                % Force failure of this clause

width_limits(37,45).                     % Limits set by user
```

The predicate **sample_points** is not defined here, since this is peripheral to our main interest. Let it suffice to say that a list of sample points along a non-branching skeleton can be derived very simply from the chain code. What possible benefits could the predicate **snake** provide?

(a) A simple extension of the ideas implicit within **snake** could be used as the basis for locating suitable robot gripping points on semi-flexible objects, such as hoses, cables, springs, etc.
(b) Since **snake** allows serpentine objects to be classified, more intelligent reconstruction of "broken" arcs could be achieved, by incorporating it into **island_hopping.**
(c) Another minor variation of **snake** would allow the measured "width" values to be compared to a set of tolerance-limit pairs, defined by some simple formula. This could have significant benefit for the robotic handling of agricultural produce, such as carrots, parsnips, bananas, etc., which despite their being highly variable in form are often tapered according to some crude formula which can be expressed in rule form. A set of rules for recognising "good" carrots is given below:

Rule 1. The length of a carrot is within the range 50–150 mm.
Rule 2. The maximum diameter of a carrot is within the range 20–30 mm.

Rule 3. The position of the maximum diameter of a carrot is within 15–30% of one of its ends.

Rule 4. A carrot is orange in colour.

Rule 5. The taper of a carrot must conform to the formula:

$X \le X_1$: $D(X) < D_2$

$X_1 \le X \le X_2$: $D_0 + D_1.X \le D(X) \le D_2 + D_3.X$

$X \ge X_2$: $D(X) < D_2 + D_3.X_2$

for some suitable values of the variables X_1, X_2, D_0, D_1, D_2, D_3.

8.5 Inspecting All ... on Every ... of a ...

Consider the task of inspecting a populated printed circuit board. There will be an *unspecified number* of integrated circuits, each with an *unknown number* of connecting leads. A (human) inspector is told that *it is necessary to inspect the solder joint on every lead of each IC on the PCB*. Given a job specification like this, we find it unremarkable that a human being is able to identify each of the ICs and then locate the solder joints associated with it. Can we specify the inspection task required of a machine in the same way?

We will see how a generic, rule-based inspection system can be built using Prolog+. Let us first develop the terminology appropriate for inspecting PCBs. Here is the top level of a program for inspecting PCBs.

```
inspect_pcb :-
    find_ic,                % Locate IC
    find_pin,               % Locate pin on that IC
    inspect_pin,            % Inspect pin found, report results to user
    fail.                   % Force back-tracking
inspect_pcb.                % Complete the inspection process
```

The predicate **inspect_pcb** operates in a perfectly standard manner: back-tracking cycles through all of the pins on all of the ICs, inspecting each one as it does so. Back-tracking is of course, effected by **fail**. This causes Prolog+ to find another pin on the IC already detected. When there are no more pins to inspect on that IC, another IC is found. Although we have not yet specified the nature of the predicates **find_ic, find_pin, inspect_pin,** we can see how easy it is to write the search strategy.

Another inspection task, from mechanical, rather than electronic, manufacturing might be stated thus:

"Every brass 6BA screw must have a washer and nut."

Yet another example is:

"All holes must be free of swarf."

The important point is that the inspector will be able to *recognise* an integrated circuit, a 6BA screw or a hole, without being told specifically what each one looks like. Other examples of this type of inspection task are:

(i) An automobile inspector may be told to check every bolt on an engine.
(ii) A building inspector may be required to check the seal around every window frame in a new house.
(iii) An inspector may be required to examine every rivet on the wing of an aeroplane.
(iv) A computer keyboard must be checked so that all of the keys have clear, legible printing.

Notice that in all of these cases, the inspection procedure may be different for certain sub-classes of objects to be inspected. For example, a car engine will probably have several different types/sizes of bolt visible.

There are three essential features of this type of inspection function:

(a) *Scanning*. For example, when inspecting a PCB, it is necessary to scan the board to locate the ICs and scan the ICs to find its solder joints.
(b) *Recognition*. A PCB inspection system has to recognise both ICs and solder joints.
(c) *Verification*. Each IC must be checked by examining each of its parts, i.e. solder joints.

Notice the hierarchical nature of this type of inspection function; the process is essentially one of tree searching. Of course, this is an ideal application for a language developed from Prolog.

Let us now turn our attention to the specific topic of selecting parts (blobs) of an image, so that detailed analysis can be performed on each one. One of the simplest object selectors is based upon **ndo**. This primitive operator shades the blobs in the input image, according to their areas, or geographic positions. The input to the following program is a binary image, containing a number of distinct globular regions. The analysis of each blob takes place as a serial process, the biggest blob being analysed first. Notice that the analysis function is specified as an argument, Z, of **do_for_all**.

```
do_for_all(Z) :-
    ndo,            % Shade blobs according to their areas. Biggest is brightest
    wri(temp1),     % Store on disc
    cycle(Z).       % Enter analysis cycle

cycle(Z) :-
    rea(temp1),     % Read image from disc
    gli(_,X),       % Find maximum intensity
    X > 0,          % Are there any non-black pixels?
    hil(X,X,0),     % Set all pixels with maximum intensity to black
    wri(temp1),     % Save result on disc
    swi,            % Return to previous image
    thr(X,X),       % Threshold it
    call(Z),        % Analyse one blob. Function is specified by Z
    !,              % Included for efficiency's sake
    cycle(Z).       % Repeat

cycle(Z).           % Terminate recursion
```

This is very simple and does not take into account the possibility that the blobs may touch the edge of the image and therefore should be examined in a different way from those which are entirely within the camera's field of view. The simplest way to deal with blobs not completely visible is to ignore them. Here is a program to remove them. Once again the input image is binary:

```
remove_incomplete_blobs :-
    edg(255,1),     % Shade border of image (1 pixel wide) white
    ndo(1,_),       % Shade blobs in order of their being found in raster scan
    thr(2,255).     % Eliminate blob of intensity 1
```

How do we cope with non-binary images? Suppose the objects to be examined are disc-like objects, such as those shown in Figure 8.19. We might, for example, be expected to read the label on each object. The trick is to create a binary image defining the limits of each blob first and then use this as a mask to select a part of the grey-scale image. Here is the program:

```
analyse_all(Z) :-
    wri(temp2)                          % Save grey-scale image on the disc
    segment_image,                      % Convert it to a binary image
    remove_incomplete_blobs,            % Remove objects on edge of image
    do_for_all((rea(temp2),min,Z)).     % Generate a mask from each disc
```

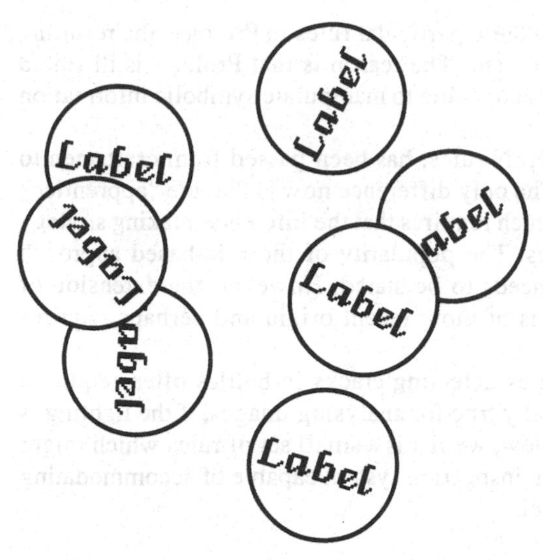

Figure 8.19. Discs with printed labels that are to be read, one at a time.

8.6 Rule-Based Inspection

In many instances, a human inspector can represent his/her knowledge about an inspection task in the form of a series of rules. We have already touched on this topic; in Chapter 2 we considered the inspection of Bakewell tarts and formulated rules for recognising "good" cakes. Amongst the other applications which are appropriate for this approach are solder joints on printed circuit boards and multi-layer semi-conductor wafers [BAR-88]. In one paper describing work on the second of these topics, the rules were of the following generic form:

"**If** a *pixel* (or a *pixel* within a certain *blob*) is on a certain *position*,
and if the *feature* value of the *pixel* (or *blob*) is within a certain *parameter* range,
then the *pixel* is considered a certain *conclusion* and should be treated by an *action*."

Ejiri and his colleagues [EJI-89] devised a set of rules for recognising faults on semi-conductor wafers based on the above paradigm. Here are two sample rules:

"**If** a pixel is on layer i,
and if the brightness of the pixel is **not** greater than the minimum allowable threshold and less than the maximum allowable threshold $t_{i,max}$
then the pixel is a part of a defect candidate and should be flagged."

and

"**If** a pixel is on the bonding pad,
and if the effective actual bonding area S is less than r% of the designed area S_d,
then the pixel is regarded as a portion of a defective pad."

While it is a simple matter to represent these particular rules in Prolog+, the resulting program will probably not be very efficient. The reason is that Prolog+ is ill suited to defining pixel-level operations; it is better able to manipulate symbolic information about images, or regions of interest.

Knowledge, in the form of IF...THEN rules, has been passed from craftsmen to their apprentices over the centuries. The only difference now is that the "apprentice" in our case is an inspection machine which requires that the inference-making strategy be expressed in terms of formal rules. The popularity of the rule-based approach when training human beings hardly needs to be stated. However, the extension of this idea to train machine inspectors is of more recent origin and perhaps requires further illustration.

Even seemingly simple tasks such as detecting cracks in bottles often require a set of rules [BAT-85a]. This is especially true for analysing images, if the lighting is not set by the inspection machine. Below, we list is a small set of rules which might conceivably form the basis of a tolerant inspection system capable of accommodating wide changes in the ambient light level.

Rule 1

If back-lighting is used, % This is a test, not a command
and if the background intensity is bright, % Clearly, this is a test
and if there is a dark line, % So is this
then there is a crack present and the bottle should be rejected. % Deduction

Rule 2

If dark-field illumination is used, % This is a test, not a command
and if the background intensity is dark
and if there is a bright line,
then there is a crack present and the bottle should be rejected.

Rule 3

If front illumination is used, % This is a test, not a command
and if the background intensity is dark,
and if there is a bright line,
then there is a crack present and the bottle should be rejected.

Rule 4

If front illumination is used, % This is a test, not a command
and if the background intensity is bright,
and if there is a dark line,
then there is a crack present and the bottle should be rejected.

Rule 5

If the maximum intensity is less than 50, % Level 50 is very dark
and if the camera aperture is not fully open,
then there is not enough light and camera aperture should be opened (a little).

Rule 6

If the maximum intensity is less than 50, % Level 50 is very dark
and if the camera aperture is fully open,
then there is not enough light and the lights should be switched on.

Rule 7

If the maximum intensity is 255, % Level 255 is saturated white level
and if the camera_aperture is not fully closed, % This is a test
then there is too much light and the size of the camera aperture should be reduced.

Rule 8

If the maximum intensity is 255,
and if the camera aperture is fully closed, % This is a test
then there is optical overload and the machine should warn the user.

Even without bothering to explain details, the following Prolog+ representation of these rules is self-evident.

```
rule(1,crack,reject) :-
    back_illumination,                          % This is a test, not a command
    dark_line.

rule(2,crack,reject) :-
    dark_field_illumination,
    bright_line.

rule(3,crack,reject) :-
    front_illumination,
    bright_line.

rule(4,crack,reject) :-
    front_illumination,
    dark_line.

rule(5,not_enough_light, camera_aperture(open)) :-
    gli(_X),
    X ≤ 50,
    not(camera_aperture(fully_open)).           % This is a test, not a command

rule(6,not_enough_light, lights(on)) :-
    gli(_X),
    X ≤ 50,
    camera_aperture(fully_open).                % This is a test, not a command

rule(7,too_much_light, camera_aperture(close)) :-
    gli(_X),
    X = 255,
    not(camera_aperture(fully_closed)).
```

```
rule(8,optical_overload, warning(optical_overload)) :-
        gli(_X),
        X = 255,
        camera_aperture(fully_closed).
```

Clearly, we can take this example very much further, if necessary, simply by coding a lot more rules. However, we have enough detail here for our present purposes.

The first point to make is that it is a very simple matter to code rules like these using Prolog+. In order to assist in this process, the rules could be expressed in the form of a restricted sub-set of English, based upon Definite Clause Grammars (see Chapter 4).

The second point is that we have not specified how, nor in what order, these rules will be used. The familiar Prolog convention is, of course, that the database is searched from top to bottom. In our particular example, the rules are searched in the order

```
rule(1,_,_), rule(2,_,_),..., rule(8,_,_).
```

Clearly, this is not the best order, since we need to adjust the lighting/camera before we can properly recognise a crack. The most obvious solution is to adjust the order in which the rules appear in the database. Rules 5–8 seem to be obvious candidates for being placed towards the top of the database.

So far, we have implicitly assumed that there will be an implicit recycling, through database reference and then acting on the information found. For example, if rule 7 "fires", the program should make the necessary adjustment to the camera aperture and then repeat the process. (It may then be necessary to readjust the camera again.) These arguments suggest the use of

(a) Back-tracking to search through the set of rules
(b) Recursion to repeat the process over and over again

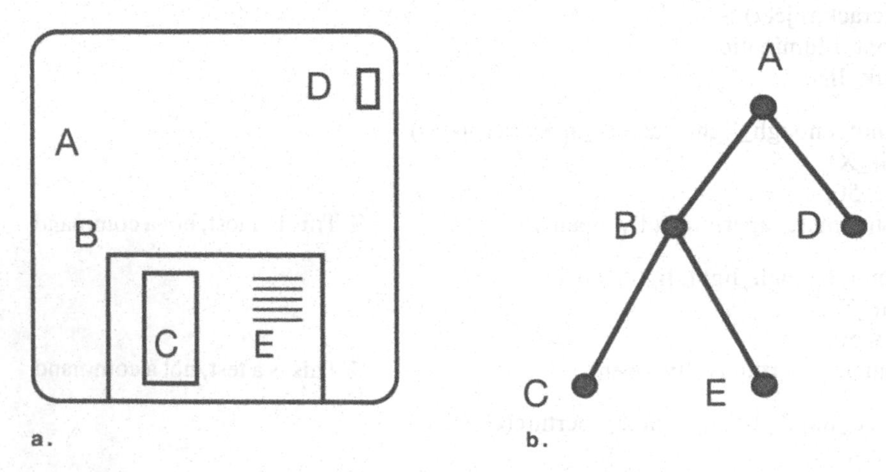

Figure 8.20. (a) Sketch of a computer floppy disc. (b) Tree showing the relationship between its parts. Feature A is taken to be a rectangle and E represents printing.

Here is a simple control program, which combines these two flow-control mechanisms in a fairly obvious way.

```
go :-
      rule(N,State,Action),
      advise_user(N,State),
      do(Action),
      go.
```

Readers who are familiar with Artificial Intelligence techniques will recognise that we are forcing Prolog to simulate *forward chaining*. It is of interest to note that LPA MacProlog contains facilities for incorporating forward chaining, as well as backward chaining.

8.7 Declarative Programming of Vision Systems

Our goal in this section is to develop ideas which the author hopes might eventually lead to a rather simpler method of programming industrial vision systems than has been available hitherto. Although that goal is still a long way off, we can take our first few tentative steps. We shall discuss the possibility of using a restricted subset of English to explain what objects/images look like. The result will be a series of simple sentences, which can be parsed using a Definite Clause Grammar (DCG, see Chapter 4), to generate a Prolog+ program for recognising objects. The long-term objective is to provide a much simpler method of programming an industrial vision system, so that people who are not experienced in the use of computers can use it, with a minimum of training. However, the author does not envisage that it will ever be possible to perform the programming task entirely in English, nor any other human language, since many of the "descriptions" required to teach/specify a given visual inspection function are essentially non-verbal in nature. For example, some "descriptions" require pointing with a cursor, comparison with a pictorial reference image, large amounts of numeric data, sketches, diagrams, etc.

Figure 8.20 shows a diagrammatic representation of a computer floppy disc. This will form a useful basis for our discussions of declarative programming.

We shall assume that a "clean" binary image like this can been created using some suitable image processing algorithm, possibly of the general form described in Chapter 5. However, these details are not important for the moment. In the particular instance of computer floppy discs, a suitable Prolog+ program for creating a "line drawing" might be as follows:

```
create_line_drawing :-
      frz,              % Digitise an image
      mdf(5),           % Noise removal filter
      sed,              % Edge detector
      thr(8),           % Convert to binary image form
      ndo,              % Shade blobs according to their areas
```

```
    thr(245),              % Keep only big blobs
    mdl.                   % Generate the skeleton
```

A person can easily describe the object in Figure 8.20, using a series of very simple English sentences, such as those given below:

1. The outer_edge is called A.   % This sentence is not essential
2. A is a rectangle.
3. B is a rectangle.
4. B is inside A.
5. C is a rectangle.
6. C is inside B.
7. E is some printing.
8. E is inside B.
9. D is a rectangle.
10. D is inside A.

Later, we shall present a program which can convert sentences like these into the following so-called *list format*.

1. [the, outer_edge, is, called, a].
2. [a, is, a, rectangle].
3. [b, is, a, rectangle].
4. [b, is, inside, a].
5. [c, is, a, rectangle].
6. [c, is, inside, b].
7. [e, is, some, printing].
8. [e, is, inside, b].
9. [d, is , a, rectangle].
10. [d, is inside, a].

These sentences have been deliberately phrased in such a way that formalisation, using Definite Clause Grammars (DCGs), will be straightforward. In Section 4.5, we saw how this could be done. Using a program very similar to the one presented there, these sentences could be converted into the following set of facts:

```
outer_edge('A').    % Derived from sentence 1
rectangle('A').     % Derived from sentence 2
rectangle('B').     % Derived from sentence 3
inside('B','A').    % Derived from sentence 4
rectangle('C').     % Derived from sentence 5
inside('C','B').    % Derived from sentence 6
printing('E').      % Derived from sentence 7
inside('E','B').    % Derived from sentence 8
rectangle('D').     % Derived from sentence 9
inside('D','A').    % Derived from sentence 10
```

The next question we have to consider is how we can combine this set of facts into a Prolog+ program, which can recognise the image of a floppy disc? Clearly, a predicate for recognising floppy discs would need to verify that all of the above conditions are satisfied. *Note:* the author, not a program, wrote the following piece of code.

```
recognise(floppy_disc) :-
    outer_edge(A),     % Find the outer edge, call it A
    rectangle(A),      % Check that A is a rectangle
    rectangle(B),      % Find a rectangle. Call it B
    inside(B,A),       % Check that B is inside A
    rectangle(C),      % Find a rectangle. Call it C
    inside(C,B),       % Check that C is inside B
    printing(E),       % Find some printing. Call it E
    inside(E,B),       % Check that E is inside B
    rectangle(D),      % Find a rectangle. Call it D
    inside(D,A).       % Check that D is inside A
```

Clearly, it is a trivial task for somebody who is familiar with Prolog programming to write a relation such as **recognise(floppy_disc)** directly, but our task in this section is to explore the possibilities of making even this level of program writing unnecessary. In fact, **recognise(floppy_disc)** can be generated by parsing free-format English sentences. (The program is listed later.)

There are several important points that are worth making here:

(i) **recognise(floppy_disc)** uses variables, whereas the set of facts uses constants. Notice, for example, that the fourth fact in the above set is

inside('B','A')

and contains quotes around both A and B. (A and B are constants.) On the other hand, the fourth sub-goal in the body of **recognise(floppy_disc)** is

inside(B,A)

and contains no quotes. (A and B are variables.)

(ii) The set of facts might conveniently be stored in the database using a clause of the form

```
stuffing([
        outer_edge('A'),
        rectangle('A'),
        rectangle('B'),
        inside('B','A'),
        rectangle('C'),
        inside('C','B'),
        printing('E'),
        inside('E','B'),
        rectangle('D'),
        inside('D','A') ])
```

(iii) An alternative form of expression might be developed around the use of Prolog operators, although this is rather less satisfactory than normal English. The user might describe the floppy disc in the following way. (The equivalent DCG sentences are included, as comments, for the purposes of comparison.)

```
'A' outer_edge.        % [the, outer_edge, is, called, a]
    'A' rectangle.     % [a, is, a, rectangle],
    'B' rectangle.     % [b, is, a, rectangle]
    'B' inside 'A'.    % [b, is, inside, a]
    'C' rectangle('C'). % [c, is, a, rectangle]
    'C' inside 'B'.    % [c, is, inside, b]
    'E' printing.      % [e, is, some, printing']
    'E' inside 'B'.    % [e, is, inside', b]
    'D' rectangle.     % [d, is , a, rectangle]
    'D' inside 'A'.    % [d, is inside, a]
```

Using Prolog operators, a person who has received a small amount of training could write complete programs. Of course, a little syntactic "sugar" might be added to assist this. For example, it is a trivial matter to define "**&**" to represent goal conjunction (,) and "**if**" to represent the symbol ":-". Here is the resulting simplified program.

```
recognise(floppy_disc) if
    A outer_edge &
    A rectangle &
    B rectangle &
    B inside A &
    C rectangle &
    C inside B &
    E printing &
    E inside B &
    D rectangle &
    D inside A.
```

Thus, there are three different levels of declarative programming by which a person might describe the objects that the vision system can expect to see:

(i) Direct programming in Prolog+

(ii) Using Prolog+ operators

(iii) Using English-like sentences and a DCG parser to generate the Prolog+ program automatically

These are, of course, listed in order of *decreasing* programmer skill required; the third item listed would, if it were possible, permit people with quite low skill levels to program vision systems.

8.8 Feature Recognition

We must now discuss the "low-level" feature recognition predicates, such as **rectangle, inside, printing, outer_edge**, since these are clearly of crucial importance. A little thought will verify that these define *recognition* functions. This has significant implications for their implementation, since certain types of algorithm are required for *recognition*, as distinct from *verification*..

An algorithm for recognising rectangles whose sides are parallel to the sides of the image will be defined in Prolog+ later (Section 8.8), as will another algorithm for testing whether one object is inside another. Finding the outer edge is quite straightforward and is also algorithmic:

```
outer_edge(A) :-
    bed,              % Binary edge detector
    ndo(1,_),         % Shade blobs according to the order they are found
    thr(1,1),         % Isolate the first one found
    wri(A).           % Save image in disc file A
```

How can we recognise printing? (It should be understood that our task is not to read the text but simply to locate the presence of printing in an image, wherever it may occur.) This is a little difficult and raises an important point: the process for recognising printing is *heuristic*, not algorithmic in nature. That is, it usually achieves the desired result, but we cannot guarantee it. Of course, the trick for recognising printing is to avoid using the geometric test of character structure, but instead uses an ad hoc technique for recognising it as a texture pattern. This informal approach will be familiar to readers with experience in applying pattern recognition to practical problems; the procedure is rule-based, although we do not, at this stage, propose to define the rules in detail.

The obvious operators for recognising printing are the *image filters* (notably N-tuple operators) and *morphological operators*, which automatically scan the whole image [SCH-89]. However, it is unlikely that we will ever be able to find just one filter that will identify printing, whilst rejecting everything else. It is much more probable that the outputs from a set of n filters, f1,...,fn, must be applied to the image, in order to locate any printing. These filter outputs must then be combined using a set of logical rules which can be expressed in Prolog+ in a straightforward way.

```
/* __________To detect extra fonts and type sizes simply add more clauses to
          "filters_to_apply" and "how_to_combine".___________ */

recognise(printing) :-
    thr(0),                  % Make the image all white
    wri(temp),               % Save white image on disc
    zer,                     % Make image all black
    wri(result),             % Save black image on disc
    filters_to_apply(L),     % Find which filters to apply
    filter_and_combine(L),   % Now apply them
    fail.

recognise(printing) :-
    rea(result).             % Read result from disc
```

```
filter_and_combine([]) :-
        rea(temp),                          % Read "temp" image from disc
        rea(result),                        % Read "result" image from disc
        max.                                % OR white regions in two images

filter_and_combine([X|Y]) :-
        rea(image),                         % Read original image from disc
        apply_filter(X),                    % Apply filter X. O/P is binary
        rea(temp),                          % Get intermediate result image
        how_to_combine(X,Z),                % Find how to combine filter X O/Ps
        call(Z),
        filter_and_combine(Y).              % Apply other filters

% ________________________Sample of database______________________

filters_to_apply([f7,f4,f12,f14).           % Apply four filters this time
.................                           % Lots more clauses like these
filters_to_apply([f3,f5,f14).               % Apply three filters this time
how_to_combine(1,min).                      % min ANDs two images together
how_to_combine(1,max).                      % max ORs two images together
how_to_combine(1,(sof,neg,son,min,neg)).    % Compound logical operator
how_to_combine(1,min).
................                            % Lots more clauses like these
how_to_combine(15,min).
how_to_combine(15,max).
how_to_combine(15,min).
```

In practice, a great many more types of image feature will have to be accommodated, compared to those mentioned so far. A list of additional feature types will be presented later.

Next, we describe a program which encompasses these ideas. Although its scope is still trivial, the program is capable of being expanded easily to detect a greater range of features than is included here.

8.9 Program for Natural Language Declarative Programming

The program consists of three distinct modules, containing the following top-level predicates:

(i) **describe_image**

(ii) **object_recognition**

(iii) **run_program**

The program structure is shown in Figure 8.21. A Data Window[10] is used for communication.

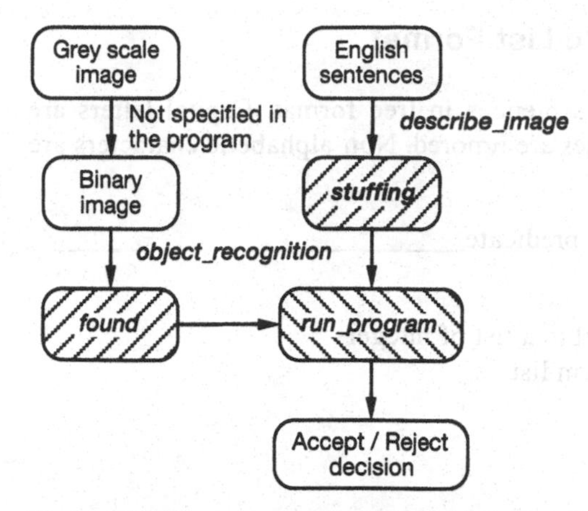

Figure 8.21. Program structure for Natural Language declarative programming. The program consists of three modules: **describe_image, object_recognition, run_program. found** and **stuffing** represent data-transfer relations, generated by **object_recognition** and **describe_image** respectively.

Data Window

```
% Seed clause, to ensure that Prolog knows where to put new clauses

stuffing.

% This is a typical clause asserted by "describe_image"

stuffing([
        true,
        is_a('Var_outer_edge', circle),
        is_a('Var_second_edge', circle),
        contains('Var_outer_edge', 'Var_second_edge'),
        is_a('Var_inner_most_edge',rectangle),
        contains('Var_second_edge', 'Var_inner_most_edge') ]).

% Seed clause, to ensure that Prolog knows where to put new clauses

known.

% This clause was asserted by "describe_image"

known(['Var_inner_most_edge', 'Var_second_edge', 'Var_outer_edge']).

% This clause was asserted by "object_recognition"

found.

% The three following clauses were asserted by "object_recognition"

found(circle, [255, 255, 184], 255).
found(circle, [255, 255, 225], 254).
found(rectangle, [210, 193, 99, 146], 253).
```

Converting Normal English into List Format

This module allows the user to type sentences in free format. Capital letters are converted to lower case. Multiple spaces are ignored. Non-alphabetic characters are ignored, but a warning is issued.

```
% ________________________Top-level predicate_________________________

english_to_list(X,L) :-
      name(X,L1),              % Convert to a list of integers
      transf(L1,[],[],L),      % Tranform list
      !.

% Terminate recursion

transf([],A,B,C) :-
      name(E,A),               % Convert last bit back into character form
      append(B,[E],C),         % Complete the list
      writenl(C).

% Multiple space – if we get multiple spaces, eliminate one of them

transf([32,32|B],C,D,E) :- transf([32|B],C,D,E).

% Single space – end the word

transf([32|B],C,D,E) :-
      name(G,C),               % Convert list of integers back into a word
      append(D,[G],F),         % Add it the list of words
      transf(B,[],F,E).        % Repeat

% Lower case alphabetic charaters

transf([A|B],C,D,E) :-
      A ≥ 97,                  % Integer representation of letter "a"
      A ≤ 122,                 % Integer representation of letter "z"
      append(C,[A],F),         % Add to list
      transf(B,F,D,E).         % Repeat

% Convert upper case letters to lower case

transf([A|B],C,D,E) :-
      A ≥ 65,                  % Integer representation of letter "A"
      A ≤ 90,                  % Integer representation of letter "Z"
      Z is A + 32,             % Convert to lower case
      append(C,[Z],F),         % Add to list
      transf(B,F,D,E).         % Repeat

% Eliminate non-alphabetic characters. Give the user a warning.

transf([A|B],C,D,E) :-
      message(['WARNING: Non-alphabetic character in the input string']),
      transf(B,C,D,E).
```

Sample Query Using "english_to_list"

The query

english_to_list('The Cat sAt* oN the 123344+++++MAt',Z)

instantiates Z to the list

[the, cat, sat, on, the, mat]

Describing Images in List Format English

```
% ______________________Top-level predicate______________________

describe_image :-
    clearup,                        % Initialise the database
    run_parser.                     % Operate the parser

% ______________________Running the parser______________________

run_parser :-
    prompt_read(['Type a sentence in English'],X),
    english_to_list(X,Y),           % Convert to list format
    not(X = end),                   % User has finished entering data
    phrase(sentence,Y),             % Parser
    writenl(Y),                     % Keep a record of the sentences entered
    !,                              % Included merely for the sake of efficiency
    run_parser.                     % Repeat for next sentence

run_parser :-
    writenl('Now we can run the program.'). % Finished entering data

% __________Structure the data for storage in "stuffing"______________

save_info(X,Y,Z) :-
    known_stuff(X,X1),              % Convert X to a variable if it is not known
    def(Y),                         % Verify that Y is known
    known_stuff(Z,Z1),              % Convert Z to a variable if it is not known
    Q =.. [Y1,X1,Z1],               % Create a term from its component parts
    stuffing(R),                    % Get previous goal list
    append(R,[Q],S),                % Append new goal to it
    assert(stuffing(S)),            % Save revised goal list
    retract(stuffing(R)).           % Delete old goal list

% We know something about objects of type X, so keep X as a constant

known_stuff(X,X) :-
    def(X).                         % X is defined somehere in the program
```

```
% We know nothing about objects of type X, so make X a variable,Y

known_stuff(X,Y) :-
      name(X,Z),                              % Construct ASCI string from X
      name(Y,[86, 97, 114, 95|Z]),            % Add "Var_" to it, at front end
      (        (known(Q),                     % Get list of "unknown" atoms
                      not(member(Y,Q)),       % Is new one member of that set?
                      assert(known([Y|Q])),   % Update list of unknown atoms
                      retract(known(Q))) ;    % Delete old list
      true).                                  % Force this to succeed

% ______________DCG Rules for recognising sentences__________________

/* Rule 1 recognises sentences of the form

[the, ship, is, a, hulk]                          */

sentence -->
      article, [X], is_verb, article, [Y] ,
      {save_info(X,is_a,Y)}.

/* Rule 2 recognises sentences of the form

[the, ship, is, called, fanny_adams]              */

sentence -->
      article, [X], is_verb, called, article, [Y],
      {save_info(X,named,Y)}.

/* Rule 3 recognises sentences of the form

[the, car, is, inside, the, garage]               */

sentence -->
      article, [X], is_verb, [Y], article, [Z],
      {member(Y, [above,below, inside, under]), save_info(X,Y,Z)}.

/* Rule 4 recognises sentences of the form

[the, cupboard, contains, a, bucket]              */

sentence -->
      article, [X], [Y], article, [Z],
      {member(Y, [touches, contains, encloses, 'lies inside']), save_info(X,Y,Z)}.

% __________________Definitions of secondary items_________________

article --> [] | [a] | [an] | [the] | [some].

is_verb --> [is] | [are].

called --> [called] | [termed] | ['known as'].
```

```
% _______________Tidy up the database for a new run________________

clearup :-
    retractall(known(_)),
    assert(known([])),
    retractall(stuffing(_)),
    assert(stuffing([true])).
```

Recognising Objects in Images

```
% ______________________Top-level predicate______________________

object_recognition :-
    # rea(original),              % Read the image from disc file
    # ndo(3,_),                   % Shade the blobs in it
    # wri(temp2),                 % Save the shaded image for further
                                    processing
    retractall(found(_,_,_)),     % Clear the database
    run_object_recognition.       % Find what objects it contains

% ___________Second-level predicate for object recognition____________

run_object_recognition :-
    # rea(temp2),                 % Read the shaded image from disc
    # gli(_,X),                   % Find intensity of the brightest blob in it
    X > 0,                        % Is maximum intensity > 0?
    # hil(X,X,0),                 % Delete brightest blob
    # wri(temp2),                 % Save the image of remaining blobs
    # swi,                        % Switch images
    # thr(X,X),                   % Isolate the brightest blob
    remember(blob_identity,X),    % Remember how to get this blob again
    # wri(temp1),                 % Save image containing just this one blob
    recognise,                    % What type of object is it?
    !,                            % Included only to increase efficiency
    run_object_recognition.       % Repeat for remaining blobs

run_object_recognition :-
    writenl('Finished').          % All blobs have been analysed

% ____________Third-level predicate for object recognition___________

recognise :-
    writenl('Trying to fit a rectangle'), % Tell user what we are doing
    rectangle.                    % Test for rectangle

recognise :-
    writenl('Trying to fit a circle'),    % Tell user what we are doing
    circle.                       % Test for circle
```

```
recognise :-
    writenl('Trying to fit a triangle'),      % Tell user what we are doing
    triangle.                                 % Test for triangle. Not defined here

recognise :-
    writenl('Trying to fit a polygon'),       % Tell user what we are doing
    polygon.                                  % Test for polygon. Not defined here

% Add similar clauses here for recognising other types of objects

recognise :- writenl('Figure was not recognised').

% _____Recognising rectangles whose sides are at 0° and 90°__________

rectangle :-
    rectangle(A,B,C,D),                       % Call main recognition clause
    write('Rectangle found! Parameters: '),   % Message for the user
    writenl([A,B,C,D]),                       % Message for the user
    recall(blob_identity,X),                  % See "run_object_recognition"
    assert(found(rectangle,[A,B,C,D],X)).     % See "Data Window"

rectangle :-
    writenl('Rectangle NOT found!'),          % Message for the user
    !,
    fail.                                     % Forced failure

rectangle(A,C,E,F) :-
    # rea(temp1),                             % Read image of just one blob
    # blb,                                    % Make the figure solid
    # bed,                                    % Binary edge detector
    # cwp(N),                                 % Count white points
    # dim(A,B,C,D),                           % Find parameters of min-area
                                              %   rectangle (MAR)
    E is B – A,                               % Calculate the length of one of its
                                              %   sides
    F is D – C,                               % Calculate the length of one of its
                                              %   sides
    # cpy,                                    % Copy image
    Z is ((2*(E + F)/N) – 1),                 % Calculate goodness of fit
    abs(Z,Y),                                 % Modulus
    !,                                        % No back-tracking
    Y < 0.100.                                % Is the MAR sufficiently close to
                                              %   figure?

% ________________Recognising circles____________________

circle :-
    circle(A,B,C),                            % Call main recognition clause
    write('Circle found! Parameters: '),
    writenl([A,B,C]),                         % Message for the user
```

```
    recall(blob_identity,X),             % See "run_object_recognition"
    assert(found(circle,[A,B,C],X)). % See "Data Window"

circle :-
    writenl('Circle NOT found!'),        % See "Data Window"
    !,
    fail.                                % Forced failure

circle(X,Y,Z) :-
    # rea(temp1),                        % Read image of just one blob
    # blb,                               % Make the figure solid
    # cpy,                               % Copy image
    # cgr(X,Y),                          % Find centroid
    # hic(X,Y),                          % Intensity cone
    # min,                               % Mask cone with figure
    # hil(0,0,255),                      % Set background to white
    # gli(Z,_),                          % Find intensity of point furthest from centroid
    Z1 is Z*1.05,                        % Find threshold par. for slightly smaller circle
    int(Z1,Z2),                          % Convert to integer
    # swi,                               % Switch images
    # thr(Z,Z2),                         % Threshold creates annulus from cone
    # neg,                               % Annulus becomes black, background is white
    # rea(temp1),                        % Read image of the blobs again
    # min,                               % Operate mask
    # edg(0,4),                          % Eliminate image border
    # cwp(N),                            % Count white pixels
    !,                                   % Avoid back-tracking
    N = 0.                               % Is edge entirely within annulus?

% Add further clauses here for recognising triangles, polygons, etc.
```

Verification Stage

```
% ____________Run the program written by describe_image _____________

run_program :-
    stuffing(L1),                        % Consult database for list of goals
    known(L2),                           % Get list of "unknown" atoms (i.e. variables)
    tohollow(L1,L3,L2),                  % Convert variables to hollow form
    run_program(L3).                     % Now run the program

% Terminate recursion

run_program([]).                         % Stop when the goal list is empty
```

```
% ____________Runs the program, one term at a time____________

run_program([A|B]) :-
    (       (call(A),                         % Try to satisfy one goal and ...
                write('The goal "'),          % ... report the result
                write(A),
                writenl('" was proved')) ;
    (       write('The goal "'),
                write(A),
                writenl('" failed'),
                fail)   ),
    run_program(B).                           % Repeat for rest of goal list

% ________________What to do if the object is a circle________________

is_a(A,circle) :- found(circle,_,A).

% ________________What to do if the object is a rectangle____________

is_a(A,rectangle) :- found(rectangle,_,A).

% ___Add similar clauses for other figures, e.g. triangles, polygons, etc.___

% ________________Check that object A contains object B________________

contains(A,B) :-
    not( A = B),
    # rea(original),
    # thr(1,255),
    # ndo(3,_),
    # wri(b),
    # thr(A,A),
    # blb,
    # wri(temp2),
    # rea(b),
    # thr(B,B),
    # blb,
    # cpy,
    # rea(temp2),
    # swi,
    # sub,
    # thr(255,255),
    # cwp(Z),
    !,
    Z = 0.

inside(A,B) :-
    contains(B,A).
```

```
% _________Other relations – incomplete list. Also see Appendix II_________

above(A,B) :-
        not(A = B),
        location(A,Xa,Ya),
        location(B,Xb,Yb),
        !,
        Ya > Yb.

below(A,B) :-
        above(B,A).

left(A,B) :-
        not(A = B),
        location(A,Xa,Ya),
        location(B,Xb,Yb),
        !,
        Xa < Xb.

right(A,B) :-
        left(B,A).

bigger(A,B) :-
        not(A = B),
        size(A,Sa),
        size(B,Sb),
        Sa > Sb.

smaller(A,B) :-
        bigger(B,A).

larger(A,B) :-
        bigger(A,B).
```

A Simple Example

The compound goal

```
object_recognition,
describe_image,
run_program.
```

was examined. The binary input image, stored in file **original**, is shown in Figure 8.22.

Sub-Goal: object_recognition

The following facts were asserted into the database. (See the section above entitled *Data Window*.)

```
found(circle, [255, 255, 184], 255).
found(circle, [255, 255, 225], 254).
found(rectangle, [210, 193, 99, 146], 253).
```

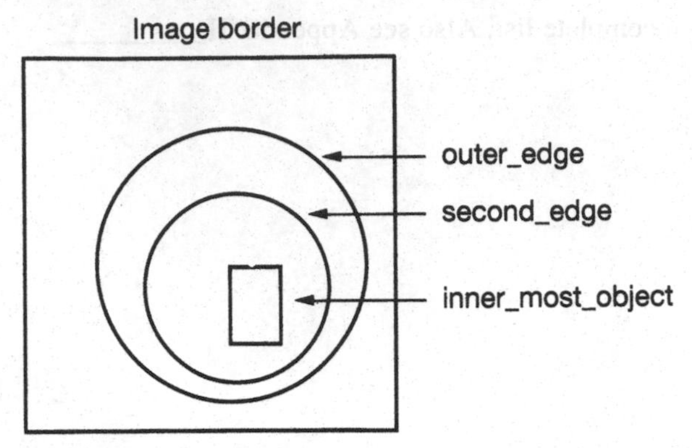

Figure 8.22. Test image, used to demonstrate Natural Language declarative programming.

Sub-Goal: describe_image

The following message was generated in the default output window. (The first six lines merely echo the user input sentences.)

```
[the,outer_edge,is,a,circle]
[the,second_edge,is a,circle]
[the,outer_edge,contains,the, second_edge]
[the,inner_most_edge, is, a, rectangle]
[the,second_edge,contains,the,inner_most_edge]
Now we can run the program.
№1    yes
```

The following goal list was then generated.

```
[ true,
is_a('Var_outer_edge', circle),
is_a('Var_second_edge', circle),
contains('Var_outer_edge', 'Var_second_edge'),
is_a('Var_inner_most_edge',rectangle),
contains('Var_second_edge', 'Var_inner_most_edge') ]
```

Using conventional Prolog syntax and the same variable names, we might write the recognition program thus:

```
run_program :-
    is_a(Var_outer_edge, circle),                       % Instantiates
                                                          "Var_outer_edge"
    is_a(Var_second_edge, circle),                      % Instantiates
                                                          "Var_second_edge"
    contains(Var_outer_edge, Var_second_edge),          % Test two circles
    is_a(Var_inner_most_edge',rectangle),               % Instantiates
                                                          "Var_inner_most_edge"
    contains(Var_second_edge,Var_inner_most_edge).% Test circle and rectangle
```

Sub-Goal: run_program

The following output data were generated in the Default Output Window:

The goal "true" was proved
The goal "is_a(255, circle)" was proved
The goal "is_a(255, circle)" was proved
The goal "contains(255, 255)" failed
The goal "is_a(254, circle)" was proved
The goal "contains(255, 254)" was proved
The goal "is_a(253, rectangle)" was proved
The goal "contains(254, 253)" was proved
№1 yes

8.10 Extending the Idea

There are three principal ways in which the very naive approach outlined above can be extended:

(a) Synonyms and antonyms
(b) Additional syntax structures
(c) Additional image features

Of course, (b) represents an addition to the syntactic analysis ability of our program, while (a) and (c) effectively extend its vocabulary. In Section 8.7, we listed a range of image features. Recognition programs have not yet been written for all of the features listed there, although there seems to be little difficulty in doing this for most of them.

Synonyms and Antonyms

Here is a small thesaurus containing a number of useful synonyms and antonyms. While the list is not complete, it does show how the expressional power of Prolog+ can be put to good use to extend the vocabulary of a language. It is, of course, conceptually easy, if a little tedious, to extend this list further.

```
inside(X,Y) :- contains(Y,X).
encloses(X,Y) :- contains(X,Y).
includes(X,Y) :- contains(X,Y).
beneath(X,Y) :- above(Y,X).
below(X,Y) :- above(Y,X).
over(X,Y) :- above(X,Y).
under(X,Y) :- above(Y,X).
above(X,Y) :-significantly_higher(X,Y).
right(X,Y) :- left(Y,X).
```

```
beside(X,Y) :-
        left(X,Y),
        not(significantly_higher(X,Y)),
        not(significantly_higher(Y,X)).

beside(X,Y) :-
       left(Y,X),
       not(significantly_higher(X,Y)),
       not(significantly_higher(Y,X)).

alongside(X,Y) :- beside(X,Y).
bigger(X,Y) :- significantly_bigger(X,Y).
larger(X,Y) :- bigger(X,Y).
significantly_larger(X,Y) :- significantly_bigger(X,Y).
smaller(X,Y) :- bigger(Y,X).
significantly_smaller(X,Y) :- significantly_bigger(Y,X).

about_same_size(X,Y) :-
       not(significantly_bigger(X,Y)),
       not(significantly_bigger(Y,X)).

wider(X,Y) :- significantly_wider(X,Y).
narrower(X,Y) :- wider(Y,X).
narrower(X,Y) :- significantly_narrower(X,Y).
significantly_narrower(X,Y) :- significantly_wider(Y,X).
fatter(X,Y) :- wider(X,Y).
fatter(X,Y) :- significantly_fatter(X,Y).
significantly_fatter(X,Y) :- significantly_wider(X,Y).
thinner(X,Y) :- wider(Y,X).
thinner(X,Y) :- significantly_thinner(X,Y).
significantly_thinner(X,Y) :- significantly_wider(Y,X).
taller(X,Y) :- significantly_taller(X,Y).
shorter(X,Y) :- taller(Y,X).
shorter(X,Y) :- significantly_shorter(X,Y).
significantly_shorter(X,Y) :- significantly_taller(Y,X).
darker(X,Y) :- significantly_darker(X,Y).
brighter(X,Y) :- darker(Y,X).

about_same_brightness(X,Y) :-
       not(significantly_darker(X,Y)),
       not(significantly_darker(Y,X)).

significantly_brighter(X,Y) :- significantly_darker(Y,X).
```

Additional Syntax Structures

It is, of course, very difficult to anticipate all of the possible sentence structures that might be used in even a limited dialogue like the one in which our program will participate. Research is under way that will, it is hoped, eventually lead to a *complete* definition, in Prolog, of the structure of the English language [GAZ-89]. Our intention here is, of course, much less ambitious.

The images with which our vision system is concerned are essentially uncomplicated. If they were not, we could not expect a person to describe them. Moreover, if an image contains a lot of different objects, people instinctively revert to describing the texture as "wood-grain", "machined surface" or "printing". (Printing may be regarded as a type of texture, if we are not trying to understand it.) In addition, we may use collective terms, for example "a group of spots", "several irregular blobs", "a number of streaks", "five parallel diagonal lines", etc. Clearly, we must eventually extend our program to encompass such phrases as these. Using group concepts, like these, simplifies the nature of the images.

Comparative statements are also important. A simple comparison can be handled using the following DCG rule:

```
sentence ⟶
      article, [X], [is], [Y], [than], article, [Z],
      {def(Y),save_info(X,Y,Z)}.
```

This rule translates the sentence

[the, sun, is, brighter, than, a, candle]

into the clause

brighter(sun, candle)

which is asserted by **save_info** into the database. In a similar way, superlatives can be understood, using a rule like this:

```
sentence ⟶
      article, [X], [is], comparator, [Y1], [than], article, [Z],
      {concat('significantly_',Y1,Y2), def(Y2), save_info(X,Y2,Z)}.
```

This rule translates both of the following sentences

[the, sun, is, **much**, brighter, than, a, candle]

and

[the, sun, is, **significantly**, brighter, than, a, candle]

into

significantly_brighter(sun, candle).

The sentences

[thing_a, is, concentric, with, thing_b]

and

[thing_a, and, thing_b, are concentric]

have the same meaning and should be translated into the same object clause. An extension of this idea is that we should be able to handle *lists* in English:

[thing_a, thing_b, thing_c and, thing_d, are concentric]

So far, our program does not do this. In a similar way, the sentences

[object1, object2, and, object3, are, squares]

[object1, object2, and, object3, are, all, squares]

and

[the, following, are, squares, object1, object2 and object3]

should result in the same clause being generated and asserted in the database.

Another type of sentence that our program does not yet accommodate is the following:

[circle, 'A', is, inside, circle, 'B']

Notice that we may not have previously defined 'A' or 'B' to be circles; our existing program would have to use three separate sentences:

['A', is, a, circle]
['B', is, a, circle]
['A', and, 'B', are concentric]

Obviously, there are many other ways to express the same idea:

[circle, 'B', encloses, circle, 'A']

[the, circle, called,'A', is, inside, circle, 'B']

['A' is a, circle, and, is, inside, 'B', which, is, also, a, circle]

[objects, 'A', and, 'B', are, both, circles, and, 'A' is, inside, the latter]

Additional Image Features

Obviously, the list of features to be detected should include at least those items listed below:

rectangle, arbitrary orientation
rectangle, normal[11] orientation
square, arbitrary orientation
square, normal orientation
parallelogram
trapezium
circle

annulus
ellipse
ovoid
irregular blob (this is, of course, the default type for blobs)
convex blob
triangle
polygon
smooth arc
circular arc
polygonal arc
spot
streak
edge
wedge
lines
parallel lines
lines at right angles
holes (also called *lakes*)
indentations (also called *bays*)
printing
corner
points of high curvature
limb ends on a skeleton
joints on a skeleton
texture

How are we likely to recognise these and similar image features? Some of them are recognisable by the Library predicates (see Appendix II), although there is, in many cases, no simple answer, since there are many different ways to detect even a simple figure, such as a circle or square. The method that is "best" in a given situation depends upon a variety of circumstances. For example, there are many manifestations of a circle (see Figure 8.23). Note:

(i) The ratio of the area to the square of the perimeter of a circle is $1/(4\pi)$. This forms the basis of a test, which though simple to implement, cannot be used in certain situations, for example when the circle is partially occluded.

(ii) Three arbitrarily chosen points on the circumference of a circle may be used to estimate the position of its centre and radius. These three parameters may then be used to generate a circle which is then compared to other points in the image. In this way, a simple test may be devised to verify that other points from the edge of the same image object lie on the circle.

(iii) The procedure described in (ii) may be extended to cover those situations in which the circle is incomplete.

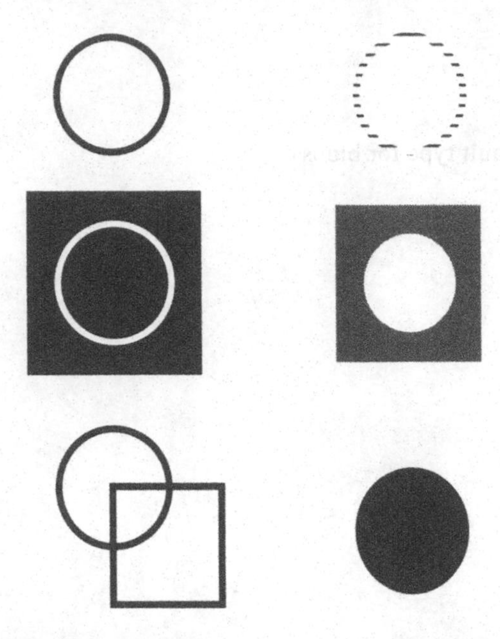

Figure 8.23. Different manifestations of a circle. Human beings refer to all of these figures as "a circle". It is likely that a Prolog+ program would require six different definitions to cover all of these cases.

(iv) Circle position and radius parameters may be estimated by a least squares fit. (This permits the use of many more data points, with greater scatter than permitted by method (ii).) The fact that a blob is a circle may be established by calculating the goodness of fit. This permits the concept of a "circle" to be given an arbitrarily sloppy definition.

(v) The Hough and Radon transforms have both been be modified to detect circles.

(vi) The minimum-area rectangle that can be drawn around a complete circle has unity aspect ratio. The difference between this rectangle and the circle contains three convex regions, each with an area of $((1 - \pi/4)/4)$ of that of the minimum-area rectangle. (This is a heuristic, not an algorithmic test, but might be useful, for example, when the circle is partially occluded by an opaque object.)

(vii) Find the centre of the top-most row which lies in the supposed circle. Repeat this for the bottom row. Construct a line which joins these two points. This provides an estimate of the centre of the circle, while the separation of the two points is an estimate of the diameter. Continue as in (i).

In addition, circles may be solid, or merely represented as outlines. When there is some overlapping with other objects, the edge of the circle may be completely hidden, or it may still be visible. (Compare Figure 8.19 with Figure 7.19.) The point is that detecting a circle may be less than straightforward, since many different tests for circularity might need to be applied, and we have to assume, here, that fast hardware is available to evaluate these tests within a reasonably short time. The same is true of other figures, such as squares, rectangles, triangles, etc.

8.11 Concluding Remarks

While we could write rules to recognise each of those sentences listed in the preceding section, it is clearly impractical to try and anticipate all possibilities; as a rule of thumb, we may assume that about two DCG rules are needed for each new type of sentence.

How can we limit the complexity of our program without imposing severe restraint on the person who will try to use English language programming for the vision system?

There are several ways that we can suggest:

(a) Forbid the use of certain words, such as **and, which** and **that**, since these are invariably associated with complicated sentence structures.

(b) Give the user a few sample sentences as an abbreviated form of training. For example, we might tell him/her to use simple sentences exemplified by those listed below

['A', is, a, circle]

['B', is, a, circle]

['A', and, 'B', are concentric]

(c) Provide (immediate) feedback to advise the user that the sentences being entered are not accepted by the parser.

Despite these comments, it should be possible to extend the program, without great effort, so that it can inspect a significant variety of industrial artifacts (see Table 8.2).

The task of recognising a banana is interesting, because it emphasises the *heuristic* nature of our approach to natural language programming. The use of natural language programming is inextricably associated with heuristic recognition procedures; we are not in the business of trying to use English to replace C, Fortran, Pascal or even Prolog. There is a saying in English that we should always use appropriate "horses for courses". The importance of this will become clear, as we discuss our final program, this time for recognising bananas.

Table 8.2. Products suitable for inspection using DCG programming techniques.

Product	Remarks
Button battery	Seen from above, this consists of several concentric circles
Aerosol can	Seen from above, this consists of several concentric circles
Reel of tape	Consists of several concentric circles
Bakewell tart	A trivial modification of Figure 8.23
Coffee mug	Both side and top views would be accommodated separately
Floppy disc (rear)	A trivial extension of the example discussed above
Cartons	With simple printing. Resemble floppy discs
Banana	Use colour, length and width
Electrical socket	Front view has three or four rectangles and two circles
In-flight meal tray	Assume that objects are separate from each other
Gift packs	Gift packs of assorted objects in pockets on a tray
Chocolates	Similar to gift packs

Here is a definition of a banana which recognises an infinity of non-bananas, although in everyday life, very few objects pass this test for banana-ness:

```
object_is (banana) :-
    colour(yellow),          % Yellow
    spotty,                  % Spotty
    length(X)
    X ≥ 50,                  % Length ≥ 50 mm
    X ≤ 200,                 % Length ≤ 200 mm
    width(Y)
    Y ≥ 20,                  % Width ≥ 20 mm
    Y ≤ 50,                  % Width ≤ 50 mm
```

The ideal use for natural language declarative programming is in defining objects in this manner. Just in case you don't remember what a banana looks like, here is its description in DCG format.

```
[a, banana, is, yellow]
[a, banana, has, spots]
[the, length, of, a, banana, is, greater, than, 50, mm]
[the, length, of, a, banana, is, less, than, 200, mm]
[the, width, of, a, banana, is, greater, than, 20, mm]
[the, width, of, a, banana, is, less, than, 50, mm]
```

I hope that, by now, you are convinced that, though not perfect, such a description is *sufficient for most purposes.*[12] (The words in italics are the key to accepting these ideas.) If you can admit this, then I hope you will also appreciate that natural language is a satisfactory programming medium.

So far we have only considered single-level, (non-hierarchical) scene descriptions. There is no provision in **describe_image** and **run_program** for defining object hierarchies. Such a facility is clearly needed if we are to describe really complex images.

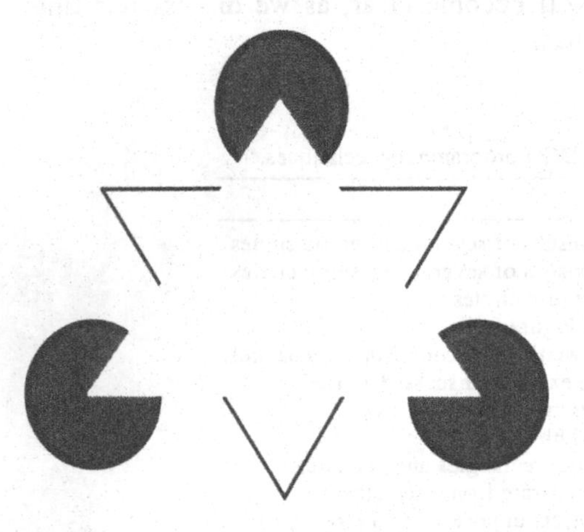

Figure 8.24. Kanizsa's triangle. Can a Prolog+ program reconstruct the "invisible" white triangle?

Although this would require only a relatively minor change to our program, we must remember that a "naive user" is intended to program the vision system using natural language. There is obviously a limit to the extent that a person, who does regularly exercise his/her mind in such activities as computer programming (whether that be in an imperative or declarative language), can think in a hierarchical manner. Indeed, it is the very discipline of using formal languages (including programming languages, flow charts and block diagrams) that permits a systems engineer to represent and think about very complex machines. By reverting to the use of natural language, we are implicitly discarding these tools. We are all willing to accept that mathematical notation is used, in preference to English, German or Russian, as a language to express certain ideas in calculus, trigonometry and algebra. In the same way, we must realise that formal programming languages will always have their place in programming vision systems; we cannot expect to use natural languages to do everything!

Some people claim that ultimately vision systems will need to cope with awkward situations, like that illustrated in Figure 8.24, which is, of course, well known from psychology textbooks. Of course, this will be very difficult to achieve. A question remains: can we achieve anything *useful*?

Notes

1. A small amount of randomness in stitch length, or slight deviations from a straight line, might be introduced to simulate a hand-made effect.
2. AVS, Inc., 640 Taft Street, Minneapolis, MN 55413, U.S.A.
3. The criteria for inspecting food products are often expressed as rules – for human inspectors to follow.
4. Although we have not discussed the control of a camera and lens in detail, this is straightforward in a language with the expressional power of Prolog+. Extra hardware is needed.
5. The user can employ the pull-down menus while the **yesno** dialogue box is visible.
6. The **balloon** predicate defined later does essentially the same thing.
7. In the plant dissection application, there is a similar situation where the stem is the major feature to be analysed. The same remarks apply to this type of image.
8. Mr I.P. Harris.
9. This can be defined as a VCS macro. The Autoview image processor possessed a primitive operator for performing this function.
10. In LPA MacProlog, it is necessary to define a window as being of type "Data Window", so that the programmer can see the results of the operations **assert** and **retract**. Moreover, it is necessary to place a "seed" in such a window, in order to ensure that **assert** knows where to place new clauses. Hence, if MacProlog executes

    ```
    assert(new_clause(qwerty,zxcvb)
    ```

    the new clause will only be placed in the Data Window if there already exists a clause with the same name. (The new clause will still be asserted but will not be visible to the programmer in the Data Window.) One convenient way to ensure that new clauses are visible is to place a "seed" clause with the same name but with a different arity in the Data Window. Such a "seed" clause might be

    ```
    new_clause.
    ```

11. With its sides at 0° and 90°.
12. Name five objects, other than bananas, that satisfy this test. Failed? Then I have proved the point!

CHAPTER 9

MULTI-PROCESSING AND OTHER IDEAS FOR THE FUTURE

"What's past is Prolog."
The Tempest, William Shakespeare

9.1 Models of Multi-Processor Systems

It has been implicit throughout the earlier chapters of this monograph that Prolog+ is used to control a single image processor. However, there are many occasions when a single view of an object, or scene, does not provide sufficient data to enable a sensible decision to be made about it. Consider the system shown in Figure 9.1, in which there are several cameras, viewing a very complex object. Each of the cameras is connected to its own dedicated image processor. These image processors must, of course, be coordinated together. For example, they must be synchronised, to capture images and to begin processing at the appropriate moment. In addition, their outputs must be combined, even if this is as simple a process as logically ANDing them together. Of course, it is an unnecessary restriction to expect that each slave processor will always run the same program; their actions need to be coordinated at a much higher conceptual level. Can this be achieved by some suitable extension to Prolog+, or is the framework that we have described restrictive in some way? Does Prolog+ need to be modified in order to build intelligent multi-processor systems?

One of the simplest multi-processor system organisations uses what might be termed the *broadcast mode* of issuing commands to the slaves; the master simultaneously addresses a group of slaves and gives them all the same task to do. The slaves then perform their tasks in parallel. If the slaves need to communicate with each other, as they perform their allotted task, they may do so directly, or via the master; the difference is merely an implementation detail and need not concern us

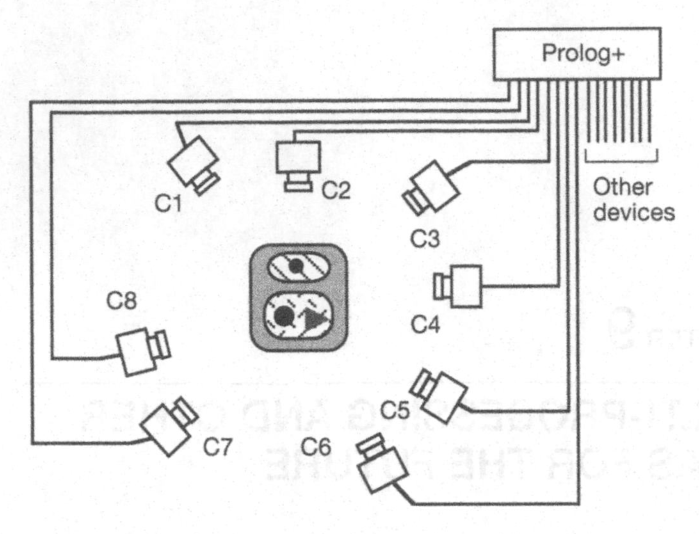

Figure 9.1. Several cameras viewing a very complex object.

here. The broadcast mode is clearly far less versatile than that in which the slaves operate concurrently, on different tasks.

In the *concurrent mode* of operation, the master will probably assign each slave a different task, at different moments in time. The master may wish to take advantage of the slaves' varying abilities, so that each slave can be given tasks that it is particularly well suited to accomplish. The essence of the concurrent mode is that the slaves can be assigned different tasks to perform at different times. Hence, the master must establish a communication protocol whereby the slaves can report back, giving the results of their work. One possible scheme is that the master accepts the request to talk to a slave as soon as it is received. When a master is already talking to a slave and another request for attention is received, the master can take one of two actions: either attend to the second slave at once, or finish the conversation with the first one and then attend to the second. The master may switch between these options, depending upon the priorities which the slaves are given. Of course, the assignment of priority is determined solely by the master and may be a dynamic process.

In a concurrent, multi-processor, robot-vision system, the slaves are auxiliary processors, which might typically be either vision systems or robot controllers. Since the master cannot reasonably be expected to exercise real-time supervision and monitoring of every minute detail of the operation of all of the slaves, they must be trusted to operate semi-autonomously, with only very occasional "high-level" commands from the master processor.

The data highway between a slave and the master is usually quiet, since the slaves are likely to spend quite a long time completing their assigned tasks. The silence on the data highway is broken only when the master gives a slave a command; when a slave requests permission to address the master; or when a slave is returning the results of its calculations to the master. When its assigned task is complete, a slave begs permission to address the master. In the discussion that follows, we shall

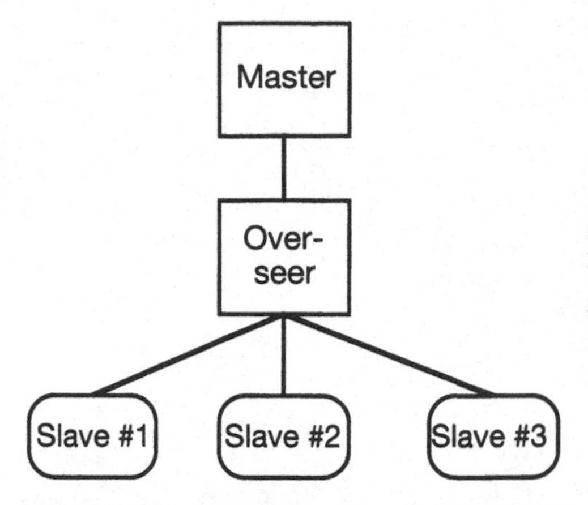

Figure 9.2. The role of the over-seer is to act as an intermediary between the master and the slaves.

assume that the master has accepted the obligation to give immediate attention, if at all possible, to each request for a dialogue from one of the slaves. This means that the master must suspend whatever it is doing and take part in a dialogue with a slave. Of course, that dialogue might alter the internal state (the Prolog+ database), so that future actions of the master might be different from what they would have been, had the conversation not taken place.

Since Prolog+ is not particularly well equipped to handle real-time activities, it may be necessary to provide it with some assistance, in the form of an over-seer which controls the slaves and which is the only subordinate unit permitted to talk directly to the master (see Figure 9.2). The over-seer is envisaged as being implemented by a processor running a program in a conventional imperative language. However, we shall see that Prolog+ can be made aware of external events using a serial input port to control the spawning of new goals. The spawning process might be likened to that performed by a conventional interrupt handler.

9.2 Handling Interrupts in Real Time

Prolog systems often provide only rudimentary means of monitoring external events in real time. LPA MacProlog, for example, permits the user to halt a program, by simply typing

COMMAND/(period).

This is the only form of interrupt that is currently available in this particular dialect of Prolog.

In MacProlog, it is possible to redefine what action should follow the receipt of an interrupt, by redefining a special built-in predicate, called **'<INTERRUPT>'**, in the following way:

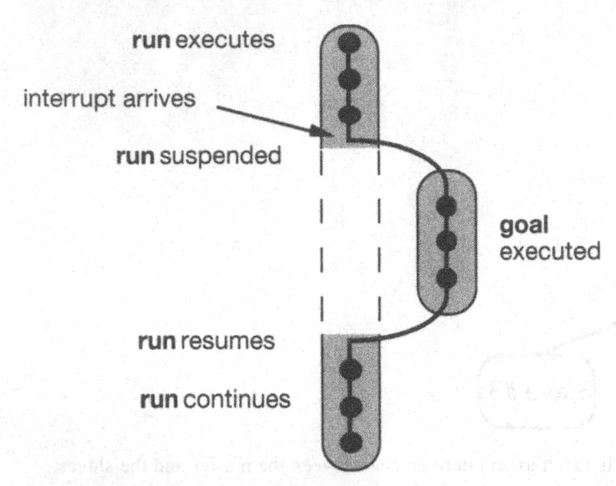

Figure 9.3. Using a meta-interpreter to poll an input port regularly.

'<INTERRUPT>'(_) :- goal.

This specifies that when the **COMMAND/(period)** interrupt arrives, the process that is currently running will be suspended and **goal** will be satisfied. Now, although this interrupt is initiated from the keyboard, some simple hardware could be built which could convert, say, a TTL logic signal into one on the Macintosh ADB bus. (This is the Macintosh keyboard/mouse bus.) In this way, it would be possible to obtain a mechanism which makes Prolog aware of external events, such as a signal from a slave processor that it needs attention. However, this is not a totally satisfactory solution, since Prolog does not handle interrupts easily or quickly. Let us develop a different approach, using what is called a *meta-interpreter*.

A meta-interpreter is a Prolog+ program which defines what action is needed to run a Prolog+ program [STE-86]. The idea of a meta-interpreter is important for our present discussion, since it is possible to write a Prolog+ program which regularly polls an input port. In this way, it is possible to simulate an interrupt handling mechanism, so that when some data arrive on an input port, a new goal is inserted into the program that is running (Figure 9.3).

Meta-Interpreter Program Listing

Here is the listing of a simple meta-interpreter program, which has been modified for the purpose described above. (Notice that the program does not have the ability to run programs which contain a **cut** (!). This will be discussed again below.)

```
% Headless clause to set-up the I/O port

:-    remember(switch2,off),                      % Set up user-defined switch
      seropen(modem),                             % Open the serial port
      serconfig(modem,both,9600,8,none,1),        % Configure the serial port
      remember(count,0).                          % Initialise the counter
```

```
% Modify the default action for non-defined relations

'<ERROR>'(2,A) :-
    swrite(A),
    swritenl(' was not defined, so it failed by default.'),
    fail.

% The meta-interpreter, contains four clauses, dealing with four types of goal

% Clause 1: What to do when the goal is "true"

interpret(true,_) :- !.

% Clause 2: What to do when a compound goal (A,B) is to be interpreted

interpret((A,B),G) :-
    !,
    interpret(A,G),                      % Apply A to the interpreter
    interpret(B,G).                      % Apply B to the interpreter

% Clause 3: What to do when the goal is a simple one in an interpreted window

interpret(A,G) :-
    idef(A),                             % Is A in an interpreted window?
    clause(A,B),                         % Find the body of A
    swrite('To prove'),                  % Optional message for the user
    swrite(A),
    swrite('we must first prove {'),
    swrite(B),
    swritenl('}'),
    interpret(B,G),                      % Now interpret B
    swrite(A),
    swritenl('is true'),
    interrupt_handler(G).                % See if there has been any input recently

% Clause 4: What to do when the goal is either a system or user-defined predicate

interpret(A,G) :-
    swrite('Compiled or system goal'),   % Optional message for the user
    swritenl(A),
    call(A),                             % Execute goal A
    interrupt_handler(G).                % Has there been any input recently?

% Versions of write and writenl that can be switched on/off using switch2

swrite(A) :-
    recall(switch2,on),
    write(A),
    !.

swrite(_) :- !.
```

```
swritenl(A) :-
        recall(switch2,on),
        writenl(A),
        !.

swritenl(_) :- !.

% Find out whether there has been any input recently

interrupt_handler(G) :-
        writenl(modem,'Send data'),   % OK to send data (message to peripheral
                                      %    device)
        put(modem,12),                % Output LF character
        serstatus(modem,in,X),        % Find state of the input buffer
        not(X = 0),                   % Is buffer empty or not?
        get(modem,Y),                 % Find first character in the buffer
        write('Character: '),         % Message for the user
        writenl(Y),
        call(G),                      % Initiate the interrupt process, G
        !.                            % Prevent resatisfaction of "interrupt_handler"

interrupt_handler(G):- !.             % No input arrived; don't initiate interrupt
                                      %    process, G.
```

A Simple Program to Illustrate the Meta-Interpreter

Notice that **go** does not contain a **cut** (**!**).

```
go :-
        member(X,[1,2,3]),            % See note1
        member(Y,[a,b,c,d]),
        writenl([X,Y]),
        fail.

go.
```

Output from the Query "go"

```
[1, a]
[1, b]
[1, c]
[2, a]
[2, b]
[2, c]
№1        yes
```

Output from the Goal "interpret(go,writenl('New process spawned'))"

Notice that switch2 was off.

```
[1, a]
[1, b]
Character: 103              % Character "g" arrived on the input port
New process spawned         % New goal initiated by the meta-interpreter
[1, c]
[2, a]
Character: 109              % Character "m" arrived on the input port
New process spawned         % New goal initiated by the meta-interpreter
[2, b]
[2, c]
Nº1    yes
```

For readers who are particularly interested to discover more about the meta-interepreter, here is the output generated by the program when switch2 is on:

```
To prove go we must first prove
    {member(_1204, [1, 2]),member(_1217, [a, b, c]),writenl([_1204, _1217]),FAIL}
Compiled/system goal member(1, [1, 2])
Compiled/system goal member(a, [a, b, c])
[1, a]
Compiled/system goal writenl([1, a])
FAIL() was not defined, so it failed by default.
Compiled/system goal member(b, [a, b, c])
[1, b]
Compiled/system goal writenl([1, b])
FAIL() was not defined, so it failed by default.
Compiled/system goal member(c, [a, b, c])
[1, c]
Compiled/system goal writenl([1, c])
FAIL() was not defined, so it failed by default.
Compiled/system goal member(2, [1, 2])
Compiled/system goal member(a, [a, b, c])
[2, a]
Compiled/system goal writenl([2, a])
FAIL() was not defined, so it failed by default.
Compiled/system goal member(b, [a, b, c])
[2, b]
Compiled/system goal writenl([2, b])
FAIL() was not defined, so it failed by default.
Compiled/system goal member(c, [a, b, c])
[2, c]
Compiled/system goal writenl([2, c])
FAIL() was not defined, so it failed by default.
To prove go we must first prove {true}
```

go is true
№1 yes

It is clear that a meta-interpreter provides a useful alternative/addition to the standard **TRACE** facility.

Using the Meta-Interpreter for Real-Time Control

As we have shown, the goal **interpret(run,goal)** will try to satisfy **run** in the normal manner. While there are no data appearing on the (modem) input port, Prolog+ will try to satisfy this goal. However, when the first character arrives on the input port, the meta-interpreter suspends action on **run** and immediately tries to satisfy the secondary goal **goal**. Clearly, the interaction between **run** and **goal** may take a variety of forms:

(i) **goal** may be totally transparent to **run**. That is, **goal** succeeds and **run** resumes with only the delay in its completion to indicate that the interrupt had any effect.

(ii) **goal** may add, modify or delete parts of the database which **run** is using. Of course, the future action of **run** may be quite different from that which would have occurred, had there been no interrupt.

(iii) **goal** may take different courses of action, depending upon where the attempt to satisfy **run** has reached. This may be entirely passive on the part of **run**, or it may deliberately deposit messages for **goal** to read when the latter is triggered into action by the interrupt.

We conclude this section with a brief discussion of the problems associated with including **!** (**cut**) in the goal **run**. The simple meta-interpreter listed above will correctly prove **!** as it moves forward, but it will not inhibit back-tracking. Sterling & Shapiro [STE-86] explain why it is difficult to write a meta-interpreter which can correctly operate with **!** and it unnecessary to repeat their arguments here. They then explain that it is possible to write a meta-interpreter which can correctly operate with **!** included in **run**, if there is a facility known as **ancestor cut**. (Unfortunately, LPA MacProlog does not possess ancestor cut.) The ancestor cut is needed in the definition of **interpret**, not **run**. It is then possible to include **!** in **run**.

We can avoid the problem associated with including **!** in the goal run, in several ways:

(a) Rewrite **run**, avoiding the use of **!** altogether. (**!** is, in any case, not accepted by "purist" Prolog programmers, who regard it as an abomination, since it violates the true spirit of the language. They therefore avoid it at all costs.)

(b) Put the lower-level predicates which **run** uses and which include **!** in compiled windows. Such predicates will not be interpreted; they are evaluated as normal, while **interpret** generates a simple output message.

(c) Write a more complex version of **interpret** which uses **assert** and **retract** to insert/delete **!** in its own code, not that of **run**.

9.3 Spawning Goals

Let us now explain what tasks **run** and **goal** might perform. Clearly, in the master-slave arrangement, **run** would act as the master and **goal** part of the slave over-seer.[2] (The over-seer is a hardware–software sub-system (Figure 9.2), which knows the wishes of the master and intelligently assigns tasks to the secondary processors, i.e. the slaves.) As we explained earlier, the last mentioned devices might perform either image processing or device control tasks. Each time an interrupt occurs (assuming that the interrupt switch is on), a new Prolog+ goal is initiated. This is the process we shall call **spawning.**

Spawning may be likened to normal goal invocation in Prolog, except that the new goal is triggered by an external event, not by the program reaching a line of code specifying what action is to be taken next. Moreover, the newly spawned goal is independent of the process that was running just before the interrupt occurred. Spawning is a model of a natural human action. To understand this, consider the following situation. A person P1 is sitting at a desk writing some notes. This is the task we shall call G1 and may be regarded as being equivalent to running Prolog+ with goal G1. A second person, P2, arrives and interrupts the writing. G1 stops. P2 is equivalent to an auxiliary processor which wishes to communicate with Prolog+. A conversation, or some other action, might then take place, involving interaction between P1 and P2. This is a goal G2, which is quite separate from G1. Afterwards, P1 will resume the initial task (task G1), but what he writes might be altered by the conversation G2. That is, P1's memory (the Prolog+ database) might be changed during the conversation.

Let us now discuss a number of predicates for spawning new goals when interrupts arrive.

Spawning Goals on Interrupt

In this section, we shall define a number of predicates for controlling what happens when an interrupt occurs.

```
% Clear the interrupt flag (The state of the flag is known only to Prolog+)

clear_interrupt :- remember(interrupt,off).

% Set the interrupt flag

set_interrupt :- remember(interrupt,on).

% Test the interrupt flag

test_interrupt :- recall(interrupt,on).

% Define what action is to follow an interrupt occurring in the future

interrupt_spawn(G) :-
    asserta(goal(G)),
    set_interrupt.
```

```
% Satisfy A, but suspend the proof-finding process if an interrupt occurs

int_call(A) :-
        test_interrupt ,            % Is the interrupt on?
        goal(B),                    % Consult database
        interpret(A,B),             % Satisfy A. Interrupt will trigger B

% Satisfy A without being concerned about interrupts

int_call(A) :- call(A).
```

Process Control Functions

The process control functions operate at a higher level than those just described. They are not listed here because their precise form depends upon the nature and I/O format of the slave processors.

The predicate **start_processor(A,B)** has four clauses:

(a) **start_processor(A,B)** initiates process B on the slave processor A. Prolog+ does not wait for the result of process B before continuing with the evaluation of the next goal. Notice that slave processor A may not be able to perform the assigned task, B. In this event, **start_processor** will still succeed, but the fact that this is so is indicated by the result that is returned by the predicate **status_processor**.

(b) **start_processor(all,initialise)** orders all slave processors to perform the initialisation function that is appropriate for them. Notice that the resulting action might be different for each slave. All slaves must respond immediately to this command. Notice that **start_processor(all,initialise)** always succeeds and that Prolog+ does not wait for the initialisation process to be completed.

(c) **start_processor(all,G,L)** forces all processors that can perform task G to do so. The list of identities of the processors that attempt to perform G is given by L. Notice that **start_processor(all,G,L)** always succeeds but that L may be instantiated to the empty list []. Again, Prolog+ does not wait for the task G to be completed.

(d) **start_processor(any,G,N)** finds one processor that can perform task G. The identity of the processor that attempts to perform G is given by N. Notice that **start_processor(any,G,N)** always succeeds but that N may be instantiated to the value **none**.

The predicate **status_processor(A,B)** instantiates B to the state of the slave processor A. (A is a scalar quantity.) B may assume one of three values:

[idle]
[running]
[results | Results_list], where Results_list is a list of the results available

The related predicate **test_processors(A)** instantiates A to a list describing the state of all of the slave processors. A typical list might be:

[[idle], [results| Results_list 1], [results| Results_list 2], [running], [idle], ...]

Both **status_processor** and **test_processors** have the effect of clearing **Results_list**. As a result, two successive calls may well yield different values:

```
status_processor(6,A),     % Instantiates A to [results, [1,2,3,4]] and clears results
status_processor(6,B),     % Instantiates A to [idle]
A = B                      % Fails because A is NOT equal to B
```

9.4 Mail Boxes

Mail boxes are widely used to control complex multi-processing systems. The concept is explained in Figure 9.4. The contents of a mail box can be set from one process, for example the master, and sensed by another, such as a slave.[3] Data transfer can also take place in the reverse direction. Both the idea and use of a mail box are similar to that of a property in LPA MacProlog, except that the latter can be only be accessed from within a Prolog program and not by an external device. This fact can be put to good use, as we hope to show.

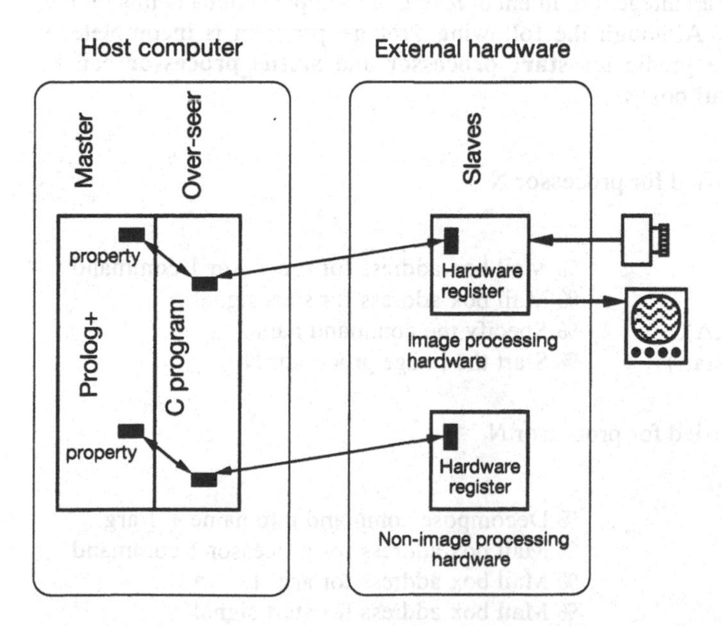

Figure 9.4. Mail boxes being used to control slave devices. Since the role of the over-seer is simply to free the master from the demands of exercising real-time control over large number of slaves, it is transparent to both the Prolog+ programmer and to the slaves. In other words, the over-seer is an implementation detail which does not alter the program flow, except to provide greater speed.

First, let us explain the idea and use of properties in MacProlog. There are two built-in predicates **remember** and **recall**, which enable the contents of a memory register to be set and sensed. These are used in the following way:

remember(R,V) % Sets the contents of register R to the value of V

recall(R,V) % Instantiates V to value stored in register R

(LPA MacProlog also allows the use of the three-argument property-handling predicates, **set_prop** and **get_prop**, but these are not relevant to our discussion here.) Now, suppose that (some of) the registers available for use by **remember and recall** could also be sensed and set by external devices, outside the Prolog host. We would then have the basis for an effective means of controlling external devices and slave processors from within a Prolog+ program. It would be a relatively straightforward matter to build a set of memory-mapped hardware registers, which could be accessed from the main bus of the Prolog+ host computer and from external processors (see Figure 9.5). Two predicates are all that is needed to form the basic inter-processor communication path:

(a) **set_mail_box(A,B)** places the data item B in mail box A.

(b) **get_mail_box(A,B)** instantiates B to the contents of mail box A.

In the program that follows, it is assumed that the Jth mail box associated with the Ith slave processor is numbered (8*I + J). It is also assumed that the mail boxes are able to represent either an integer argument or result, or a simple command mnemonic, such as **neg**, **thr**, etc. Although the following Prolog+ program is incomplete, it does indicate how the predicates **start_processor** and **status_processor** can be implemented using mail boxes.

```
% No arguments specified for processor N

start_processor(N,A) :-
    D is 8*N,                   % Mail box address for processor 1 command
    I is D + 5,                 % Mail box address for start signal
    set_mail_box(D,A),          % Specify the command name
    set_mail_box(I,start).      % Start the image processor N

% One argument specified for processor N

start_processor(N,A) :-
    A =.. [B,A1],               % Decompose command into name + 1 arg.
    D is 8*N,                   % Mail box address for processor 1 command
    E is D + 1,                 % Mail box address for arg. 1
    I is D + 5,                 % Mail box address for start signal
    set_mail_box(D,B),          % Specify the command name
    set_mail_box(E,A1),         % Specify the first parameter value
    set_mail_box(I,start).      % Start the image processor N
```

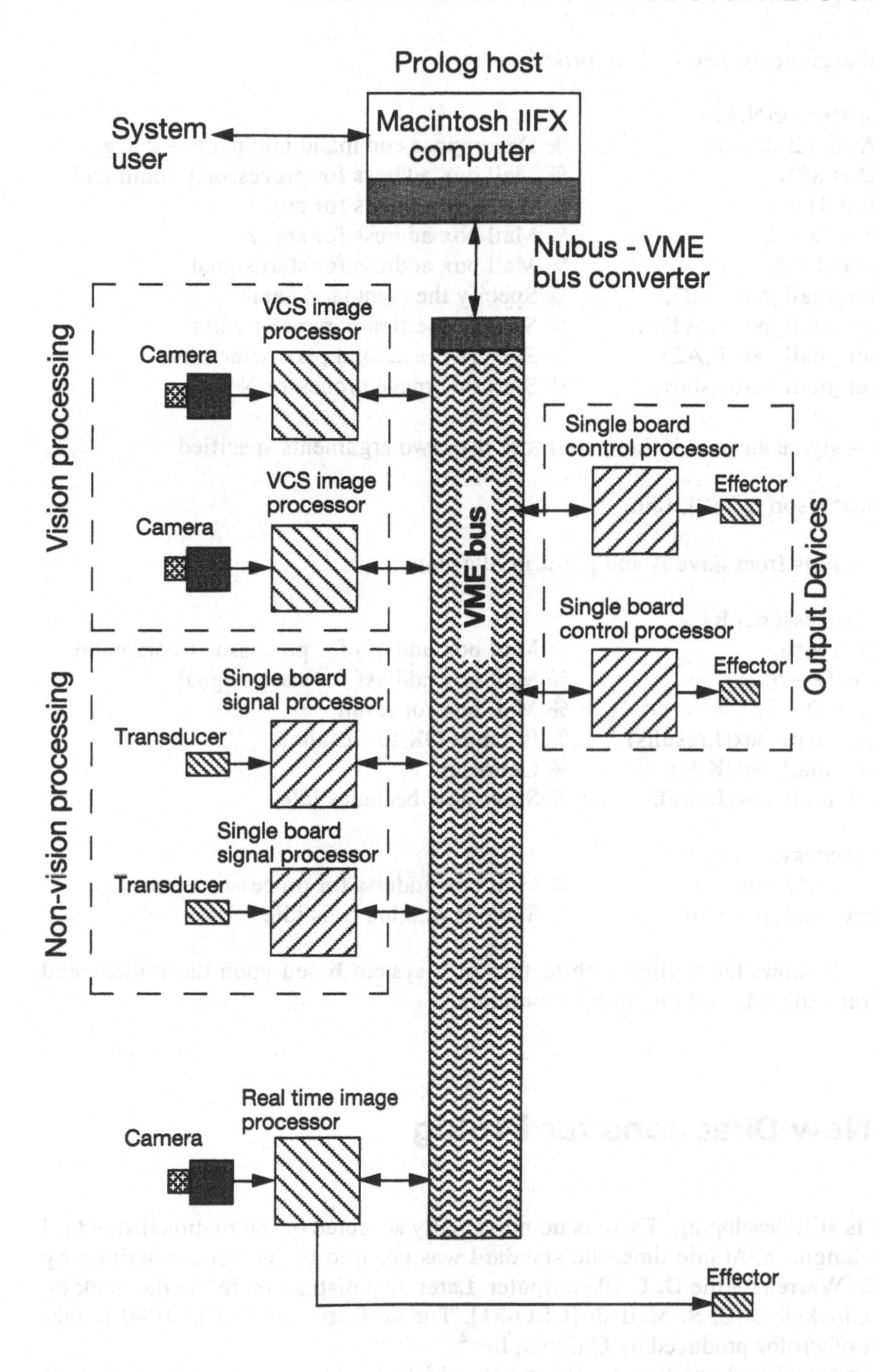

Figure 9.5. Using mail boxes and a hardware bus-adaptor to enable a Prolog+ system to control several slaves which operate on a different protocol. This system is currently being built by the author in his laboratory.

```
% Two arguments specified for processor N

start_processor(N,A) :-
      A =.. [B,A1,A2],          % Decompose command into name + 2 arg.
      D is 8*N,                 % Mail box address for processor 1 command
      E is D + 1,               % Mail box address for arg. 1
      F is D + 2,               % Mail box address for arg. 2
      I is D + 5,               % Mail box address for start signal
      set_mail_box(D,B),        % Specify the command name
      set_mail_box(E,A1),       % Specify the first parameter value
      set_mail_box(F,A2),       % Specify the second parameter value
      set_mail_box(I,start).    % Start the image processor N

% Fail – signal an error if there are more than two arguments specified

start_processor(_,_) :- !, fail.

% Get results from slave A and put it in idle state

status_processor(A,R) :-
      D is A*8,                 % Mail box address for processor 1 command
      J is D + 6,               % Mail box address for status signal
      K is D + 7,               % Mail box for result
      get_mail_box(J,results)   % Is status OK for results?
      get_mail_box(K,R),        % Get result
      set_mail_box(J,idle).     % Slave now becomes idle

status_processor(A,B) :-
      C is A*8 + 6,             % Mail box address for processor command
      get_mail_box(C,B)         % Slave is running or is idle
```

Figure 9.5 shows the outline architecture of a system based upon these ideas and which the author is building in his laboratory.

9.5 New Directions for Prolog

Prolog is still developing. There is no universally accepted or international standard for the language. At one time, the standard was taken to be the version written by D. H. D. Warren for the DEC 10 computer. Later, that distinction fell to the book by W. F. Clocksin & C. S. Mellish [CLO-81]. The de facto standard in 1990 is that version of Prolog produced by Quintus, Inc.[4].

There are several exciting developments which should make Prolog+ even more powerful in the near future. Some of these already exist in a commercial product, while others are merely promised. As has been stated earlier, LPA MacProlog already supports:

(a) Graphics
(b) Forward chaining
(c) Object oriented programming
(d) An expert system tool-kit

although Prolog+ does not yet make full use of these facilities. It should be understood that there is no fundamental problem to inhibit this development, merely the lack of engineering effort available.

Within the very near future it is confidently expected that LPA MacProlog or other dialects of the language will also support:

(i) Data arrays
(ii) Arithmetic formulae
(iii) An improved interface to conventional languages (e.g. C and Pascal)
(iv) An interface to a spreadsheet program (e.g. Excel)
(v) An interface to a database program and a query language (e.g. HyperCard, SQL)
(vi) Interfacing software for semi-autonomous image processing, signal processing, data acquisition and other hardware
(vii) Interfacing software for hyper-media. (The present, rather modest speech synthesis interface is only just the beginning of this)
(viii) Real-time control facilities (e.g. a more powerful interrupt handler)

Clearly a flood of new facilities like these presents a challenge to the software designer and it may well take some considerable time, possibly years, before all of these new ideas can be incorporated into a well integrated, fully engineered package. What we have begun in the present version of Prolog+ is just the beginning!

9.6 Communicating Sequential Prolog

There have been several attempts to adapt Prolog so that it can make good use of a parallel/concurrent processor as its host. One of these variants is called *Communicating Sequential Prolog* (CS-Prolog)[8] and it was developed specifically to operate upon a network of transputers [FUT-89]. At first sight, CS-Prolog seems to offer considerable scope for extending the ideas implicit in Prolog+.

In order to understand the nature of CS-Prolog, consider a set of standard Prolog programs, each running on a different processor (transputer). These programs are assumed to be able to communicate with each other, by passing messages between them. (We have already seen something very similar to this in Section 9.4, where we discussed mail boxes as a means of getting a single Prolog+ system to control a number of slave image processors.)

A Prolog goal, being evaluated by an intepreter/compiler, will be called a *process*. Any number of processes can be evaluated concurrently in CS-Prolog. It is important to appreciate that these processes operate independently, except that, from time to time, a process can pass a message to another process, or perhaps to several processes. Synchronisation between a pair of cooperating processes is achieved in the following way. The execution of a particular process (P_j) is suspended if a message ($M_{i,j}$) is expected from another process (P_i) but has not yet arrived. When P_i does eventually send $M_{i,j}$, the execution of process P_j will resume. However, the execution of P_i will not be delayed, if P_j has not anticipated the arrival of $M_{i,j}$ from P_i. Messages provide the only mechanism, in CS-Prolog, for communicating between processes. Other possible mechanisms, for example, shared variables and accessing dynamically defined databases, are not allowed.

CS-Prolog has several built-in predicates, which enable it to handle messages. One of these, is **send(A,B)** and sends the message A to all of those processes named in the list B. Obviously, A must not contain any uninstantiated variables.

In addition, there are two other important predicates **wait_for(M)** and **wait_for_dnd(M)**. These differ in the way they work on back-tracking. We shall explain **wait_for** first. Let us suppose that process, p1, has already sent a message, using

send('This is a message for processor number 2',[p2])

and that process p2 contains the following goal, in which Z is uninstantiated.

wait_for(Z)

This goal will immediately succeed and Z will become instantiated to

'This is a message for processor number 2'.

If, on the other hand, p2 tried to satisfy

wait_for(Z)

before p1 had encountered

send('This is a message for processor number 2',[p2]),

process p2 would be suspended until the message arrived. If the first message sent is not unifiable with the argument of **wait_for**, the receiving process (p2) will remain suspended until one that does match is received. Table 9.1 explains the sequence of actions in response to an unwanted message (**'Message for p2'**) and a wanted one (**my_message**).

Notice that **wait_for** is deterministic; it simply waits for the first matching message sent to it. On the other hand, **wait_for_dnd** is not deterministic; in the event of

Table 9.1

Process p1	Process p2	Process p3
send('Message for p2',[p2,p4,p9])	[p2 is running and so it ignores message from p1]	—
—	wait_for(my_message)	—
—	[p2 is suspended]	—
—	—	send(my_message,[p2,p6])
—	[p2 resumes]	—

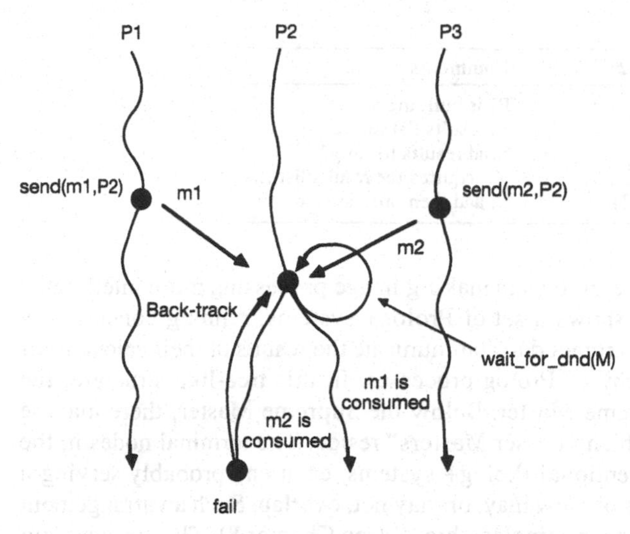

Figure 9.6. Non-deterministic flow in a multi-stream Prolog system controlled by message passing.

back-tracking, it can pick up additional messages received from other **send** procedures (see Figure 9.6).

Of course, the messages passed between processors can be a bit more interesting that those that we have indicated so far. For example, p1 can build a goal for processor p2 to run (Table 9.2). Another interesting possibility is that the message contains a results list (Table 9.3). Finally, notice when the second argument of **send** is uninstantiated, the message is sent to all other processors. Hence, the goal

send(Q,_)

broadcasts the Q to all processes.

In addition to **wait_for, wait_for_dnd** and **send,** there are several other predicates peculiar to CS-Prolog:

run(G) Runs the goal G on a special processor, called the "root" processor.[5]

new(G,N,T) Creates a process with goal G and name N on processor T.

delete_process(P) Delete process(es) unifiable with P. Deleted processes are restored when back-tracking occurs in the caller process.

In addition, there is a mechanism for communicating time signals from process to process, but this need not concern us here.

Table 9.2

Process p1	**Process p2**	**Comments**
Instantiate A, B & C	—	P1 is building a ...
M = .. [A,B,C],	—	... goal for p2 to run
send(M,[p2]),	—	Send goal as a message to p2
—	wait_for(Z),	p2 acquires the new goal ...
—	call(Z),	... and then runs it

Table 9.3

Process p1	Process p2	Comments
process1(A,B,C)	—	P1 is building a ...
M = [A,B,C],	—	... results list for p2
send(M,[p2]),	—	Send results list to p2
—	wait_for(Z),	p2 acquires the results list ...
—	process2(Z),	... and then processes it

How can these ideas be useful to us in making image processing more intelligent? Consider Figure 9.7, which shows a set of Prolog+ systems, running concurrently (on different processors) and arranged to communicate the results of their calculations *upwards* , through a hierarchy of Prolog processes. In this tree-like structure, the "root" processor is the Supreme Master. Below the Supreme Master, there may be several different levels in which "Lesser Masters" reside. The terminal nodes in the tree are intended to be conventional Prolog+ systems, each one probably serving a different camera. Their fields of view may, or may not, overlap. Such an arrangement is, of course, ideal for inspecting complex objects (see Chapter 8). The arrangement shown in Figure 9.7, though as yet untested, seems to hold considerable promise for the future. Let us now list the advantages that this system appears to offer.

The *lowest level* in the hierarchy is a set of standard Prolog+ systems. Each one, acting on its own, provides a powerful medium for expressing intelligent image processing procedures.

The network organisation shown in Figure 9.7 suggests a natural way in which the ideas expounded in this book can be extended, seemingly without limit.

Several Prolog+ systems can operate simultaneously on different views of an object. These views may overlap; they may be disparate; or they may represent different degrees of zoom. The same computing architecture can handle all of these.

The structure is highly systematic and message passing follows exactly the same protocol, throughout the tree.

Messages passing between processors (particularly in those levels of the tree close to the root) can represent very abstract ideas. Processors further away from the root will probably rely upon more concrete concepts. There is a smooth, "seamless" transition from one level of abstraction to another.

The hierarchical architecture allows the expressional power of Prolog to be put to good use in a consistent, well-structured manner.

The system provides a consistent means of processing images and controlling external devices, such as lights, robot, manufacturing machines, etc.

The operating speed (of inferential decision-making processes) can be increased, virtually indefinitely, simply by adding more processors.

The system can be tailored easily to suit the demands of a given application.

The system can be made to degrade gracefully should one unit fail. Self-testing is also straightforward. The homogenous structure makes servicing and the provision of spares straightforward.

Finally, it is interesting to ponder whether a "standard" Prolog+ system could simulate a tree-like structure like the one shown in Figure 9.7. A Prolog+ system

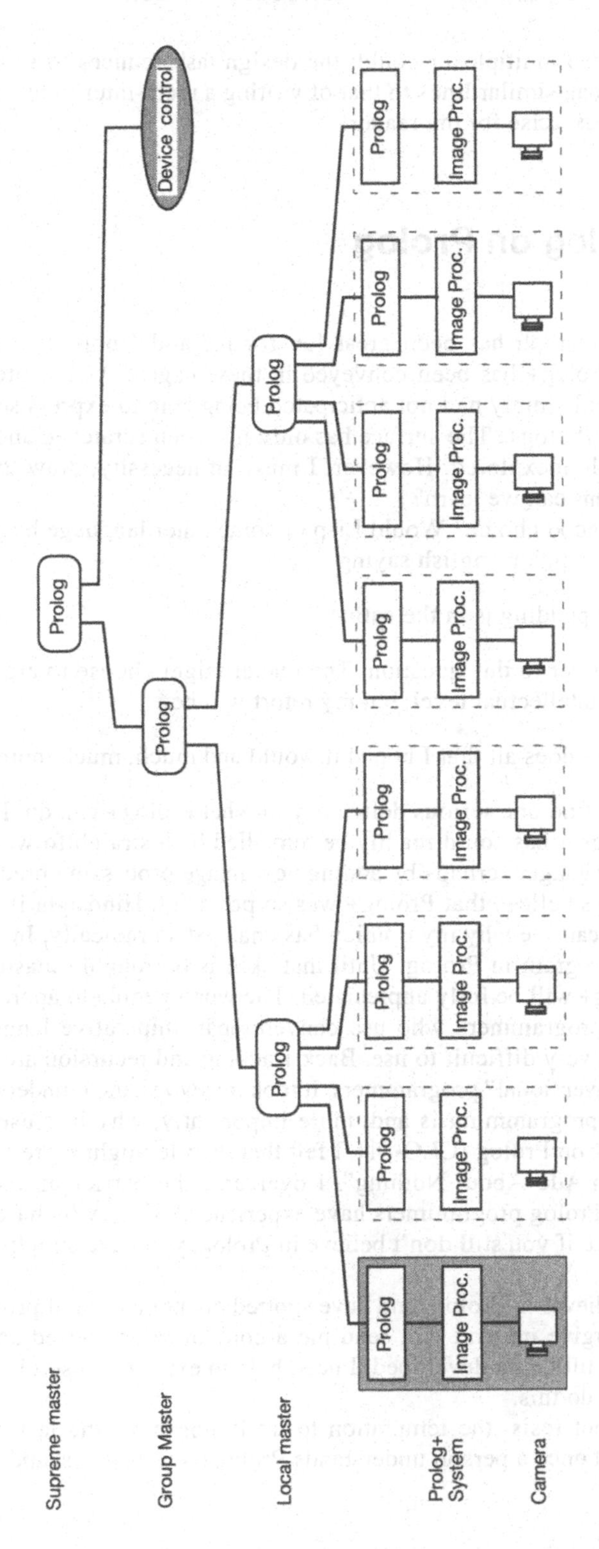

Figure 9.7. Proposed architecture for the next range of intelligent image processing systems.

with an input video multiplexer could; the design task reduces to standard Prolog programming, along similar lines to that of writing a meta-interpreter (Section 9.2). This is left as an exercise for the reader!

9.7 An Epilog on Prolog

Writing this monograph has been great fun for me and I hope that some of my enthusiasm for Prolog+ has been conveyed in these pages! As I wrote, many new ideas came to me; I simply had not anticipated being able to express so many, very different ideas in Prolog+. The surface has only just been scratched and there many exciting ideas still unexplored. However, I must, of necessity, draw this book to a close. What lessons can we learn?

Was Prolog a good choice? Would Lisp or some other language have been more appropriate? The popular English saying

"The proof of the pudding is in the eating"

represents my answer to this question. The reader might choose to argue this point at a much higher intellectual level, but my retort will be

"It works! Prolog+ does all that I hoped it would and much, much more."

I cannot honestly find one serious deficiency in what Prolog+ can do. I believe that what shortcomings it has could readily be remedied by a straightforward extension of the existing language, perhaps by adding new image processing predicates.

I did not always believe that Prolog+ was so powerful. Hindsight is wonderfully exact, and now I can see why my opinion has changed so radically. In short, I have learned how to program in Prolog. Until that skill is thoroughly mastered, neither Prolog nor Prolog+ will be fully appreciated. The worst people to appreciate Prolog are experienced programmers who use conventional, imperative languages. They find its structures very difficult to use. Back-tracking and recursion are not familiar tools to most "conventional" programmers. It took me two years to understand exactly what declarative programming is and, more importantly, why it is useful. When I first read *the book* on Prolog [CLO-81], I felt that its title might more appropriately have been "Much Ado About Nothing". I overcame the barrier of understanding, which almost all Prolog programmers have experienced, simply by hard work. I am sorry to report that, if you still don't believe in Prolog, you have simply got to study it even more!

If you are a believer in Prolog, and have spotted my poor style of programming, I hope you will forgive me and will send me a copy of your revised and improved code. Prolog+ is still being developed. I need help to extend its usefulness further – please help me to do this.

Finally, I cannot resist the temptation to let Prolog have the last word, which clearly shows that once a person understands Prolog there is no escape!

```
go :-
    yesno(['Do you understand this program?']),
    next_action.

go :-
    study(prolog),
    go.

next_action :-
    recommend(friends, prolog),
    use(prolog).

next_action :-
    use(prolog),
    recommend(friends, prolog).
```

Notes

1. The list/set membership predicate in LPA MacProlog is **on**. Throughout this monograph we have used **member** instead, since this more accurately reflects the function that we wish to perform. This also avoids ambiguity with the meaning "is on top of", which is useful when we are discussing tasks such as stacking.
2. It may be necessary or desirable for hardware reasons, to split the task of over-seer between the predicate **goal** and a program running on a separate intermediate-level processor. However, this is a detail which need not concern us at this point.
3. The over-seer in Figure 9.4 is effectively transparent to the master.
4. Quintus Inc., 1310 Villa Street, Mountain View, CA 94041, also market LPA MacProlog in the U.S.A.
5. The "root" processor is the only one connected to the host computer.

Appendix I

VCS Command Repertoire

The following is a complete list of the commands available in the commercial VCS software package at the time of writing (March 1990).[1] Operations which are not useful within the context of Prolog+ are indicated by italics.

Image Arithmetic

acn	Add a positive, or negative, constant to each intensity
mcn	Multiply each pixel intensity by a constant
abs	"Absolute value" i.e. fold intensity scale about half-white
din	Double all intensities
hin	Halve all intensities
sqr	Square intensities
exp	Exponentiate intensities
log	Logarithm of intensities
add	Add current and alternate images
sub	Subtract alternate image from current image
mul	Multiply current image by alternate image
div	Divide current image by alternate image
dif	Absolute difference. Subtract alternate and current images. Ignore sign

Binary Image Operations

bed	Edge detector, for binary images
blb	Fill all holes ("lakes")
ccc	Draw the circumcircle around a blob-like object
chu	Draw the convex hull around a blob-like object
exw	Expand white regions (add one pixel all way around)
skw	Shrink white regions (remove one pixel all way around)
dil	Dilate image. Expand white regions along given directions
ero	Erode image. Shrink white regions along given directions
egr	Edge grow. Extend arcs
mid	Mid-point skeleton. Crude, fast approximation to the true skeleton
mdl	Medial line transformation. Draw skeleton of a blob-like object
cny	Preserve pixels critical for connectivity. Remove those which are not
cnw	Count white neighbours in each 3 x 3 neighbourhood
rnc	Run-code; code image in a series of black and white runs
irc	Regenerate image from run-code

Extracting Features from a Binary Image

dim	Find dimensions of a blob
dms	Find dimensions of a sparsely scanned blob
blp	Find dimensions, area and centroid of each blob
eul	Euler number. Number of blobs minus number of holes
bve	Find object's extremities along a given vector
cgr	Find coordinates of the centroid of a blob-like object
cgs	As cgr, in a sparsely scanned image
lmi	Locate centroid, area, angle of minimum moment of inertia
med	Median coordinates of the image (may be grey or binary)
ilb	*Initialise learn buffer for learn/recognise functions*
lrn	*Learn a shape and store the result in a buffer*
rec	*Recognise a shape by comparing with others stored by lrn*
ndo	Shade distinct objects
fcc	Chain code (Freeman code)
ifc	Reconstruct image from the code chain code
mat	*Binary match*

Extract Features from a Grey-Scale Image

avr	Average intensity
gav	Average and standard deviation of intensites
gli	Intensity limits
med	Median coordinates
cur	*Cursor-controlled investigation of grey-level values*
kur	*Like CUR but uses a mouse instead of a cursor*
ppi	Print point intensities of pixels in a given rectangle
pri	Plot intensity along a given row
lgt	Store the intensities along a line
vgt	Store grey-levels along a vector
hgi	Intensity histogram
hgc	Cumulative histogram
hge	Equalisation histogram
hpl	Plot graph of contents of a given buffer
hpi	Plot intensity histogram of image (= HGI + HPL)
hpc	Plot cumulative histogram of image (= HGC + HPL)
hti	Tabulate intensity histogram
htc	Tabulate cumulative histogram
hnt	Find the Nth trough in a given buffer

Filters (for Binary or Grey-Scale Images)

con	General linear 3×3 convolution operator

cnr	General radial 3 × 3 convolution operator
hpf	High-pass filter
lpf	Low-pass filter. Equivalent to **con(2,3,2,3,5,3,2,3,2)**
bfl	Noise removal for a binary image
brm	Remove isolated black pixels from a binary image
wrm	Remove isolated white pixels from a binary image
mdf	Find Nth largest intensity in each 3 × 3 neighbourhood (N=5, median filter)
raf	Local average filter; replace each pixel with mean intensity in M × N neighbourhood
lnb	Largest neighbour; replace each pixel with maximum intensity in 3 × 3 neighbourhood
red	Roberts edge detector
sed	Sobel edge detector
gra	Crude, simple edge detector
lgr	Largest gradient of each 3 × 3 neighbourhood
bgr	Biggest gradient; largest 4-neighbour, or direction code
hgr	Horizontal gradient
vgr	Vertical gradient
drg	Derive radial gradient relative to a given address
dbn	Direction of brightest neighbour
rid	Rank intensity direction; direction of Nth brightest neighbour in 3 × 3 neighbourhood
gfa	Grass fire transform

Geometric Operations

aad	Aspect adjust: compensate for aspect ratio of camera
ang	Calculate angle and length of vector given start and end points
ctr	Cartesian-to-radial transformation
rtc	Radial-to-Cartesian transformation
tbt	Top-to-bottom transformation (i.e. reflect image about its middle row)
lrt	Left-to-right transformation (i.e. reflect image about its middle column)
yxt	Y-to-X transformation (i.e. reflect image about the diagonal, y = x)
pex	Picture expand: double image size
psq	Picture squeeze: halve image size
psh	Picture shift
psw	Picture shift, with wraparound
roa	Rotate current image anticlockwise 90°
roc	Rotate current image clockwise 90°
tur	Turn current image through a given angle about a given position

Graphics Functions

zer	Fill the current image with black
set	Fill the current image with white

gry	Fill the current image with a given grey-level
edg	Set the border of the current image to given grey-level
pfx	Fix intensity of a given point at a given grey-level
vpl	Vector plot; draw a straight line at a given grey-level
box	Fill a given rectangle with a given grey-level
cir	Fill a given circle with a given grey-level
lab	Write text on the image
wgx	Wedge: intensity varies across the screen
lic	Linear intensity cone
hic	Hemispherical intensity cone
dra	*Draw curves on framestore image using mouse*
gcs	*Graphics colour select*
gof	*Graphics mode off*
gon	*Graphics mode on*
gpf	*Graphics protection off*
gpn	*Graphics protection on*

Grey-Scale Functions

heq	Histogram equalise over given intensity range
lge	Linear grey-scale expansion within given limits
hil	Highlight pixels in a given range; set them to given grey level
rin	Integrate from left to right within each row
csh	Column shift; repeat the right-most column across the window
rox	Row extend; set each pixel to the highest intensity found to the left along that row
lfx	Line fix: set row of pixels to the values contained in a buffer
max	Maximum; set each pixel in window to higher of current or alternative
min	Minimum; set each pixel in window to lower of current or alternative
neg	Negate
psa	Picture shift and accumulate if a given grey level matched
thr	Threshold pixels inside a given intensity range to white; outside to black
fgr	Fix grey levels inside given grey range and set the rest to black
pct	Percentage threshold; set threshold to produce given percent of black
tlu	Threshold look-up; threshold images in real time under user control
dlt	Delta code

Input/Output Functions

ctm	Camera to monitor continuous frame grab (Stop with **frz**)
frz	Freeze frame in the framestore after continuous grab (**ctm**)
ifs	Input to framestore; grab frame and input to current image
cap	Capture frame when an external event occurs
fil	Select video input filter
gai	Select video input gain

mod Select framestore mode
off Select video input off-set
lnt *Load negative transform to input LUT; negate incoming image*
stl *Set threshold; set input LUT threshold limits to give binary from grab*
tlu *Threshold lookup table; adjust input LUT threshold limits during grab*
rfs Read from framestore and store in current memory image
ofs Output current memory image to framestore (usually automatic)
wri Write whole of current image to disc (even if window is on)
rea Read image from disc (file created by previous **wri**)
sab Save buffer of any type or length to disc file
lob Load buffer from disc file (usually created by **sab**)

Binary Image Processing

and AND corresponding pixels in current and alternate images
ior OR corresponding pixels in current and alternate images
xor Exclusive OR of corresponding pixels in current and alternate images
not Logical inverse
bis Set specified bits in each pixel intensity
bic Clear specified bits in each pixel intensity
bif Complement specified bits in each pixel intensity
sca Clear less significant bits in each pixel

LUT Lookup Table Functions

tra Transform each pixel in current image using a given LUT
slt Linear LUT
lnt Load negative transform to input LUT; negate image
stl *Set threshold lookup for live grab to produce a binary image*
tlu *Threshold LUT: adjust input LUT threshold limits during grab*
slu Select input lookup table number from several previously loaded
wlu *Write a given input LUT to the framestore from a given buffer*

Miscellaneous Functions and Utilities

ini Initialise/reset whole system
fsi *Framestore initialise: reset all LUTS and special modes*
sri *Set framestore region of interest to image processing functions*
fsm *FrameStore mode on; memory image mode off*
mim *Memory image mode on; framestore mode off*
rco Region copy; copy from one region of the framestore to another
cpy Copy current image to alternate image
pag *Page switch; select alternate framestore page*

psl	*Page select; select page 1 or 2 of the framestore (MIM only)*
don	*Turn automatic display of current image on*
dof	*Turn automatic display of current image off*
son	Switching on
sof	Switching off
swi	Switch current and alternate images (so does <CR>)
tab	Tabulate contents of a given buffer on the terminal
pon	*Printing on; automatic printing of pixels in a selected area*
pof	*Printing off; turn off automatic printing of selected area*
lsz	Select label size for **lab, hpl,** etc.

Window-Related Functions

won	Limit processing to defined window
wof	Process over whole image
swc	Set window coordinates/assign window number
gwc	Get window coordinates; display current window number and its coordinates
mwc	*Move window coordinates; set position and size of window interactively*
mwk	*Like mwc but using a mouse instead of cursor keys*
wgr	Write to selected window from a region of the same size
wcw	Write a region of image from the window
wrw	Write the current window to a specified disc file
rew	Read new image from disc into the current window (use only after **wrw**)
wfx	Set pixels in a given rectangle to values from given buffer
wgt	Store pixels in a given rectangle in a given buffer
ron	*Relative coordinate mode on (relate coordinates to top-left of window)*
rof	*Relative coordinate mode off*
aon	Absolute addressing mode on (i.e. same as rof)

Notes

1. The author is grateful to the directors of Vision Dynamics Ltd. for their permission to reproduce this list, which is based upon information to be found in the manual of the VCS image processing software. New commands are being added continually and the author has access to certain commands not yet available in the commercial version of the software. A few of these are mentioned in the main text.

Appendix II

Library Predicates

The Library Predicates can be divided into the following major categories:

List processing and arithmetic functions
Standard images stored on disc and test images
Self documentation
Stack (Last-in-first-out store)
System control
Speech synthesis
Journal control
Lighting control
Device control
Camera control
LPA MacProlog default controls
Operators

The more important Library predicates are listed below. In addition, nearly all of the predicates discussed listed elsewhere in this monograph are held in the Library, although they are not listed here. Since the Library is continually being extended, the following list does not constitute a "finished product".

Predicate and arguments	Operation/test
List Processing and Arithmetic Functions	
create_integer_set(A,B,C,D)	Match D to the set formed from the union of C and the set of integers in the range [A,B]
keep_even_terms(A,B)	Match B to the list containing only the even terms in the "input" list A
keep_odd_terms(A,B)	Match B to the list containing only the odd terms in the "input" list A
keep_constants(A,B)	Match B to the list containing only the defined elements in the "input" list A
keep_constants(A,B)	Match B to the list containing only the variables in the "input" list A
shrink_list(A,B,C)	Match C to the string formed by concatenating the (string) elements contained in A
list_print(A)	Perform a "pretty print" routine upon the input list A
list_threshold(A,B,C)	Match C to that list consisting of 1s and 0s formed by thresholding the list B at level A
list_run_lengths(A,B)	Match B to the list formed by finding runs of identical elements in list A
list_smooth1(A,B)	Match B to the list formed by applying a three-point averaging operator to the numeric elements in the list A

Predicate and arguments	Operation/test
list_smooth2(A,B)	Match B to the list formed by applying the **middle_value** operator to the numeric elements in the list A
list_max(A,B)	Match B to the maximum value of the numeric elements in A
list_min(A,B)	Match B to the minimum value of the numeric elements in A
list_sum(A,B)	Match B to the sum of the numeric elements in A
list_range(A,B)	Match B to the difference between the maximum and minimum values of the numeric elements in A
list_mean(A,B)	Match B to the mean of the numeric elements in A
list_variance(A,B)	Match B to the variance of the numeric elements in A
middle_value(A,B,C,D)	Match D to the median value of the numbers A,B and C
strip(A,B,C)	Match C to the list formed by stripping the first B elements from the "input" list A
cull(A,B,C)	Match C to the set obtained by culling the first occurrence of A from the list B

Standard Images Stored on Disc and Test Images

face	Read the standard "face" image from disc
standard_binary_image	Read the standard binary image from disc
standard_industrial_image	Read the standard industrial image from disc
piece_parts	Read the piece parts image from disc
con_rod	Read the con-rod silhouette image from disc
staircase	Generate an image with a "staircase" intensity pattern (used to set up monitor)
test_pattern	Generate a test pattern
grid	Generate a grid pattern
random_image(A)	Generate random image

Self Documentation

update_self_documentation	Update the self-documentation database[1]
get_help(A)	Find "help" information about topic A
document_it	List the set of Library predicates
predicate_classes(A)	Match A to the set of Library predicate classes
class_contains(A,B)	Match B to list of Library predicates in class A
window_gen_times_and_dates	List times and dates when each of the program windows was last edited

Stack (Last-In-First-Out Store)

lifo_pop(A,B)	Match B to the result of popping stack A
lifo_push(A,B)	Push B onto stack A
lifo_clear(A)	Clear/create a new image stack called A
lifo_kill(A)	Destroy stack A and delete all data stored on it
lifo_read(A)	Read the top element from stack A. Do not alter the stack
image_lifo_pop(A)	Pop an image from image stack A
image_lifo_push(A)	Push an image onto image stack A
image_lifo_clear(A)	Clear/create a new image stack called A
image_lifo_kill(A)	Destroy image stack A and delete all images stored on it
image_lifo_read(A)	Read an image from image stack A. Do not alter the stack

Predicate and arguments	Operation/test
System Control[2]	
initialise_system	Restart the system by resynchronising Prolog and the image processor
purge	Clear all data from the I/O channel (i.e. between Prolog and the VCS image processor. (Used to tidy up after a keyboard interrupt)
transparent_mode	Switch transparent/interactive mode on
load_macro_library	Load the VCS macro library into RAM
load_vcs_program(A)	Load a VCS program into RAM[3]
run_vcs_program(A)	Run the VCS program currently in its memory
add_process_command	Tell Prolog+ that there is a new VCS processing command available
add_analyse_command	Tell Prolog+ that there is a new VCS analysis command available
extend_menu	Extends one of the pull-down menus. Uses a convenient interactive dialogue
Speech Synthesis	
speak(A)	Utter the phrase defined by A
speak_list(A)	Utter the phrases defined by the list A
switch(speech,A)	Switch speech synthesiser on/off
say_time	Tell the time
say_date	Say the date
cut	Use **!,cut** in lieu of !
repeats	Use in lieu of **repeat**
fails	Use in lieu of **fail**
Journal Control	
switch(journal,A)	Switch the journal recorder on/off
switch(banner,A)	Switch banner display of current IP command on/off
Lighting control	
light(A,B)	Set lamp A to brightness level B
laser(A)	Switch the laser on/off
back_light(A)	Switch back-lighting unit on/off
fibre_optic(A)	Switch the fibre-optic light source on/off
ultra_violet(A)	Switch the ultraviolet light source on/off
lighting_state(A)	Match A to the lighting state vector
projector(A)	Switch the slide projector on/off
step_projector(A)	Step the slide projector forward/backward
Device Control	
home	Send the robot to its home position
calibrate_axes	Determine mapping between robot and image processor coordinate axes
convert_axes(A,B)	Convert between robot and image processor coordinate axes
grasp	Operate robot gripper
release	Release robot gripper
move_to(A)	Move a multi-axis robot to a point in space defined by list A
nudge(A)	Move robot by an incremental amount, defined by A

Predicate and arguments	Operation/test
robot_at(A)	Where is the robot now?
solenoid(A,B)	Switch solenoid A on/off
output_port(A,B)	Set the output port A to the bit pattern defined by B
input_port(A,B)	Instantiate B to the state of the input port B
Camera Control	
select_camera(A)	Choose a camera
zoom(A)	Set zoom of selected camera
focus(A)	Set focus of selected camera
aperture(A)	Set aperture of selected camera
focus(A)	Set focus of selected camera
pan(A)	Set pan of selected camera
tilt(A)	Set tilt of selected camera
camera_state(A)	Find camera state-vector
filter(A)	Rotate filter turret so that colour filter A is in front of the lens
LPA MacProlog Default Controls	
'<ABOUT>'(A)	Give user a bit of information about Prolog+
'<ERROR>'(A)	Undefined predicates fail
'<INTERRUPT>'(A)	Abort the current program, when a keyboard interrupt arrives. Disable the robot and any other dangerous devices (e.g. laser). Purge the I/O port
'<DEBUG>'(A)	Modified debugger which uses speech
Operators	
# (prefix)	Primitive image processor control operator. Used in VSP
@ (prefix)	Alternative to #
¶ (prefix)	Primitive device control operator. Used in VSP
• (infix)	Repeat goal a given number of times
if (infix)	Use in lieu of :-
& (infix)	AND, use in lieu of ','
or (infix)	OR, use in lieu of ';'

Notes

1. The Library predicates each have descriptions in a defined format and are embedded within the comments just above the program code. When update_self_documentation is invoked, these desriptions are extracted from the comments and are stored in regular Prolog+ facts. document_it and get_help both scan this database, in order to find the required information.
2. These predicates are specific to the particular implementation consisting of the VCS image processor linked to LPA MacProlog.
3. Both macros and programs are available can be written in VCS. In effect, these allow the user to extend the VCS command repertoire.

Appendix III

Very Simple Prolog+

Author: Bruce Batchelor, University of Wales College of Cardiff, P.O. Box 904, Cardiff CF1 3YH, Wales, U.K. Telephone: Int + 222 874390. Fax: Int + 222 874192.

Notice: Copies of all or part of this program, for non-commercial use, may be made freely and without payment, provided that the author is informed in writing. A licence for commercial development of this software may be obtained by writing to the author at the above address.

System configuration: The hardware comprises a VCS512 framestore, plugged into a VME-bus computer, running the OS/9 operating system. Overall control is exercised by a Macintosh II computer running LPA MacProlog, version 2.5. There is no robot or other electro-mechanical device connected.

Cable details: Linking the Macintosh II computer to the VME/OS-9 System:

Mini DIN connector	25-way D-type connector
3 (transmit data)	2 (receive data)
4 and 8 (strapped together)	7 (ground)
5 (receive data)	3 (transmit data)
—	4 and 5 (strapped together)

Features: The software listed below shows how a simple Prolog+ system may be constructed.

(i) The # operator is used to control the VCS image processor. (An alternative symbol, @, may be used instead. This is provided because @ is easier to find on the Macintosh keyboard. However, # is aesthetically more appealing.)

(ii) The ¶ operator is used to control external devices, such as a robot and lighting. The interface software is similar to that required for the image processor and so is not listed here.

(iii). A mechanism is provided allowing the user to define pull-down menus. Use the Extend menu item under the Utility menu, to initiate the dialogue. Each time this is done, a new fact of the form:

menu_item(Menu,Item,Command).

is asserted in the database. Subsequently, when **Item** is selected from **Menu, Command** is executed.

(iv) A transparent mode (or Interactive mode) is provided. This can be initiated using either a pull-down menu or a command key (Command/T). Since the # (or @) operator is required in programs, it was felt that users might inadvertently use it when working in Interactive mode. For this reason, Interactive mode will tolerate this operator, although it is not obligatory.

```
/* _______________The # operator for controlling VCS_______________ */

# # A :- # A.                        % Allow # to be used in interactive mode

# @ A :- # A.                        % Allow @ to be used in interactive mode

# end_of_file :- # swi.              % Press RETURN in interactive mode to
switch                                 images

# A :-
      A =.. [P|Q],                   % Construct list from VSP command
      atom(P),                       % Make sure that P is an atom
      consts([P|Q],C),               % C is list of constants in [P|Q]
      D =.. C,                       % Construct command from list of constants
                                       in A
      varsin([P|Q],E),               % E is list of variables in [P|Q]
      !,
      banner(vcs(D,E),['Current operation:',D]).   % Command VCS to do it

/* ________Alternative to the # operator – either symbol may be used__________ */

@ X :- # X.

/* ___________The ¶ operator for controlling external devices_____________ */

¶ A :-
      A =.. [P|Q],
      atom(P),
      consts([P|Q],C),
      D =.. C,
      varsin([P|Q],E),
      !,
      device_control(D,E).           % "device_control" closely resembles "vcs"

/* _______________Deleting variables from a list_______________ */

consts([],[]).

consts([A|B],[A|C]) :-
      not(var(A)),
      consts(B,C).
```

```
consts([A|B],C) :-
        consts(B,C).

/* ________________Define the pull-down menus________________________ */

start :-
        kill_menu('Fonts'),
        kill_menu('Utility'),
        kill_menu('Grey'),
        kill_menu('Binary'),
        kill_menu('Anal'),
        kill_menu('Device'),

        install_menu('Utility',['Extend menu', 'Initialise system', 'Interactive mode/T', 'Switch
images/H','-']),
        install_menu('Grey',[ ]),
        install_menu('Binary',[ ]),
        install_menu('Anal',[ ]),
        install_menu('Device',[ ]),
        build_menus.

/* ___________Initialise the data link to the VCS image processor_____________ */

initialise :-
        wkill('Output data to VCS'),
        wcreate('Output data to VCS',0,40,300,150,340),
        seropen(modem),
        serconfig(modem,both,9600,8,none,1),        % Search for "?" (query)
        banner(skip(modem,63),'(1) Press RESET on the VCS512 framestore (2) Press
                             RESET on the master processor (3) Wait 30 seconds '),
        write(modem,0),
        nl(modem),
        skip(modem,93),                             % Search for key character "]" (square
                                                      bracket)
        skip(modem,58),                             % Search for key character ":" (colon) */
        date_and_time,
        nl(modem),
        skip(modem,36),                             % Search for key character "$" (dollar)
        write(modem,host),
        nl(modem),
        skip(modem,36),                             % Search for key character "$" (dollar)
        write(modem,vcs),
        nl(modem),
        skip(modem,58),                             % Search for key character ":" (colon)
        vcs(prolog,_),                              % Switch VCS into mode suitable for Prolog
        nl,
```

```
        write('Prolog+ is now running'),
        nl,
        initialise_vcs.

/* ____________What to do under the "Utility" menu____________________ */

'Utility'('Extend menu') :-
        scroll_menu(['Which menu do you wish to extend?'],

        ['Utility','Grey','Binary','Anal','Device'],[ ],Z),
        [Y] = Z,
        prompt_read(['What is the name of the item you wish to add to the'",Y,'"
                                                        menu?'],X),
        prompt_read(['What goal do you wish to be associated with the item '",X,
                                                        '" in menu"',Y,'" ?'],W),
        nl,write([W,X,Y]),nl,
        extend_menu(Y,[X]),
        assert(menu_item(Y,X,W)),
        date(Year, Month, Day),
        time(Hour,Min,_),
        stringof(['Extension',Min,'-',Hour,'-', Day, ' -', Month, '-', Year], File),
        save(File),
        nl,
        write('Program saved in file:'),
        write(File),
        nl.

'Utility'('Initialise system') :- initialise.          % Initialise VCS-Prolog data link

'Utility'('Interactive mode') :- interactive_mode.    % Enter Transparent mode

'Utility'('Switch images') :- swi.                     % Switch images

'Utility'(X) :-
        menu_item('Utility',X,Y),                      % Find out what to do ...
        call(Y).                                       % ...and do it

'Utility'(X) :- not_found(X).

/* ________________What to do under the other menus___________________ */

'Grey'(X) :- menu_item('Grey',X,Y), call(Y).
'Grey'(X) :- not_found(X).

'Binary'(X) :- menu_item('Binary',X,Y), call(Y).
'Binary'(X) :- not_found(X).

'Anal'(X) :- menu_item('Anal',X,Y), call(Y).
'Anal'(X) :- not_found(X).
```

```
'Device'(X) :- menu_item('Device',X,Y), call(Y).
'Device'(X) :- not_found(X).

not_found(X) :-
        message(['Menu item not found – check the contents of "Data window"']).

/* ______________________Initialise VCS___________________________ */

initialise_vcs :-
        wshow('Output data to VCS'),
        # swc(1,0,0,512,512),                  % 512*512 pixels/image
        # swc(2,128,128,256,256),              % 256*256 pixels/image
        # swc(3,192,192,128,128),              % 128*128 pixels/image
        # won(1),                              % Default option is 512*512 pixels
        # son,                                 % Retain current image
        # com.

/* __________________Output the date and time to OS/9_________________ */

date_and_time :-
        date(Z,M,D),
        Y is Z -1900,
        write(modem,Y),
        write(modem,'/'),
        write(modem,M),
        write(modem,'/'),
        write(modem,D),
        write(modem,' '),
        write(modem,'9:00 am'),                % Simplified date – always 09.00
        nl(modem).

/* ___________This is the predicate responsible for all I/O to VCS___________ */

vcs(A,B) :-
        seropen(modem),                        % Open modem port
        serconfig(modem,both,9600,8,none,1),   % Configure it
        write(modem,A),                        % Send command to VCS
        nl(modem),
        skip(modem,10),                        % Look for RETURN character
        read(modem,X),                         % Instantiate X to incoming data
        X = [D|B],                             % Get error no. (D) and values (B)
        skip(modem,58),                        % Look for ":" character
        tell('Output data to VCS'),            % Update journal
        Y =.. [A],
        write(Y),
        write(','),
        told,
```

```
        !,
        check_error(D),
        tell('Output data to VCS'),
        write('  /* Value(s) returned: '),
        pp1(B),
        write(' */'),
        nl,
        told,
        !.

/* _________Pretty print routine for outputting lists to the modem port___________ */

pp([]) :-
        nl(modem).

pp([A|B]) :-
        write(modem,' '),
        write(modem,A),
        pp(B).

/* ________Pretty print routine for printing lists with spaces as separators________ */

pp1([]).

pp1([A|B]) :-
        write(' '),
        write(A),
        write(' '),
        pp1(B).

/* ______________________Error check for VCS______________________ */

check_error(0) :-!.                          % Error zero signifies no error at all

check_error(A) :-
        message(['Image processor has signalled an error, number ',A]),
        !,
        fail.

/* _______________________Interactive mode_______________________ */

interactive_mode :-
        prompt_read(['Please specify just one image processing goal'],X),
        not(X = end),                        % Terminate interactive session
        # X,
        !,
        interactive_mode.

interactive_mode.
```

```
/* ________________________Building menus___________________________*/

build_menus :-
        menu_item(A,B,C),                   % Consult database
        extend_menu(A,[B]),                 % Extend menu A by adding [B]
        fail.                               % Consult all facts in database

build_menus.

/* ____________Typical image processing operator definitions____________*/

swi :- # swi.

contrast_enhance :-
        # cpy,                              % Image processing command
        # gli(X,Y),                         % Image processing command
        Z is 0 – X,
        # acn(Z),                           % Image processing command
        W is Y – X,
        # mcn(255,W).                       % Image processing command

/* ______________________Repeating commands_______________________*/

0•G.                                        % Terminate recursion

N•G :-
        call(G),                            % Satisfy goal G
        M is N – 1,
        !,
        M•G.                                % Recurse

/* ________________Dummy to show the use of ¶ operator_________________*/

device_control(A,[B]) :- prompt_read([A],B).   % User simulates robot

/* ___________________________Operators______________________________*/

op(900, xfx, •).                            % Repetition operator
op(900, fy, #).                             % Image processing
op(900, fy, @).                             % Alternative image processing
op(900, fx, ¶).                             % Device control

/* ___________________________Database ____________________________*/

menu_item.                                  % Dummy – ensures new facts are inserted
                                              into correct place
```

```
% Facts put into database by invoking "Utility"("Extend menu")

menu_item('Grey', 'Blur - mild', # lpf).
menu_item('Grey', 'Blur - severe', # raf(15, 15)).
menu_item('Grey', 'Negate', # neg).
menu_item('Grey', 'Threshold - fixed', # thr(128, 255)).
menu_item('Grey', 'Threshold - variable', (prompt_read(['Specify the threshold
                                        parameter'], _1),# thr(_1, 255))).
menu_item('Grey', 'Contrast enhance', contrast_enhance).
menu_item('Grey', 'Histogram equalisation', # heq(0, 255)).
menu_item('Grey', 'High pass filter', (# raf(11, 11),# sub)).
menu_item('Utility', 'Resolution: Full', # won(1)).
menu_item('Utility', 'Resolution: Half', # won(2)).
menu_item('Utility', 'Resolution: Quarter', # won(3)).
menu_item('Anal', 'Area', (# cwp(_1),nl,write('Area = '),write(_1),nl)).
menu_item('Binary', 'Edge detector', # bed).
menu_item('Binary', 'Expand white regions', # exw).
menu_item('Binary', 'Shrink white regions', # skw).
menu_item('Binary', 'Convex hull', # chu).
menu_item('Binary', 'Fill holes', # blb).
menu_item('Binary', 'Biggest blob', (# ndo(3, _1),# gli(_2, _3),# thr(_3, _3))).
menu_item('Anal', 'Count blobs', (# ndo(3, _1),nl,write('There are '),write(_1),write('separate
blobs'),nl)).
menu_item('Device', 'Initialise robot', message(['The robot is being intialised now'])).

% Alternative definition when the VCS image processor is hosted on the Macintosh computer

vcs(A,C) :-
        call_c(A,[B|C]),
        !,
        check_error(B).
```

Appendix IV

An Alternative Implementation

Vision system: Intelligent camera

Lens:	Any standard C mount lens may be fitted
Sensor:	CCD
Size:	Length 310 mm (excluding lens) Width 110 mm Height 60 mm
Weight:	2.0 kg (excluding lens)
Power:	20 watts
Resolution:	256×256 pixels
Intensity:	256 grey levels
Storage:	Five image planes
Control lines:	Eight input/eight output

Prolog: LPA MacProlog, version 2.5 or later

(Host computer: Macintosh, with ≥ 1 MB RAM)

Prolog+ Software:

Control software interfacing Prolog to intelligent camera
Pull-down menus (Can be extended by the user)
Speech synthesiser
Library (Similar to the list in Appendix II)
Demonstrations
- Identifying a non-picture playing card
- Picking up a con-rod
- Stacking blocks
- Dissecting a plant
- Burglar alarm

VCS simulator[1]

Manufacturer: Image Inspection Ltd,
Unit 7, First Quarter,
Blenheim Road,
Surrey KT19 9QN,
United Kingdom.
Tel: Int+ 372 726150
Fax: Int+ 372 726276

Typical execution times: See Table 3.1

Notes

1. The simulator enables many of the programs in this book to be run without any modification whatsoever. In some cases however, it is not possible to simulate a VCS operation precisely; an audible warning is given that the simulation is approximate and the program continues running. In a few instances, where there is no simulation possible, the software warns the user and then the program halts.

References and Further Reading

Vision

ARB-87 M. A. Arbib & A. R. Hanson, *Vision Brain and Cooperative Computation*, MIT Press, Cambridge, MA, 1987.

GOM-88 E. H. Gombrich, *Art and Illusion*, Phaidon Press, Oxford, 1988.

GRE-90 R. L. Gregory, "Perception: where art and science meet", *Proc. Royal Society of Arts*, vol. CXXXVIII, no. 5406, May 1990, pp. 399 – 403.

HOF-85 D. Hofstadter, *Metamagical Themas*, Penguin Books, London, 1985.

MAR-82 D. C. Marr, *Vision*, Freeman, San Francisco, 1982.

PYL-88 Z. W. Pylyshyn, *Computational Processes in Human Vision*, Ablex, Norwood, New Jersey, 1988.

Prolog

BRA-86 I. Bratko, *Prolog Programming for Artifical Intelligence*, Addison-Wesley, Wokingham, 1986.

COE-88 H. Coelho & J. C. Cotta, *Prolog by Example*, Springer-Verlag, Berlin, 1988.

CLO-81 W. F. Clocksin & C. S. Mellish, *Programming in Prolog*, Springer-Verlag, Berlin, 1981.

CSP [8]CS-Prolog is a joint product of Multilogic Computing Ltd., Brainware Gmbh and Densitron Computers Ltd. The last mentioned may be contacted at Unit 4, Airport trading Estate, Biggin Hill, Kent, TN16 3BW, U.K.

FOR-89 N. Ford, *Prolog Programming*, John Wiley, Chichester, 1989.

FUT-89 I. Futo, & P. Kascuk, CS-Prolog on multi-transputer systems, *Microprocessors and Microsystems*, vol. 13, no. 2, 1989, pp. 103–112.

GAZ-89 G. Gazdar & C. Mellish, *Natural Language Processing in Prolog*, Addison-Wesley, Wokingham, 1989.

KLU-85 F. Kluzniak & S. Szpakowicz, *Prolog for Programmers*, Academic Press, London, 1985.

MAR-86 C. Marcus, *Prolog Programming*, Addison-Wesley, Wokingham, 1986.

MER-89 D. Merritt, *Building Expert Sytems in Prolog*, Springer-Verlag, New York, 1989.

ROS-89 P. Ross, *Advanced Prolog*, Addison-Wesley, Wokingham, 1989.

STE-86 L. Sterling & E. Shapiro, *The Art of Prolog*, MIT Press, Cambridge, MA, 1986.

Image Processing (General)

BAL-82 D. H. Ballard & C. M. Brown, *Computer Vision*, Prentice-Hall, New Jersey, 1982.

BAT-85a B. G. Batchelor, Principles of digital image processing, in *Automated Visual Inspection* (B. G. Batchelor, D. A. Hill, & D. C. Hodgson, eds), IFS (Publications), Bedford and North Holland, Amsterdam, 1985.

BOL-81 L. Bolc & Z. Kulpa, *Digital Image Processing Systems*, Springer-Verlag, Berlin, 1981.

CAS-78 K. R. Castleman, *Digital Image Processing*, Prentice-Hall, Englewood Cliffs, 1978.

FU-77 K. S. Fu (ed.), *Syntactic Pattern Recognition Applications*, Springer-Verlag, Berlin and New York, 1977.

GON-77 R. C. Gonzalez & P. Wintz, *Digital Image Processing*, Addison-Wesley, Cambridge, MA, 1977.

HAL-79 E. L. Hall, *Computer Image Processing and Recognition*, Academic Press, New York, 1979.

HOR-86 B. K. P. Horn, *Robot Vision*, MIT Press, Cambridge, MA, 1986

JOY-85 Joyce Loebl Co., *Image Analysis: Principles and Practice*, Gateshead, 1985.

PAV-77 Th. Pavlidis, *Structural Pattern Recognition*, Springer-Verlag, Berlin and New York, 1977.

PRA-78 W. K. Pratt, *Digital Image Processing*, John Wiley, New York, 1978.

PUG-83 A. Pugh, *Robot Vision*, IFS (Publications), Bedford, England and Springer-Verlag, Berlin and New York, 1983.

ROS-82 A. Rosenfeld & A. C. Kak, *Digital Picture Processing*, (2nd edition), Academic Press, New York, 1982.

SCH-89 R. J. Schalkoff, *Digital Image Processing and Computer Vision*, Wiley, New York, 1989.

Industrial Applications of Image Processing

BAR-88 S.L. Bartlett et al., "Automated solder joint inspection", *Trans. I.E.E.E. on Pattern Analysis and Machine Intelligence, PAMI* vol. 1, no. 10, 1988, pp. 31 – 42.

BAT-78 B. G. Batchelor, *Pattern Recognition, Ideas in Practice*, Plenum, London, 1978.

BAT-85a B. G. Batchelor, D. A. Hill, & D. C. Hodgson (eds), *Automated Visual Inspection*, IFS (Publications), Bedford, and North-Holland, Amsterdam, 1985.

BAT-85b B.G. Batchelor and A.K. Steel, A flexible inspection cell, *Proc. Int. Conf. on Automation in Manufacturing, Part 4: Automated Vision Systems*, Singapore, September 1985, pp.108 – 134. Singapore Exhibition Services Pte. Also in *Proc. 5th Int. Conf. on Robot Vision and Sensory Control*, Amsterdam, October 1985, pp. 449 – 468. pub. by I.F.S. (Publications), Bedford.

BAT-87 B. G. Batchelor & G. E. Foot, Image acquisition and processing techniques for industrial inspection, in I. B. Turksen, *Computer Integrated Manufacturing*, Springer-Verlag, Berlin, 1987.

BOW-85 C. C. Bowman & B. G. Batchelor, Automated visual inspection, in R. S. Sharpe (ed.), *Nondestructive Testing*, vol. 8, pp. 361 – 444. (Contains an extensive bibliography.)

BOY-88 R. D. Boyle & R. C. Thomas, *Computer Vision, a First Course*, Blackwell, Oxford, 1988.

BRE-81 J. Breschi, *Automated Inspection Systems for Industry*, IFS (Publications) Ltd., Bedford. Originally published in German, R. Oldenburg Verlag, Germany, 1979.

BRO-86 A. Browne & L. Norton-Wayne, *Vision and Information Processing for Automation*, Plenum Press, New York, 1986.

CHI-82 R. T. Chin & C. A. Harlow, Automated Visual Inspection: A Survey, *IEEE Trans. on Pattern Analysis and Machine Intelligence*, vol. 4, no. 6, 1982

DOD-79 G. C. Dodd & L. Rossol, *Computer Vision and Sensor-based Robots*, Plenum, New York, 1979.

EJI-89 M. Ejiri, H. Yoda & H. Sakou, Knowledge directed inspection for complex multilayered patterns, *Machine Vision and Applications*, vol. 2, 1989, pp. 155 – 166.

HOL-84 J. Hollingum, *Machine Vision, the Eyes of Automation*, IFS (Publications), Bedford, 1984.

HOR-86 B. K. P. Horn, *Robot Vision*, MIT Press, Cambridge, MA, and McGraw-Hill, New York, 1986.

PAR-78 J. R. Parks, Industrial sensory devices, in B. G. Batchelor (ed.), *Pattern Recognition – Ideas in Practice*, Plenum Press, 1978.

ZIM-82 N. J. Zimmerman, *Robot Vision in Holland*, DEB Publishers, Pijnacker, 1982.

ZIM-83 N. J. Zimmerman & A. Oosterlinck, *Industrial Applications of Image Analysis*, DEB Publishers, Pijnacker, 1983.

Interactive Image Processing

BAT-79 B. G. Batchelor, Interactive image analysis as a prototyping tool for industrial inspection, *Computers and Digital Techniques*. vol. 2, pp. 61 – 69, 1979.

BAT-80 B. G. Batchelor, P. J. Brumfitt & B. D. V. Smith, A command language for interactive image analysis, *Proc. IEE*. vol. 127(E), 1980. pp. 203 – 218.

BAT-82 B. G. Batchelor, D. H. Mott, G. J. Page & D. N. Upcott, The Autoview interactive image processing facility, in N. B. Jones (ed.), *Digital Signal Processing*. Peter Perigrinus, London, 1982. pp. 319 – 351.

BAT-91 B. G. Batchelor & F. M. Waltz, *Interactive Image Processing*, Springer-Verlag, in Press.

VCS *VCS Manual*, Vision Dynamics Ltd, Times House, Marlowes, Hemel Hempstead.

Image Processing and Prolog

BAT-86a B. G. Batchelor, Merging the Autoview image processing language with Prolog, *Image and Vision Computing*, vol. 4., no. 4, November 1986, pp.189 – 196.

BAT-86b B. G. Batchelor, An appraisal of artificial intelligence techniques for industrial vision systems, *Proc. Int. Conf. on Automated Optical Inspection*, Innsbruck, Austria, April, 1986, *Society of Photo-optical Instrumentation Engineers*, vol. 654, 1986, pp. 250 – 263.

BAT-87a B. G. Batchelor, An interactive AI language for image processing and robotics, Proc. SPIE conf. on *Intelligent Robots and Computer Vision: Sixth in a Series*, Cambridge, MA, U.S.A., Nov. 1987, vol. 848, pp. 600 – 607.

BAT-87b B. G. Batchelor, An interactive environment for developing intelligent image processing algorithms for inspection, *Proc. SPIE Conf. on Automated Inspection and High Speed Vision Architectures*, Cambridge MA, Nov. 1987, vol. 849. Also published in *Proc. S.P.I.E.* vol. 848, 1987, pp. 100 – 108.

BAT-87c B. G. Batchelor, Industrial vision systems for inspection and manufacturing, in *Systems and Control Encyclopaedia*, Pergamon, 1987, pp. 2473 – 2479.

BAT-88a B. G. Batchelor, F.M. Waltz, & M.A. Snyder, A design methodology for industrial vision systems, *Proc SPIE conf.*, Dearborn, June 1988, vol. 959, pp. 126 – 145.

BAT-88b B. G. Batchelor, I.P. Harris, J.R. Marchant & R.D. Whitfield, Automatic dissection of plantlets, *Proc. SPIE on Automated Inspection and High Speed Vision Architectures II*, Cambridge MA, Nov. 1988, vol. 1004, pp. 159 – 167.

BAT-88c B. G. Batchelor, I.P. Harris, Waltz & M.A. Snyder, Applications of the ProVision language in robot vision, *Proc SPIE conf.*, Cambridge, MA, June 1988, vol. 1002, pp. 473 – 481.

BAT-89a B. G. Batchelor,Integrating vision and AI for inspection and robot control, *Proc. IEEE Conf. on Image Processing*, Singapore, September 1989, pp. 146 – 151.

BAT-89b B. G. Batchelor, Integrating vision and AI for industrial applications, *Proc. Int. SPIE Conf. on Intelligent Robots and Computer Vision VIII: Systems and Applications*, Philadelphia, Pennsylvania, Nov. 1989, pp. 168 – 173.

BAT-89c B. G. Batchelor, A Prolog lighting advisor, *Proc. Int. SPIE Conf. on Intelligent Robots and Computer Vision VIII: Systems and Applications*, Philadelphia, Pennsylvania, Nov. 1989, vol. 1193, pp. 295 – 302.

BAT-89d B. G. Batchelor & G. E. Foot, Image acquisition and processing techniques for Industrial inspection, *Proc. Nato Advanced Study Institute, Computer Integrated Manufacturing, Istanbul*, 1987, pp. 505 – 553, 1989.

BRU-84 P. J. Brumfitt, Environments for image processing algorithm development, *Image and Vision Computing*, vol. 2, no. 4, 1984, pp.198 – 203.

FAI R. Fairwood, University of Guildford, U.K., private communication

GAR-89 B. Garner, B. Cheng & D. Lui, An interactive concept classifier for scene analysis, *Proc. Int. SPIE Conf. on Intelligent Robots and Computer Vision VIII: Systems and Applications*, Philadelphia, Pennsylvania, Nov. 1989, vol. 1193, pp. 123 – 130.

HAN H. Hanakalahti, Technical Research Centre of Finland, private communication.

MOT-85 D. H. Mott, A development tool for AI in vision, *Sensor Review*, vol. 5, no. 1, Jan. 1985, pp. 29 – 32.

POW-88 D. Powers, Prolog simulator for studying visual learning, *Proc. Int. SPIE Conf. on Intelligent Robots and Computer Vision VII*, Cambridge, MA, Nov. 1988, vol. 1002, pp. 482 – 489.

SNY-88 M.A. Snyder, F. M. Waltz & B. G. Batchelor, A real time inspection algorithm development station for use in a symbolic, *Proc SPIE on Automated Inspection and High Speed Vision Architectures II*, Cambridge MA, Nov. 1988, vol. 1004, pp. 176 – 183.

TOL J. Tolman, University of Bradford, U.K., private communication.

SUBJECT INDEX

VCS commands and Prolog+ predicates are set in bold type.